America
the
Poisoned

America the Poisoned

Lewis Regenstein

How Deadly Chemicals are Destroying Our Environment, Our Wildlife, Ourselves and — How We Can Survive!

ACROPOLIS BOOKS LTD.

Washington, D.C.

Dedication

This book is dedicated to pesticide victims everywhere, and to those fighting to save them, indeed everyone, from the deadly poisons that the chemical industry and our own government have allowed to contaminate us all.

ACROPOLIS BOOKS LTD.
Colortone Building, 2400 17th St., N.W.
Washington, D.C. 20009

Printed in the United States of America by
COLORTONE PRESS, Creative Graphics, Inc.
Washington, D.C. 20009

Library of Congress Cataloging in Publication Data

Regenstein, Lewis.
America the poisoned.

Includes bibliographical references and index,
1. Environmental health—United States. 2. Environmentally induced diseases—United States. 3. Carcinogenesis. 4. Abnormalities, Human—Etiology. I. Title.
RA565.R38   363.1′72′0973   82-1813
ISBN 0-87491-486-8   AACR2

Photo Credits:
Front Cover, USDA Photo
Back Cover, top to bottom: USDA Photo, National 4-H Council, USDA Photo, USDA Photo
Spine, National 4-H Council.

Acknowledgments

This book could not have been written without the support of The Fund for Animals. I particularly appreciate the encouragement of its president Cleveland Amory, and director Marian Probst.

My thanks also to the many people who helped supply invaluable information and assistance, including Glenn Chase, Paul Cheng, Melanie Cook, Doris Dixon, Larry Miller, Sharon Frazier, Phil Keisling, Ben Lieber, Lucille, Nevin, and Jimmy Mendenhall, Veronica de Mello, Angela Miller, The Mitchell Millers, Elliott Norse, Eric Jansson of Friends of the Earth; Jacob Scherr of the Natural Resources Defense Council; the staff of the Environmental Defense Fund; and last but not least to my wife Janice, for putting up for years, though not always cheerfully, with my clippings, scribblings, and boxes of material scattered all over the house.

Ultimately, of course, the success and value of any book depends to a great extent on the publisher, and I cannot say enough about the enthusiasm and competence with which Acropolis Books has pursued this project. I am deeply grateful to publisher Al Hackl, Laurie Tag, Sandy Trupp, and the excellent staff at Acropolis for all they have done. The book was superbly edited by Nancy Heneson. It has been a pleasure working with all of them, and any success the book has will be due in large part to their efforts.

BATCH NO. 11-78M1477B

GROSS WT. 39.08 KILOS/86.16 LBS.   NET WT. 35 KILOS/77.16 LBS.   2.838 CU. FT.

MONTROSE CHEMICAL CORPORATION OF CALIFORNIA
TORRANCE, CALIFORNIA

EPA EST. NO. 8775-CA-1

INSECTICIDE, DDT 75% WATER DISPERSIBLE POWDER. SPECIFICATION NO. PHS/NCDC-1-102A

CAUTION: HARMFUL IF SWALLOWED. ABSORBED
THROUGH THE SKIN WHEN IN SOLUTION. AVOID
BREATHING DUST OR SPRAY MIST. AVOID
CONTAMINATION OF FEED AND FOODSTUFFS.

STOW AWAY FROM BOILERS AND BULKHEADS
IN A COOL PLACE.

UNITED STATES OF AMERICA

UNITED STATES OF AMERICA
SHIPPING MARKS:
MINISTRY OF HEALTH
B.P.K.L.N. - D JAWA TIMUR
J1. JOHAR 24
SURABAYA, INDONESIA
PORT OF ENTRY: TG. PERAK, SURABAYA
PIO/C 497-0239-5-6440536
MALARIA CONTROL LOAN 497-U-034
P.O. WPN-F-Z0329-1

DDT was generally banned in the U.S in 1972 because of its potent ability to cause cancer, and its devastating impact on birds, fish, and other wildlife. By this time, it was being found in virtually all human tissue and breast milk samples, as well as in food, air, and water throughout the country. Yet, over 40 million pounds a year of DDT have continued to be manufactured in the U.S. and sold abroad, much of it distributed by the U.S. Agency for International Development. Since food crops grown abroad for export to the U.S. are often contaminated with the pesticide, DDT remains a part of the American diet. *Photo courtesy of Pat Goudvis, 1978.*

Contents

Chapter 2 Pesticides: Feeding The Insects, Poisoning The People.......................... 79

Chapter 3 Toxic Wastes: 80 Million Tons A Year............. 136

Part II   243

The Impending Disaster

Part III   279

Chemicals That Might Kill Us: Case Studies Of The Major Toxics Scandals

The decimation of our nation's symbol, the American bald eagle, by pesticides aptly symbolizes the devastation these chemicals have wrought on our environment, our wildlife and ourselves. *Photo courtesy of Lois Constantine.*

Introduction

This book tells the story of the major environmental and health crisis of our times: the pervasive presence in our society of chemicals that may be killing and disabling millions of Americans.

The U.S. Environmental Protection Agency (EPA) has called this toxic chemical contamination "the most grievous error in judgment we as a nation have ever made," describing it as "one of the most serious problems our nation has ever faced." Indeed, the dimensions of this situation are truly incredible.

By the time the government got around to generally banning some of the most deadly pesticides—several of which cause cancer and/or birth defects in animals at the lowest doses tested—these toxins were being found in the tissues of 99 percent of all Americans tested, as well as in our food, air and water.

Mothers' milk has become so contaminated with banned chemicals that it would be illegal to sell it in supermarkets. In fact, even before birth, infants are surrounded by toxic chemicals and continue to be exposed throughout life. One of the most toxic compounds—PCB's—has been so heavily and widely used that it is found not only in virtually all human tissue and breast milk samples, but may also be present in the flesh of all living creatures.

The results of our constant contact with such chemicals are now becoming apparent. One in four Americans can expect to get cancer. This means that of the Americans now alive, over 56 million will contract the disease, which kills over 1,000 of us every day. The chances of getting cancer by the age of 74 are almost one in three.

Chemicals that cause sterility are regularly found in human semen samples, and studies show that the sperm counts of American males appear to have dropped 30 to 40 percent in the last 30 years, with toxic chemicals held largely to blame.

The health of some 35 million Americans is in jeopardy because of polluted air, a U.S. government study shows; scientists estimate that such pollution may be causing some 200,000 deaths a year. Top U.S. officials have warned the air pollution resulting from our current energy policies could help warm the atmosphere enough to melt the polar icecaps and flood New York, Los Angeles and other coastal cities world-wide. Thus, the crippling of our environmental and health protection laws by the Reagan administration could have catastrophic consequences, costing the lives of literally millions of Americans.

Hundreds of thousands of children—perhaps millions—may contract cancer just from exposure to sleeping garments treated with a flame-retardent chemical known as Tris, which were worn by some 60 million younsters. After being banned for sale in the U.S., millions of these garments were shipped abroad to be sold in foreign countries, despite a plea from Congress to President Carter to halt the exports. And massive amounts of banned chemicals are regularly—and legally—shipped abroad, where they poison huge numbers of people in foreign countries and return to us as residues on bananas, coffee, tea, sugar and other imported food.

Millions of pounds of the same compounds that composed the infamous defoliant used in Vietnam, Agent Orange, are sprayed over the U.S. each year, with predictable results. Some uses of one such herbicide were restricted after EPA admitted that it was killing unborn children in areas where it was sprayed. Yet millions of pounds of this compound are still used on such food crops as rice and on fruit orchards.

Perhaps the most shocking—and depressing—aspect of researching this book was conducting interviews with people who had received publicity in the late 1970's because they had been subjected to spraying by this herbicide and other pesticides. Of the people I was able to reach, without exception, they and/or a close family member had developed, and in some cases died of, cancer recently, or they were continuing to suffer miscarriages and other severe health disorders.

Another closely-related chemical, the nation's most widely-used herbicide, continues to kill fetuses. Following spraying episodes, 9 out of 10 women in one area of western Montana had miscarriages; in Ashford, Washington, 10 out of 12 pregnancies ended in miscarriage.

The amount of toxic wastes produced in the U.S. each year amounts to over 600 pounds for every American, and 90 percent of these deadly wastes are disposed of improperly. There are literally thousands of hazardous dump sites across the country that are as dangerous as Love Canal, or more so. At the Hooker Chemical company's Hyde Park dump near Love Canal, enough of the deadly chemical dioxin is buried to kill everyone on earth if consumed in equal amounts.

Yet the chemical industry spends millions upon millions of dollars in advertising campaigns denying that there is a problem. It even tried to stop the publication in 1962 of Rachel Carson's epic work, *Silent Spring*, suggesting that it was part of a Communist plot against America. More recently, agribusiness interests have urged the federal government not to ban a pesticide that admittedly causes cancer and sterility, but rather to let it be handled by older workers and those who do not want to have children.

Today, twenty years after *Silent Spring*, almost all of the toxic chemicals Rachel Carson discussed are still in widespread use; the few that have been restricted have often been replaced by equally or more hazardous compounds, and a thousand new and largely untested chemicals are introduced each year.

Ironically, we can grow more food for less money without relying on the massive use of pesticides. Although the use of chemical pesticides has increased ten fold in the last 30 years, crop losses to insects have doubled.

Pesticides kill off birds, frogs, predatory insects, and other beneficial creatures that help keep pests under control, while poisoning our food—and ourselves—in the process. More effective methods of insect control, such as Integrated Pest Management (IPM), have been developed that allow farmers to grow more food, more cheaply without poisoning the environment.

For example, government studies show that such systems, if used on major crops, could reduce pesticide use by 50 percent over the next 5 years, and 70-80 percent in the next decade "with no reduction in present crop yield levels." In California, mosquito control programs using IPM have cut pesticide use by 90 percent while reducing labor and material costs and virtually eliminating pesticide pollution. Similar non-toxic methods of insect control can be used by home gardeners.

It should be emphasized here that this is not just another book bemoaning the fate of humanity. For each problem discussed, a viable solution is offered, usually in the form of an alternative product or course of action that is practical, effective, and often less expensive.

We should have known better than to think we could poison our environment and our wildlife but not be affected ourselves. A book written long, long ago—centuries before *Silent Spring*—warned us that we could not kill off the birds and beasts of the earth without jeopardizing our own survival. As the Bible tells us, the fates of humans and of animals are intertwined:

> For that which befalleth the sons of men befalleth
> beasts. Even one thing befalleth them: as the one
> dieth, so dieth the other; yea, they have all one breath,
> so that a man hath no pre-eminence above a beast.
>   *(Ecclesiastes* 3:19)

In light of what we have learned in recent years, we can no longer plead ignorance. The predictable—and predicted—results of our massive and indiscriminate use of these poisons are now showing up as epidemics of cancer, miscarriages, birth defects and other health disorders.

For many of us, it is already too late, for the damage that has been done is irreversible. But for all of us, the truly tragic thing is that we are continuing down a road that we should know full well is leading us to a disaster of unprecedented proportions. And every day that passes, the problem becomes greater, as does the price we will inevitably have to pay when the day of reckoning arrives.

Part One

The Chemical Crisis We Face

Before being banned for use in Vietnam, where it was used to kill off crops and forest cover, Agent Orange had devastating effects on the environment and human health. It apparently caused miscarriages, birth defects, cancer, and liver damage among Vietnamese and Americans alike, the same effects it has when given in minute doses to test animals. Millions of pounds of the same two ingredients of this defoliant—2,4-D and 2,4,5-T—are still sprayed over huge areas of the U.S., much of it on forests and food crops, often causing similar results as in Vietnam. *Photo courtesy of Bob Rand—Agent Orange Victims International (AOVI), 1968.*

Chapter One

Herbicides: America's War Against Itself

The questions which have been raised recently concerning the hazards of 2, 4, 5—T and related chemicals ... may ultimately be regarded as portending the most horrible tragedy ever known to mankind . . In view of the potential disaster that could befall us— or conceivably has insidiously already befallen us—absolutely no delay is tolerable in the search for answers.

Senator Philip Hart,
April, 1970[1]

Death from the air: it has become a common sight in the 20th century. An aircraft sweeps in low and releases a cloud of poison, a chemical mist designed to kill living organisms. When breathed, ingested, or absorbed through the skin, the agent can sicken, disable, or even kill its victims. It can also attack the immunological and central nervous systems, cause cancer and other disorders, and kill and deform unborn children.

Throughout the 1970's and early 1980's, we have been shocked by reports of such chemical attacks by the Russians against Afghan resistance fighters, and by the Communist Vietnamese against Laotian tribespeople.[5] But most often, such scenes have quietly taken place in the United States, where insecticides and herbicides, similar or identical to the chemical warfare agents from which they were developed, have been regularly sprayed on American crops and forests, as well as on the inhabitants of these areas. An increasing number of people

21

are asking why our government permits its citizens to be treated in ways that, under the international rules of warfare, would be illegal if carried out by enemy soldiers.

Perhaps the most controversial use of pesticides involves the massive herbicide spraying of inhabited, forested areas by the U.S. Forest Service and the timber industry. The two main herbicides that have been used on U.S. forests—2, 4, 5—T and 2,4-D—are chemically similar, and are the same substances that composed Agent Orange, a 50-50 mixture of these compounds. Agent Orange was used to defoliate vast areas of Vietnam during the war there; but in 1970, its use was halted in that country after it was learned that the agent was causing a high number of miscarriages, birth defects, and even infant deaths among Vietnamese villagers. Shortly thereafter, ironically, the use of those same herbicides in U.S. forests was greatly accelerated, with results similar to those in Vietnam. While the most extreme precautions were being taken in disposing of leftover stocks of Agent Orange and even the empty containers, millions of pounds of these same chemicals were being sprayed over the United States.

By the late 1970's, forest uses alone were directly exposing some four million people a year[6] to one of these chemicals, 2,4,5-T, with many more people being exposed through contaminated food and water. The other chemical, 2,4-D, had by then become the nation's most widely used weedkiller, with a substantial proportion of the American population regularly coming into contact with it.

When 2,4,5-T or 2,4-D was sprayed from the air, a tragic, familiar and predictable pattern often emerged: pregnant women miscarried or gave birth to deformed babies. In a community in western Montana, 9 out of 10 pregnant women suffered miscarriages.[7] Around Ashford, Washington, 12 pregnancies resulted in 10 miscarriages.[8] In addition, epidemics of severe flu-like disorders swept through entire communities; pets, livestock, and wild animals were sickened and killed, and birds and other wildlife disappeared.

Yet in light of what is known about the dangers of these chemicals and the devastating health effects they have caused in test animals and production workers, there is nothing surprising about these tragedies. The most active ingredient in 2,4,5-T is a dioxin contaminant called TCDD, the most toxic chemical known to man. In laboratory animals, it causes cancer, birth defects, miscarriages, and even death at the lowest levels tested, in some cases one or a few parts per trillion.[9][10] An ounce of dioxin, if consumed in equal amounts, is thought to be sufficient to kill a million people, a drop could kill a thousand; and it is 100,000 to a million times more potent than thalidomide in causing birth defects.[11]

Although not nearly as toxic, 2,4-D also causes cancer and birth defects in laboratory animals, and appears to cause similar problems in humans. Because it is so widely used (up to 80 million pounds a year), it may cause even more damage than 2,4,5-T.

The major manufacturer of these compounds, Dow Chemical Corporation, had known for years and even decades of their hazards, as did the U.S. government. But such information was consistently suppressed, denied, and ignored in order to allow them to remain in widespread use.[12]

Finally, admitting that 2,4,5-T spraying was regularly killing unborn children, EPA placed restrictions on its use in forests, but allowed other major uses to continue, such as on rice crops and rangeland. The use of 2,4-D is still virtually unrestricted, and it remains the nation's most popular herbicide. While EPA debates what, if any, further restrictions to place on these chemicals, they continue to enter, and contaminate, our food, air and water. Meanwhile, over 230 million Americans proceed to serve as guinea pigs for the U.S. government and the chemical industry.

"The Politics of Poison"

The majestic redwood country of northern California is an area of breathtaking beauty. Many people have found it an ideal place to live and raise their families, "in the shadow of the forest, where the redwoods scrape the sky and the lumberman is king."[13] But in recent years, the residents of this area and other remote, unspoiled regions have been subjected to a "rain of death" from the skies, which has poisoned the land and its inhabitants with the most toxic chemical known to man, killing and deforming children, wiping out the wildlife and contaminating the vegetation, food and water supplies.

A good example is the northern California town of Orleans, whose 600 residents have become potential or actual victims of

America's chemical warfare against itself. Of the 30 women who are known to have become pregnant between 1976 and 1978, 19 either miscarried or gave birth to dead or deformed babies.[14] These and other severe medical problems seemd to coincide with the aerial spraying of nearby forests with 2,4,5-T. The chemical was being used by the U.S. Forest Service and timber companies to kill off unwanted vegetation such as brush, leafy plants, and unwanted hardwood trees. (Because these trees provide shade, they slow the growth of more commercially valuable trees such as redwoods and Douglas firs, which thrive in direct sunlight.)

An extraordinary number of children in this town have also been born with birth defects. The head of a local child care center in Orleans reported that one third of the children attending—5 out of 15—had cleft palates, an abnormality which often appears in the offspring of test animals that have been given dioxin.

Another example of how our own government has turned Heaven into Hell is "the hamlet of Denny, California—nineteen miles of winding mountain road from the highway—a century from the city. A kind of Shangri-La to its residents—until the U.S. Forest Service sprayed 2,4,5-T."[15] One resident of Denny, Linda Van Atta, described in 1979 how she had three miscarriages, her daughter-in-law one miscarriage, a friend one miscarriage and cancer, and another friend a baby with a hairlip.[16] Indeed, of the 20 to 25 women in Denny who were of childbearing age, at least 11 had serious reproductive or medical problems, including 8 miscarriages. Van Atta and another Trinity County woman, Pat Kneer, who suffered a miscarriage and also developed cervical cancer, blame the U.S. Forest Service for their maladies and have sued the agency for $1.5 million in damages.[17]

Another formerly healthy woman who worked as a camp cook for some forestry workers in Oregon in areas sprayed with 2,4,5-T experienced regular stomach cramps, nose bleeds, and urinated blood. After she became pregnant, she developed boils on her face, and later experienced an extremely rare condition called "mole pregnancy." As she tells it, "[w]hen I miscarried... the fetus wasn't even recognizable as a baby or anything. It was just a mess... It looked like hamburger, like chopped meat."[18]

In the controversy over 2,4,5-T, private citizens have usually had to do the job of informing the public of the dangers of pesticides because of the failure of government to act. One of the most effective recent examples of this was a highly-acclaimed documentary film, "Politics of Poison." Produced by the television station KRON in San Francisco, it described the effects of 2,4,5-T spraying on the residents of northern California. Aired on 25 April 1979, the film had a tremendous impact, not just in California but also in Washington, D.C., where it was later shown before the House Subcommittee on Oversight and Investigations. (It also forms a chapter of a recent Sierra Club book, *Who's Poisoning America.*)[19]

Accurately describing dioxin as a "fetus-deforming agent 100,000 times more powerful than thalidomide ... a synthetic chemical so powerful that an ounce could wipe out a million people," KRON presented dramatic new evidence that 2,4,5-T causes birth defects, miscarriages, and nervous disorders.

The documentary's producer and director, David Rabinovitch, and his film crew visited several of the small, remote northern California towns surrounding timber forests, and uncovered shocking evidence (much of it discussed above) that the inhabitants of these areas were suffering miscarriage and birth deformity rates between 44 and 60+ percent. Although much of this data, in and of itself, may seem just too incredible to be taken seriously, similar statistics have turned up in other areas subjected to herbicide spraying.

"Politics of Poison" also included interviews with officials of EPA and the major manufacturer of 2,4,5-T, Dow Chemical, which in 1977 alone spent $286 million fighting the government's attempt to regulate chemicals, according to the documentary. Dow spokesman Cleve Goring, calling the campaign against 2,4,5-T "chemical McCarthyism," defended the use of the chemical as:

> ...a very important symbol ... if we were to lose on this issue, it would mean that the American public has been really taken back a couple of hundred years to an era of witch-hunting, only this time the witches are chemicals, not people.[20]

Also appearing in the film was Edwin Johnson, a deputy assistant administrator of EPA in charge of pesticides, who stated: "I don't have information to demonstrate to me that there have been health problems associated with exposure to 2, 4, 5-T." Johnson, long a defender of the safety of pesticides, had given similar testimony during some 1974 Senate hearings on 2,4,5-T, where he disputed evidence of the chemical's extreme toxicity.

The film created an uproar in California, prompting over 40,000 letters from viewers demanding action. Most of the reaction was similar to that expressed by *San Francisco Examiner* columnist Bill Mandel:

> The only sensible conclusions one can draw are these: that commercial interests are spraying populated areas with herbicides considered too deadly for use as chemical weapons; that government agencies charged with the protection of the public and the environment are powerless or too cowardly to do anything about this rain of death from the skies; that health officials look everywhere for explanations except at the culprits; and that massive expenditures by the timber and chemical companies paralyze the fact-aimed opposition of scientists and residents of the affected areas.[21]

In June, 1979, two weeks after the showing of the documentary, Mendocino County voted overwhelmingly to ban aerial spraying of phenoxy herbicides, such as 2,4,5-T and 2,4-D.

EPA's response to the film was to praise it and express gratitude for the "new information" it had presented. Yet none of the information presented in this documentary should have come as a surprise to EPA or the U.S. Forest Service (USFS). Similar data had been available for decades, and citizens and conservation groups had, over the years, and to no avail, written to these agencies calling attention to the health hazards of herbicide spraying and urging that action be taken to halt it. Indeed, citizens throughout the country are continuing to fight against, and suffer from the effects of, herbicides being sprayed on their land.

The Defoliation of Oregon

Oregon is considered by many to be America's loveliest state. Its natural beauty unsurpassed, much of Oregon remains a pristine and shining jewel of forests, wildlife, lakes, and sea coast. Yet the surface beauty of much of this rugged wilderness is deceptive, for vast areas of Oregon have been poisoned, the forests contaminated, the wildlife wiped out, and the people sickened, disabled, and killed. For at least the last decade, Oregon, too, has been subjected to the rain of death from the skies.

From 1972 to 1978, some 10,000 pounds of 2,4,5-T were sprayed over 7,000 acres of Oregon's coastal forests.[22] The effects on people, animals and the environment have been devastating.

Five Rivers, Oregon is a remote, unspoiled coastal community with clean air, friendly people, and thick forests. It is also the birthplace of the state's organized resistance movement against herbicides, which are regularly and heavily applied to the area's timber stands.

In the spring of 1979, shortly after forest spraying of 2,4,5-T had been restricted, USFS sprayed 2,4-D on several hundred acres of timber around Five Rivers. Shortly thereafter, five of the seven women known to be pregnant lost their babies; all of the miscarriages occurred among women living within 1½ miles of the area sprayed. Other residents also became ill, but the USFS continued to spray herbicides around Five Rivers.[23]

On 17 July 1979, five residents of Five Rivers signed and released a letter urging that herbicide spraying be halted, and describing the community's sudden, severe health problems:

> Since May 12 [when the spraying began] the health of
> the population of our valley has undergone profound and
> disturbing changes. Chief among these has been the
> incidence of respiratory illness and intestinal disorders in
> almost every household. Several women experienced
> uterine hemorrhaging not associated with pregnancy.
> Children and adults suffered bleeding gums, bloody noses,
> and a number of women suffered from bacterial vaginal
> infections. Ten children and three adults have experienced
> an undiagnosed illness characterized chiefly by high
> fever...

There is nothing new about herbicide poisoning to the people of this area. Beginning in 1973, on a block where the roadside was regularly sprayed to kill weeds and a local spring nearby supplied the drinking water, several women started to experience uterine bleeding. Later, other residents came down with hepatitis, and an outbreak of an influenza-like illness struck the valley. Livestock began to bear deformed offspring, and chicken eggs failed to hatch. A 1976 survey of 70 people living in the neighborhood where the herbicide was sprayed revealed that almost half of the families had been sickened or seen such effects in others.

As the spraying of the area continued, it was accompanied by an epidemic of health disorders: As Phil Keisling reported in the weekly newspaper *Willamette Week*, of Portland, Oregon:

> Debbie Marano, a native eastern Oregonian, suffered her fourth miscarriage within two years. In the house next door to hers, a young man from Tennessee contracted terminal cancer; he moved out, and the woman who moved in subsequently had a miscarriage.
>
> In the next house, a 62-year old man died of pancreatic cancer; the woman who then moved in also suffered a miscarriage. When she left, the next woman miscarried as well.
>
> When Susan Parker moved out of her house to one across the road, her replacement found herself unable to conceive a second child. There were also other problems of unknown origins. Parker's husband underwent a "complete personality change," she says, adding, "It seemed with all the men in the area, their sex drive went way down."[24]

And to top things off, in the first few hours of the New Year of 1978, the home of two prominent opponents of the herbicide spraying, Steve and Carol Van Strum, was burned to the ground, killing their four children. It was commonly believed in Five Rivers that the house was deliberately set on fire because of the Van Strums' leading role in fighting against aerial spraying.

The Van Strums have continued to fight for a halt to herbicide spraying, which has continued through 1982, along with a seemingly high rate of birth defects, miscarriages, and outbreaks of meningitis, colds, and flu-like illnesses. In the spring and fall of 1981, county and state agents sprayed their property despite a highly-visible "No Spraying" sign being posted. Since then, she has had two miscarriages. At one point, Steve stopped the spraying at gunpoint, and was charged with "obstructing government administration." According to him,

"In the Five Rivers Valley, from 1979 until June 1981, only two babies have been born—and those were by mothers who were away during the first trimester. Five to ten kids would normally have been born." In addition to two older children, the Van Strums are now raising one infant—it is adopted.[25]

In nearby Allegany, Oregon, residents experienced similar health problems, including an unusually high rate of cancer. After herbicides were sprayed on nearby forests, fish died, the birds disappeared, and other wildlife was nowhere to be seen. One little girl contracted a rare blood disease, and analysis of her blood revealed the presence of 2,4-D and Silvex, another widely-used herbicide. She was hospitalized and slowly began to improve. But after she was taken home, the spraying was resumed; the child suffered a relapse and almost died.

A typical account of life in the area is provided by Eve and Vern DeRock, who lived on a farm in the village of Scottsberg near Allegany. Vern, a former lumberjack, until recently drove a logging truck for the International Paper Company after losing three fingers to a chain saw. In March of 1977, International Paper sprayed their valley with herbicides; soon afterward, fish died in the creek, and still-born fawns were found in the woods. A week later, 15 of the DeRock's pregnant cows aborted their calves. When they tried to breed their 54 heifers, only six became pregnant; five of the calves were born with grotesque deformities and died, the sixth one was born weak and sickly. The DeRocks sold off most of their cattle, but before doing so, butchered and ate several of them. All three members of the family then came down with severe stomach cramps and flu-like illnesses that could not be shaken. Vern collapsed after being exposed to a fire in a sprayed area, and later suffered a heart attack. By the fall, Eve was still experiencing flu-like symptoms and extreme weakness, and on one occasion collapsed and fell into convulsions for several days. Doctors could not determine the cause of her illness, but after undergoing exploratory surgery, she learned that she had developed chronic liver failure.

In the fall of 1978, after seeing a television news program on the human health effects of Agent Orange, the DeRocks realized that herbicides may have been causing their problems. They stopped consuming the butter and meat stored in the freezer from the cattle they had butchered, and their health began to improve. Although Vern, who was retired, was still on International Paper's payroll, both were bitter about the company's spraying of their farm. "International Paper always wanted to buy our land," she told the *Willamette Week*," but I never thought they'd try to kill us."[26]

In a December 1981 interview, Mrs. DeRock described how she and other family members who had eaten the contaminated beef were still feeling the effects of the herbicide spraying. She was continuing to suffer from brain damage, and for a long time "couldn't think forward or backwards, or complete a verbal sentence." Her nine year old granddaughter had her ovaries removed, and her brother had died two months earlier for no apparent reason. Her husband had also passed away, she told me, but she could not remember when.[27]

Elsewhere along Oregon's coastal range communities, one hears the same stories over and over. A group of women in Alsea repeatedly suffered miscarriages shortly after herbicides were sprayed in their area. In Rose Lodge, Norma MacMillan saw her husband die and her daughter fall victim to baffling and incurable illnesses less than a year after a cloud of herbicides drifted into her yard and killed her garden.

A lengthy investigation of forest spraying along Oregon's remote and lovely coastal communities, reported by Phil Keisling in a December 1979 series in *Willamette Week*, revealed in these and other stories the same shocking patterns as in other areas subjected to herbicide spraying. He found that "the similarities between most stories were striking," with the "themes that consistently recurred" being the "uncanny" disappearance and death of wildlife, the dead and deformed fetuses of animals and humans and the persistent flu-like sicknesses, respiratory disorders, and gastrointestinal illness that infected entire communities.[28]

In the rugged Oregon coastal area of Lincoln County, Drs. Renee and Chuck Stringham reported that between 1975-79, two of the babies they delivered were anencephalic, "born without brains, with heads like a frog," an extremely rare condition, giving the county an incidence rate 13 times the national average. The extraordinarily high rate of such birth defects, according to one doctor, made local women afraid to have babies, and prompted the Lincoln County Medical Society to call for a ban on the use of phenoxy herbicides until they could be shown to be safe. Sixteen out of 18 physicians in northern Lincoln County signed a petition calling for a moratorium on aerial spraying.[29]

Jack Kinman, who runs a small shop and gasoline pump on Highway 101, also owns a small plot of land settled by his grandfather in 1876. Several years ago, he let a local pest control official talk him into allowing his pasture to be sprayed after being assured that the chemical would not harm his horse. The county officer swore it was harmless and even offered to drink a cup of it (although he never did). The day after the field was sprayed, Kinman was dragging off his dead horse, the animal still foaming at the mouth. His horse is not the only animal he misses; he feels that the spraying has "killed all kinds of birds. You used to run into droves of grouse and you just don't any more. And you used to see bald eagles all the time."[30]

Additional research on the effects of herbicide spraying in Oregon turned up many other instances of people sickened and fetuses killed or damaged. Four out of 9 mothers who were exposed to spray drifts suffered miscarriages while living within the Suislaw National Forest, two of them twice, and one on 16 occasions.[31]

Several other individuals in the same area exposed to spray from Weyerhaeuser and other companies experienced uterine bleeding, severe headaches, lung congestion, difficulty in breathing, nausea, dizziness, stomach cramps, lethargy, diarrhea, bleeding noses, neuritis, acne, irritation of eyes, nose, and chest, and loss of resistance to such infectious ailments as flu, bronchitis, colds, pneumonia, and boils. Affadavits submitted in court by Oregon's Citizens Against Toxic Spray (CATS) cited these and numerous other cases of residents experiencing such severe health symptoms. In some cases, where blood tests were taken, traces of Silvex and 2,4,5-T were found.[32]

Other reports from Oregon tell of 63 horses dying, and 11 of 17 sheep producing dead or deformed lambs, after grazing on pastures sprayed with 2,4,5-T and 2,4-D. In one sprayed forest, the body count included two spotted owls—an endangered species—and a mother robin in her nest with three chicks.[33]

Most of these occurrences have been dismissed by government and industry as isolated incidents that were in no way related to herbicide use. But the pattern of aerial spraying followed by miscarriages became so predictable that a group of women in Alsea, Oregon, organized by Bonnie Hill, started keeping diaries which showed a definite correlation between the spraying and their spontaneous abortions. Nine women in the area were able to show that in the period between 1972 and 1978, the 13 miscarriages they experienced directly followed the spraying of 2,4,5-T.[34] When this data was submitted to EPA, the agency accepted it as valid and used it as a basis to later restrict some uses of the chemical.

But the spraying of similar herbicides continues, with equally tragic results.

Across the USA, A Rain of Death

Around Ashford, Washington, where the surrounding forest and the state highway are regularly sprayed with 2,4-D, a total of 12 pregnancies from January 1979 to March 1980 resulted in 10 miscarriages.[35] At a meeting with women from the community, a timber company chemist reportedly told them that "babies are replaceable," and that they should "plan their pregnancies around the spray schedule."[36]

In the Swan Valley area of western Montana, nine out of ten pregnant women—all apparently healthy—suffered miscarriages during a one year period ending in the fall of 1979, during which time 2,4-D was heavily sprayed along roadsides to kill weeds. The area's only successful birth was by a woman who left the state and returned home to have her child.[37] Six of the women living on a 15 mile stretch along Highway 83 had seven miscarriages and only one live birth in the two-and-a-half year period ending in February 1980.[38]

In Louisiana, 2,4-D spraying of roadways by state highway crews and utility companies has been linked to numerous human miscarriages as well as to grotesquely deformed offspring of horses, cows, and other livestock that graze along the roads. The problem began to attract attention after a state senator, driving behind a state highway department spray truck, inhaled the chemicals and became violently ill for two weeks. A few weeks earlier, in March 1979, farmer George Maggio spilled a small portion of 2,4-D and another chemical, carnex powder, in his lap while mixing them together. Although he washed immediately he became extremely ill and died a few days later.[39]

The spraying of 2,4,5-T and 2,4-D is not confined to the West but takes place throughout the United States and much of the underdeveloped world as well. It has become a controversial issue in Arkansas, for example, where 2,4,5-T has been used extensively on rice crops, and by the U.S. Forest Service on some 20,000 acres a year of national forests to kill off oak trees and other hardwoods that compete with more commercially valuable softwood varieties.[40]

An effort to document the harm being done by 2,4,5-T and 2,4-D has been waged for years by Erik Jansson, of Friends of the Earth (FOE) in Washington, D.C. Jansson organized a coalition of environmentalists and other organizations to fight this aerial spraying, and on 26 May 1978, in concert with Citizens Against Toxic Sprays (CATS), FOE petitioned EPA to ban 2,4,5-T, presenting lengthy and detailed evidence of its hazards to public health and the environment.

According to the affidavits and other material submitted by Jansson, 16 out of 23 women located who reported that their skin was wet by, or their gardens or water supply contaminated with, 2,4,5-T lost their babies, a miscarriage rate of 70 percent!

Among the cases listed by FOE of unborn children apparently killed by their mother's exposure to 2,4,5-T were the following:

> Six out of 8 women in Fayetteville, Arkansas lost their babies; and one child was born with a cleft head and no legs, after 2,4,5-T was sprayed on a field above a spring supplying drinking water to 20 people.

> In a small settlement in northern Wisconsin of less than 40 families, all 5 pregnant women lost their babies after spray drifted over their homes. The high rate of spontaneous abortions and stillborns was continuing in the town, and there has been a severely deformed baby born. Dead and deformed wildlife, pets, and livestock have also been found.[41]

In a 20 July 1977 letter to CATS, Gladys Pavel of Beaver, Washington described the effects of herbicide spraying in her area, including the massive loss of wildlife and domestic animals:

I first noted the effects in July 1967 after a large forested area on the North Fork of the Calahwa had been sprayed. On the river I found dead ducks, sick ducks, and dead fish. In the woods, I found dead and dying birds, squirrels, and chipmunks. Blackberry patches reeked with the smell of herbicides. About ten days after the spraying I found a dead doe and two fawns in the same area.

We lost our dog, which up to that time had been very healthy and active . . . Out of 64 chicken eggs, only 15 chicks lived; three of these are deformed.

. . . the birds are few. Two years of spraying on the berries killed the big flocks of band tail pigeons which had numbered in the thousands . . . The grouse population has diminished. Deer and elk are no longer plentiful. No longer the great numbers of ducks nesting, and the ponds are quiet of frogs singing. Wild bees and bee trees are no longer to be found.

Moths and butterflies are gone . . . Haven't seen a titmouse (small bird) for the past eight years . . . Seldom see the tiny brown wrens along the creeks or in their usual nesting areas . . . I found dead swamp robins near the house . . . In sprayed areas there is silence—no squirrels, birds, chipmunks, or rabbits. The raccoon population is practically nil compared to ten years ago. The wild river animals are no longer a threat to poultry.[42]

When Silvex (from a Bureau of Reclamation spray program) drifted onto a New Mexico farm, the rancher reported a loss of 38 out of 100 cattle; 40 percent of his pregnant cows aborted; others were born deformed; the rest of his herd had to be quarantined because of high residues of the chemical. He and his daughter suffered headaches and vomiting.

In scenic Wisconsin, the heart of America's dairyland, over a million acres are sprayed with 2,4-D each year to kill brush and weeds on farm, forest, and public utility lands. The usual results have ensued: reports of dead fish and wildlife and physical distress among humans, including birth defects, miscarriages, headaches, and nausea.[43]

Reports of excessively high rates of miscarriage, birth defects, and other reproductive disorders after areas were sprayed with 2,4,5-T came from New Zealand in 1972, Pittsville, Wisconsin in 1975, and Lincoln and Benton Counties in Oregon from 1970-1978.[44] In Colombia in 1975, 2,4,5-T was reported to be the cause of 70 miscarriages, according to Congress' General Accounting Office.[45]

The extensive spraying of forests, power line right-of-ways and other wooded areas that has taken place in recent decades also raises the prospect of the continuing spread of dioxin contamination through forest fires and the handling and burning of chopped wood or brush that has been sprayed.[46] Indeed, controlled burning has often been used with spraying of 2,4,5-T as a forest "management" technique.[47] Cooking meat and vegetables at oven temperatures can also release dioxin if the food is contaminated with 2,4,5-T, as often is the case with the meat of "game" birds and animals and livestock foraged on sprayed pastures and grazing lands.[48] Herbicides can thus continue to cause human health problems years after they have been sprayed.

A group of six children on a church camping trip learned this lesson the hard way when they had to be hospitalized in 1979 after apparently being exposed to 2,4-D. While swimming in a lake in California's Los Padres National Forest, the children's skin suddenly began to burn and itch, and they ran out of the water crying. After developing nausea and headaches, they were hospitalized and diagnosed as having experienced pesticide poisoning. A discarded sign was found in the area warning that herbicides had recently been sprayed there, and later testing showed that the lake water contained two parts per million 2,4-D.[49]

Chemical Warfare And The Resistance Movement

Many other pesticide and herbicide spraying programs take place throughout the country, often without the knowledge of the local citizens. When people do learn about what is happening, they often try to organize opposition to these programs, as has happened recently in gypsy moth suppression projects in Virginia and West Virginia, and spruce budworm control programs in Maine. Rarely, however, do local citizens have any real success in influencing the policy makers, since such decisions are usually made by industry-dominated bodies such as the U.S. Forest Service and state agricultural agencies.

Just as tobacco companies continue to deny that cigarettes cause lung cancer despite overwhelming evidence to the contrary, today the numerous accounts of herbicides causing harm to human health are dismissed as coincidental or nonexistent. For years, the government has ignored or denied these stories—first-hand accounts by people of how herbicides have harmed them or their animals—which it likes to refer to disparagingly as "anecdotal" evidence that is considered unreliable and unscientific. Yet such citizens' reports are so numerous, widespread and consistent in their accounts of their symptoms and the effects of spraying on their areas that it is impossible not to accord them a large degree of credibility.

But the government and the chemical and timber industries continue to deny that a human health problem exists except in the minds of the victims. Industry spokesmen delight in claiming that much of the opposition to 2,4,5-T stems from its being deadly to marijuana plants![50]

These stories of Americans attempting to protect themselves from the actions of their own government are reminiscent of wartime tales of citizens fighting an enemy occupation army. As Phyllis Cribby of Grants Pass, Oregon wrote in a 3 January 1980 letter to the House Subcommittee on Forests:

> This past year, residents near scheduled spray sites spent countless hours studying spray plans and units, writing protests and petitions, suggesting alternatives, and making fruitless appeals. When spray day came, some left their homes. In one area of Southwest Oregon, the residents organized site occupations, risking arrest and being sprayed in a desperate effort to protect their area and families. Their reactions are labelled "emotional," yet their concerns for human health are the same as those voiced by eminent scientists: many of them have a practical knowledge of the forest and its ecology and their concerns for it should not be taken lightly.[51]

In order to halt these herbicide spraying programs, citizens have had to resort to tactics reminiscent of the civil rights marches and demonstrations of the 1960's. In June, 1977, protesters in Lincoln County, Oregon moved into a forest owned by Publisher's Paper Company (a subsidiary of the Times-Mirror Company, owner of *The Los Angeles Times* newspaper) and succeeded in stopping a spraying program that was to be carried out by helicopters. As reported by Ronald B. Taylor in, ironically, the *Los Angeles Times*, the spraying was rescheduled for a month later, and in the early morning hours of 6 July, the group of 70 or so people returned:

> Defying the warnings, the protesters once again went into the 350-acre, brushy mountain site near Rose Lodge. During the confrontation, 20 protesters were arrested for trespassing and others were sprayed by helicopters laying down swaths of 2,4-D.[52]

(Charges against the arrested protesters were later dropped after the judge ruled that the prosecution was guilty of collusion with the paper company.)

Company officials, however, were not concerned about possible harm to the protesters, as Dow had assured them the chemicals were harmless:

> Three Publisher's Paper officials acknowledged that the demonstrators had been sprayed and that orders had been given to spray the area even if demonstrators were present. A company spokesman said the order was given only after consultation with scientific experts, including Dow scientists, who told them the material was safe and would not harm anyone, even if applied directly.[53]

Indeed, some residents of forested areas have become so frustrated with their unsuccessful attempts to halt the spraying of their areas, they have been forced to resort to the traditional American right of self-defense. In Keno, Oregon, a spraying campaign was stopped in September 1978, after three men blocked a road with their automobiles and fired a shotgun.[54]

In Humboldt County, California, according to the local sheriff's office, there have been numerous threats to shoot down herbicide-spraying helicopters.[55] Similarly, residents of the village of Honcut in Butte County, northern California, consider themselves victims of "chemical warfare" since their homes and gardens have been directly sprayed. Says one inhabitant of that town, "They're not going to take it. They'll shoot 'em out of the air if they come over. That's exactly what they said they'll do and they'll do it."[56]

In March 1980, a group of residents of Oregon's Applegate River Valley, citing studies in the scientific literature demonstrating that 2,4-D causes cancer, birth defects, and genetic changes in animals, attempted to halt the Bureau of Land Management spraying program by occupying an area about to be sprayed. Those exposed to direct spray and wind drift of the chemical reported experiencing a number of immediate and obvious health effects, including peeling of facial skin, headaches, dizziness, nausea, severe abdominal cramps, physical weakness and burning sensations of the throat and nose. Some of these symptoms persisted for weeks; no one can foretell the long-term health consequences of such exposure.[57]

In August 1980, over 30 women and children from the town of Index, Washington threw themselves across the railroad tracks in a successful effort to stop the Burlington Northern Railroad from spraying rights-of-way with 2,4-D to kill brush by the tracks. The railroad backed down—temporarily at least—after the people formed a living barrier and stopped a 2,400 gallon chemical tank car that was rolling down the rails.[58]

In a joint letter to the heads of EPA and the Federal Aviation Administration, Eric Jansson of Friends of the Earth raised the interesting prospect of a "spray them back with poison" campaign:

> I would like to suggest that if it is legal to spray people
> with poisons from aircraft and ground rigs without their
> permission, it would also be quite legal for anyone,
> including spray victims, to walk into your offices with
> pesticide cans and spray you with poisons.[59]

So far, none of these people have taken matters as far as a group of guerrillas in South America in 1976, when the Guatemalan Army of the Poor launched a midnight attack at the town of LaFlora and destroyed 22 crop-dusting planes.[60] But throughout the West, stories abound of plans by Vietnam Veterans and others to take similar action if, as expected, EPA reinstates the large scale forest spraying of 2,4,5-T.

Managing The Forests By Destroying Them

In the controversy over these herbicides, perhaps the greatest irony of all is that these chemicals, which are sprayed to speed the growth of commercial timber, may actually be damaging and destroying the very forests they are intended to help.

Congressman Jim Weaver (D—Ore), Chairman of the House Subcommittee on Forests, has warned of the dangers of using deadly chemicals to "manage" forests: "it is a very real possibility...if we put all our eggs in this phenoxy herbicide basket, 30 years and 40 years from now (we'll) wake up with a dead forest":

> We are taking an enormous gamble with our entire resource and our future timber economy, and I am appalled that the Forest Service and Bureau of Land Management continue to use herbicides under these conditions.
> ...we don't know what these plant killers, these phenoxy herbicides might be doing to the structure of the conifer, the tissues. It may be weakening them for disease 30 years down the pike. We are spraying millions and millions of acres in this country, and who knows but 20, 30, 40 years later, these trees will be enormously suscep- tible to an insect infestation and various diseases...we are changing the ecosystem which may also be weakening these trees.[61]

Forest expert William C. Denison, a professor of botany and plant pathology at Oregon State University (OSU), testified in federal court that the herbicide spraying was killing off and disrupting a necessary part of the forests' natural life-cycle. When unwanted brush, trees, and other vegetation are removed, they can no longer play their vital role of providing nitrogen for the soil and habitat for small mammals, insects, and fungi which are necessary for the maintenance of a healthy forest ecosystem.[62]

Indeed, the destruction of "weeds" that serve as natural fertilizing agents and provide nitrogen to the soil through their roots, besides representing a long-term threat to the growth of the desired trees, has forced lumber firms to drop nitrogen-rich fertilizer into such areas. These urea fertilizers are made from an increasingly scarce and expensive resource, natural gas, and cost from $50 to $75 an acre to apply.[63]

Although herbicides are intended to speed the growth of conifers, there is evidence that they have just the opposite effect. A 1961 study showed that 2,4,5-T and 2,4-D slowed the growth of conifers for years after being applied. Another study found that more than a third of the crop was destroyed after brush that had been killed by herbicides broke off under the weight of snow and crushed the young trees.[64] [65]

In a June 1979, editorial in the highly-respected magazine *BioScience,* published by the American Institute of Biological Sciences, Frank Egler writes: "I do not know of a single, comprehensive long-term research project on forest range or right-of-way vegetation that will support, condone, or even excuse the present massive indiscriminate spray programs, which are so lucrative for industry and their 'brain-washees.'"[66]

When Congressman Weaver's Forests Subcommittee held hearings in Eugene, Oregon in January 1980, various expert witnesses effectively demolished the economic justification for aerial spraying of herbicides. Included was the Northwest Coalition for Alternatives to Pesticides, which had commissioned an independent, year-long study on the economic efficacy of herbicide spraying.[67] The report's author was Dr. Jan Newton, a highly respected economist and research consultant who has done contract research for such federal agencies as the U.S. Department of Agriculture (USDA) and Health, Education, and Welfare (HEW, now HHS). She found that data demonstrating the advantages of herbicides use was virtually nonexistent; benefits were consistently exaggerated, while adverse effects were ignored or downgraded. *Not a single study could be found demonstrating increased growth of trees because of aerial application of herbicides.* "Some of the research I looked at was the worst I've ever seen in my life," she said, "if a sophomore in high school had turned in some of the papers I read, I'd flunk them." Dr. Newton testified before the Subcommittee that evidence to justify "an economic case is almost entirely nonexistent," and that "the real . . . purpose for herbicide use is to claim higher future timber yields in order to justify increased harvest(s) . . .in the present."[68]

She also stated that the U.S. Forest Service and the Oregon Department of Forestry in their January 1979 joint study had "purposefully misled the public concerning the employment impacts of herbicide use." She called their predictions of massive job losses if 2,4,5-T were banned "preposterous," and said that "there is absolutely no basis for (these) exaggerated claims about job losses due to restrictions on herbicide use."[69]

The bias of the report is hardly surprising in light of the fact that of the 14 "experts" consulted, 10 were employees of lumber companies that were using 2,4,5-T; 3 were USFS employees, which also was using 2,4,5-T; and the other was Mike Newton, a zealous defender of herbicides at OSU. Concerning the use of workers instead of herbicides to clear brush, she found that "...a great deal of misinformation has been perpetrated...It is quite possible that manual release is cost competitive with chemical treatments if all the relevant direct costs are included."[70]

Indeed, a major argument repeatedly cited by industry in its fight to retain 2,4-D and 2,4,5-T is the claim that it is much cheaper to spray poisons to get rid of unwanted vegetation than to use manual labor. Yet, the economics of this alternative method may be comparable to spraying. Data from the Willamette National Forest in Oregon show that spraying herbicides costs an average of $82.86 an acre, while manual removal, according to the Northwest Coalition for Alternatives to Pesticides, was between $70 and $110 an acre, costs which should decrease as crews gain experience and better devices are developed.[71]

Although the U.S. Forest Service has estimated that manual clearing costs would come to $376 an acre, in 1978, 3,500 acres of USFS land was hand cleared for an average of $106 an acre. Herbicide spraying cost just a few dollars less—$100 an acre.[72] A California State Department of Economic Development study reports that replacing spraying aircraft with people would raise the cost of producing lumber by less than two percent.[73] Moreover, the costs of aerial spraying will continue to rise as the price of oil climbs.

What is not in doubt is that using people instead of poisons does not contaminate the environment, uses far less gasoline and oil, and provides much-needed employment. Yet, the industry continues to resist innovative methods of managing forests.

This, then is the ultimate irony: in order to justify cutting more trees on the basis of increased future yields, the timber industry and the U.S. government are poisoning and damaging the forests in ways that probably *reduce* forest productivity over both the short and long run. Yet, because industry has such a vested interest in preserving the fallacy of herbicide benefits and thus maintaining increased timber cutting for the short term, the poisoning of the forests—and of the people—continues apace.

Science For Sale?

One reason for the lack of data on the adverse effects of herbicides and alternatives to their use is that the forest science establishment also has a vested financial interest in the continued and expanded use of such chemicals. Many of these people receive funds directly from the chemical and timber industries, and they do not bite the hand that feeds them.

A good example is the situation that prevails at Oregon State University (OSU), which has been carrying on a love affair with industry for many years. OSU is often used to give "objective, scientific" support and credibility to the claims of industry, despite the fact that the university and its faculty members receive large sums of money from the users and producers of these chemicals.

In his December 1979 exposé in the *Willamette Week*, Phil Keisling explains why it is, as one conservationist put it, "OSU researchers are supposed to dispassionately study herbicides; but it's a foregone conclusion they'll find them safe."[74] In 1978, the timber industry granted some $130,000 in the form of contracts and donations to OSU's forest research laboratory. After the School of Forestry decided to construct a new building several years ago, $150,000 was raised from the business community and other private sources. When the university recently kicked off a drive to raise money to complete a new conference center, the Weyerhaeuser lumber company was asked to contribute $45,000.

But these sums, substantial as they are, pale in comparison to the real conflict-of-interest at OSU which cannot help but erode what academic freedom and independence it has left, as well as its ability to be objective about herbicides. State law requires that a special tax be assessed on lumber according to how much is cut, with these proceeds and matching funds going to the university's forestry department. In 1978, this arrangement raised a total of $1,856,000 for the laboratory, a figure amounting to 70 percent of its budget. Therefore, any reduction in timber cutting—much of which is justified by herbicide spraying—would cut directly into the university's budget. Specifically, an 11 percent drop in timber yields—which the USFS and the Oregon Forestry Department predicted would follow a ban on 2,4,5-T—would mean a loss of $200,000 for the school.[75]

But lest stockholders feel that industry is being profligate with company money, they can rest assured that the timber giants know exactly what they are doing. Their money is well spent, and they get an excellent return on their investment. In late 1979, shortly after OSU issued a report attacking EPA's study that led to the suspension of 2,4,5-T, the chemical's main manufacturer, Dow, sent the University's foundation $5,000.

OSU's most outspoken and zealous defender of herbicides is Dr. Michael Newton, a professor of "forest ecology." He first gained notoriety in 1973 after a bizarre incident in which he wrote to the Air Force suggesting that the 2.3 million gallons of leftover Agent Orange be made available to spray U.S. forests. (The chemical, a 50-50 mixture of 2,4,5-T and 2,4-D, was banned in 1970 for use as a defoliant in Vietnam after causing numerous birth defects, miscarriages and other health problems among the inhabitants of the country: see pages 57-61). Later, it was discovered that Newton had obtained supplies of Agent Orange and had used it on 350 acres of forest land even though he had not been given a permit to do so. As a result, EPA confiscated his remaining stockpile.[76]

Newton and other OSU professors have frequently acted as apologists for herbicide use and appeared on behalf of the USFS and other pesticide users and makers at hearings and lawsuits. In 1979, a few days after EPA restricted forest spraying of 2, 4, 5-T, Newton and his wife flew at Dow Chemical's expense to its Midland, Michigan headquarters to help the firm draft statements, news releases, and other material opposing the action. Newton acknowledges having received "several thousand dollars" from Dow for consulting work. Another giant chemical firm, Monsanto, contributed $14,000 to OSU to finance research by Newton.

The university officials apparently see no improprieties or conflict-of-interest in Newton's activities, but their treatment of him contrasts sharply with that of William Denison, a botany professor who has been one of the few faculty critics of OSU's policies willing to speak out. In mid-1977, he was the only OSU faculty member to testify on behalf of CATS in its lawsuit asking for a ban on 2,4,5-T spraying, with Newton and other professors appearing on behalf of the USFS. That fall, when classes began, Denison was asked to resign.[77]

Forest Service employees who protest the use of herbicides are also shown the door. In Siskiyou County, California, three employees who refused to take part in a program to spray 2,4-D over 385 acres of timber lands were summarily fired by District Ranger George Karper.[78]

These and numerous other incidents have sent forth a clear message to scientists and bureaucrats involved in pesticide policy: the ticket to a successful career requires an uncritical view of these chemical poisons.

At other universities, particularly those that are state-run, similar situations prevail. Close financial ties to large corporations result in the neglect of projects that would benefit consumers, small farmers, agricultural workers and the environment, in favor of research supported by chemical companies and large agribusiness corporations. For example, in 1978-79, $689,000 was provided to the University of California by pesticide manufacturers. Much of the university's huge agricultural experimentation budget goes to pesticide-oriented projects. In a 1977 study, the Environmental Defense Fund found that over 90 percent of U.C.'s Cooperative Extension Service's recommendations on pest control advocated the use of chemical pesticides.[79]

Dioxin: The Most Toxic Chemical Known

In contrast to lack of analysis of the impacts of herbicides on forests, the effects of these chemicals on test animals have been extensively studied; and they have been shown to be extremely toxic not only to animals but to humans as well. [80]

There are some 75 different types of compounds called dioxin, and the most toxic is known as tetrachlorodioxin, or TCDD.[81] This is the compound that occurs in several well-known and widely used chemicals, such as the herbicides, 2, 4, 5-T (2, 4, 5-trichlorophenoxyacetic acid) and Silvex. Millions of pounds of both have been (and in some cases still are) used each year by the U.S. Forest Service, the timber industry, farmers, ranchers, and homeowners to clear away unwanted vegetation. Silvex has been used mainly as a weed killer against dandelions, chickweeds, and other weeds on home lawns, gardens, and around ornamentals, as well as in orchards and sugarcane fields.

The major manufacturers of 2,4,5-T are Monsanto, whose motto is, "without chemicals, life itself would be impossible," and Dow Chemical, which in 1977 earned $12 million from sales of the compound.[82]

Silvex and 2,4,5-T, (as well as 2,4-D) are members of a chemical family called phenoxy herbicides that are so toxic they were produced at Fort Detrick, Maryland during World War II for possible use in chemical and biological warfare. In experiments on laboratory animals, dioxin has caused death, cancer, birth defects, and other disorders in doses so small as to be almost inconceivable, in some cases a few parts per trillion. When pregnant test animals have been fed food with levels of dioxin at just one part per trillion, they have produced an unusually large number of stillborn and dying offspring.

Because the U.S. Environmental Protection Agency (EPA) refuses to ban 2,4,5-T and Silvex, millions of pounds of these chemicals continue to be sprayed on rice fields, livestock grazing lands, and other food growing and inhabited areas. Their application on forests was restricted in 1979 after their use was linked to numerous miscarriages experienced by women in sprayed areas.

The incredible toxicity of 2,4,5-T and its inevitable contaminant, dioxin, has been confirmed in numerous laboratory tests in which similar effects to those caused in humans have been induced in several species of test animals, sometimes in doses so low that special and extremely sophisticated laboratory equipment is required to detect and measure it. In May 1980, an EPA official testified in a legal proceeding that "we have not been able to establish a no-effect level (one that produces no tissue change in laboratory animals). There is reason to be concerned about exposure at any level."[83] Later, on 29 May, in a news release, EPA stated that "...TCDD, which is present in 2,4,5-T..., produces serious life-threatening effects on the fetus at minute doses including the lowest dose tested in many studies."

Toxic At Every Level Tested

Because dioxin has been found to be toxic to laboratory animals at every level tested, no matter how small, no one really knows how dangerous it really is. The amount found in 2,4,5-T can be as small as one-tenth of a part per million, which is the equivalent of several spoonsful in an olympic-size swimming pool. Government tests conducted in the 1960's using pregnant rats and mice fed 2,4,5-T produced birth defects in 40 percent of the fetuses from those mothers given the lowest dosages, and 90 to 100 percent were deformed among the rodents whose mothers had been fed the highest amounts. These results were kept secret for two years before being released in 1969. They also showed the chemical to be "unequivocally carcinogenic."[84]

Dioxin's toxicity makes it difficult to use in conducting cancer research, as it tends to kill the test animals before they can produce tumors, even when given in a few parts per trillion. But when the laboratory subjects do survive long enough to complete the experiment, dioxin causes cancer in the lowest doses imaginable.[85]

In its tenth annual report, issued in December 1979, the President's Council on Environmental Quality (CEQ) noted,

> Research since 1970 has shown TCDD (dioxin) to be one of the most toxic substances ever studied. Fetotoxic (damaging to a fetus) effects have been seen in rats at levels of exposure at least as low as 10 parts per trillion. In monkeys, toxicity to the offspring of treated animals has been seen at doses as low as 2.5 parts per trillion. TCDD has also been shown to be carcinogenic to test animals at doses as low as 2.2 parts per billion.[86]

In studies conducted by Dr. James Allen, of the University of Wisconsin Medical School, rats were fed a diet containing dioxin at levels as low as 5 parts per trillion, and a significant number developed cancerous tumors.[87]

In June 1980, two new studies conducted by the National Cancer Institute were released which also concluded that dioxin causes cancer in laboratory animals, and helped corroborate the results of Dr. Allen's research over the last 20 years. Almost half of the test group—48 out of 100—also developed toxic hepatitis.[88]

One of the pioneers in research on dioxin, Dr. Matthew Meselson of Harvard University, has described how the minutest amounts imaginable of dioxin can kill animals: "... if you feed a guinea pig one-billionth of its weight with dioxin, this will kill the guinea pig. One part per billion. Yet we do not know the sensitivity of humans."[89]

Cancer researcher Dr. Samuel Epstein of the University of Illinois, who has studied hazardous chemicals for some three decades, calls dioxin "the most toxic synthetic chemical compound known. TCDD is also the most potent known carcinogen and teratogen which ... induces cancers and injures or kills embryos [in laboratory tests] at doses as low as one-billionth of a gram per kilogram of body weight a day."[90] Dr Epstein also described how the ability of dioxin to harm a fetus is multiplied or "synergized" if 2,4,5-T is applied simultaneously, and how dioxin suppresses the body's immunological system, increasing susceptibility to bacterial infections.

In other experiments on animals done by Dow Chemical Company, the National Institutes of Health and other laboratories, dioxin and 2,4,5-T have caused lung and liver tumors and leukemia. Often, it was found difficult to induce tumors in such tests because these chemicals were so toxic they frequently killed the animals before the cancers developed.[91]

Another effect of dioxin on animals and humans is that it appears to damage the thymus gland, which plays a key role in the body's immune system, thereby lowering resistance to infection. This would cause people to experience the symptoms of common ailments and thus would not, in most cases, be attributable to dioxin. This may be the reason why there have been numerous reports of unusually large numbers of flu and other virus attacks on people living in areas that have just been sprayed, even in summer months. Dr. Granville Knight, of Santa Monica, California, found that 70 percent of his patients who contracted such viruses had measurable amounts of herbicides in their blood.[92]

Dr. Matthew Meselson's research gives an alarming indication of the widespread incidence of dioxin contamination. He has found this chemical in mothers' milk, in beef samples and in a codfish from Boston.[93] A 1973-74 EPA study found dioxin to be present in animals in an Oregon national forest sprayed with 2,4,5-T, but the agency suppressed the data until it was forced to release it during a 1977 court hearing on herbicide spraying.[94] Other scientists have found dioxin in deer and other wildlife eaten by hunters, and in cows' milk. A Florida State University study found 2,4,5-T and/or Silvex in the urine of over one third of the dormitory students and one quarter of the football team, as well as in food and city water.[95] Among the major uses still allowed for 2,4,5-T are applications to rice crops and livestock rangeland.

One of the major concerns about the presence of dioxin in food—even in tiny amounts—is that it appears to have a cumulative effect on health. In another set of experiments conducted by Dr. Allen on monkeys, it did not matter whether small doses were given over a lengthy period or large doses were added all at once to the diet; the effects were ultimately the same: once the lethal level was reached, the monkeys died. This suggests that over an extended period of time, the combined effects of exposure to even trace amounts can damage or kill whichever organism ingests or is exposed to it, including humans.

This problem is accentuated by the fact that dioxin is stored and becomes more concentrated in organisms as it moves up the food chain. This means that small amounts found on plants can build up to higher levels in the bodies of animals that consume such vegetation. Thus, humans who eat beef or venison can get a concentrated dose of dioxin that has built up over several years. Because dioxin is secreted in mothers' milk and can even cross the placenta and enter the fetus, it represents an especially serious danger to young and unborn children.

This data drawn from laboratory experiments on test animals would seem to verify the voluminous accounts and "anecdotal evidence" of herbicide spraying coinciding with the killing and deforming of fetuses and the outbreak of flu epidemics in numerous areas of the U.S. What remains unclear is why EPA still permits the widespread use of these chemicals, which may be steadily building up in our bodies and approaching lethal concentrations.

2,4-D: The Nation's Most Widely Used Herbicide

2,4-D is the nation's most popular and widely used weed killer, with 70 to 80 million pounds a year being used in 1,500 products, including Weed-B-Gon, Formula 40, Weedone, Estron, Weed-Rhap, and Brush-Rhap. It is used by home-owners, farmers, timber companies and government agencies to kill weeds and unwanted vegetation on millions of lawns and golf courses, along highways, forests, rangelands, pastures, rivers and streams and on cereal crops, sugar cane and rights-of-way.

Although it is uncertain to what extent, if any, 2,4-D (2, 4-dichlorophenoxyacetic acid) contains certain dioxin contaminants, the toxic effects of the herbicide have long been documented. In 1962, Rachel Carson described in *Silent Spring* how exposure to 2,4-D had been implicated in causing paralysis and severe neuritis, and how the chemical tended "to imitate X-rays in damaging chromosomes."[96] In 1969, the Mrak Commission, in *Pesticides and their Relationship to Environmental Health,* a report compiled for the Secretary of HEW, recommended that 2,4-D be "immediately restricted" and that human exposure to it be minimized to the greatest extent possible because of studies showing that it caused tumors and birth defects in laboratory animals.

Subsequent studies have shown that human exposure to 2, 4-D can cause neuropathy and changes in the central nervous system.[97] It has also induced birth defects and cancer of the lymph glands and mammaries in laboratory animals. In northern Sweden, forestry, sawmill and paper pulp workers regularly exposed to 2,4,5-T and 2,4-D were found to have 5 times the risk of contracting certain cancers than unexposed individuals.[98]

After reviewing all of the laboratory data on 2,4-D and another herbicide often used along with it—Picloram—Dr. Melvin D. Reuber of the National Cancer Institute (NCI) concluded that both chemicals cause cancer and reproductive problems.[99] Ruth Shearer, of NCI, also conducted a study of all of the available data on the chemical and concluded that "2,4-D, in combination with other substances, clearly promotes cancer; some evidence suggests it also may initiate it."[100]

In 1971, Dow Chemical, the major manufacturer of 2,4-D, ran a series of tests on the chemical to determine its ability to cause birth defects. Although an unusually high number of serious congenital defects were produced in the offspring of the test animals, such as skeletal deformities and water on the brain, Dow's researchers concluded that the chemical was not fetus-deforming![101]

A 1972 Russian study showed that farm animals grazed on pastures sprayed with 2,4-D experienced "spontaneous abortions, mummification of the fetuses, high sterility, a large number of stillborn lambs, and a decrease in the sexual activities of the males and the quality of the sperm." Another study showed that severe reproductive problems occurred in animals exposed to the chemical in quantities approximating food tolerance levels for humans.[102]

One reason, 2,4-D has been allowed to remain in use is that EPA has ignored and/or misinterpreted data on its ability to cause cancer. After a lengthy investigation, the Senate Sub-committee on Administrative Practice and Procedure issued a report in December 1976 stating that "a clear example of EPA's failure to evaluate data resulted in the agency's determination that there was 'sufficient' data for 'full' re-registration of the pesticide 2,4-D:"

> Yet a summary report on the study in EPA's files stated
> that there was "increased tumor formation" in the rats. An
> independent pathologist, who reviewed the raw data on
> the study at the request of subcommittee staff, concluded
> that 2,4-D "is carcinogenic (cancer-causing) in rats."[103]

In June 1981, four officials of a major testing laboratory, Industrial Bio-Test (IBT), were indicted on charges of fabricating and falsifying test results in order to prove the safety of several pesticides and other chemicals. When EPA tried to examine IBT's tests on 2,4-D, the lab admitted that its records had been shredded.[104] (See pp. 121-23)

The only mention of 2,4-D in EPA's list of suspended and cancelled pesticide is a 1967 regulation requiring that "products bearing directions for use on small grains (barley, oats, rye, or wheat) must bear the following label precaution: 'Do not forage or graze treated grain fields within 2 weeks after treatment with 2,4-D.'"[105]

White House and National Park Service (NPS) gardeners have regularly used 2,4-D each year (at a rate of 2,000 pounds in 1979) to control dandelions and other "weeds" on the White House lawn, Washington Monument grounds, the Mall, East and West Potomac Parks, and other areas frequented by the President's family and millions of tourists in Washington, as well as in other park lands.

In November 1980, the NPS announced that because of doubts about 2,4-D's safety, the agency would greatly reduce its use: it would no longer apply the chemical routinely in its 325 parks and recreation areas, but only when no alternative weed control methods were available. It remains to be seen how much of an actual reduction in use this will amount to, since government agencies have long contended that there are no effective alternatives to using 2,4-D. Dow Chemical responded to the announcement by calling the decision "distressing" and claiming that there had been "absolutely no problem" connected with 2,4-D in the three decades it has been on the market.[106]

Other government agencies, such as the U.S. Forest Service and the Bureau of Land Management (BLM), still make extensive use of 2,4-D on public timber lands to kill off vegetation that competes with such preferred species as conifers. And the Bureau of Reclamation even uses it to kill weeds in drinking water reservoirs![107]

In Louisiana, Mississippi, Florida and other southern states, 2,4-D is used to treat pastureland adjacent to crops, and in bayous and canals to kill off water hyacinths and other unwanted vegetation. Such uses pose the threat of contaminating fish, shellfish, and other aquatic creatures as well as the entire food chain.

EPA describes its studies on the chemical's cancer-causing properties as "inconclusive." But it was not until 29 April 1980 that EPA got around to asking 2,4-D's manufacturer, Dow Chemical, to provide additional information on the substance in order to determine its toxicity to humans. In the meantime, EPA stated that the chemical would not be suspended or even restricted.

In the statement released by EPA on that day, the agency cited the "chemical similarity to the dioxin-contaminated herbicides 2,4,5-T and Silvex," as another reason for its concern about 2,4-D. This fact sheet pointed out that while many of the tests conducted to determine if the chemical caused tumors, neurological damage, or birth and genetic defects were "inadequate and inconclusive," some of the results were cause for concern. For instance, at very high doses, 2,4-D caused "life-threatening birth defects (skeletal malformations) and cleft palates," and some Russian tests showed damage to unborn fetuses at much lower levels.[108]

Although 2,4-D has not been shown to be as conclusively hazardous to humans as 2,4,5-T, the toxicity of both chemicals appears to be enhanced or synergized when they are combined or used together, as they often are, and as they were in Agent Orange with such tragic results. Moreover, the fact that 2,4-D is used in such massive amounts means that it may cause far more damage to human health than 2,4,5-T, even it if proves to be far less toxic.

In January, 1981, Canada announced that it was banning or phasing out some products containing 2,4-D, in particular those found to be contaminated by certain types of dioxins. Shortly after the announcement, stores were swamped by farmers buying up available stocks while they could.[109]

The 2,4,5-T Body Count And Suspension

Attempts have been made for over a decade by conservationists, citizens groups and some government officials to have 2,4,5-T banned or severely restricted, but until recently, these efforts failed to result in significant action. Finally, in April 1978, EPA began the official public review process ("Rebuttable Presumption Against Registration," or RPAR) which precedes the final decision on whether to ban or restrict a chemical. But four years later, it is still unclear what, if any, final action will be taken in this matter.

Even under the Carter Administration, which came into office promising to protect the public from toxic chemicals, the political clout of the poison lobby remained as strong as ever, and was brought to bear on officials who tried to make much-needed policy changes. For example, Assistant Secretary of Agriculture Rupert Cutler was forced to weaken a quite modest restriction he had placed on the spraying of 2,4,5-T by the U.S. Forest Service. On 11 August 1978, he announced that the chemical could no longer be sprayed from the air in the national forests closer than a quarter mile from streams or a mile from homes.[110] One purpose of this sensible restriction was to keep 2,4,5-T from contaminating streams, lakes, drinking water, and people. However, after the Department was pressured by Oregon Governor Bob Straub, several congressmen, the Forest Service, and the timber industry, Cutler was forced to back down a few weeks later and reduce the buffer zone for streams by 85 percent, to 200 feet. This modification effectively nullified much of the impact of the ruling, since it is well known that pilots have extreme difficulty in maintaining 200 foot buffers and preventing drifts, especially in high or variable wind conditions.

By 1979, the evidence of 2,4,5-T's toxicity to humans, combined with the pressure from citizens and environmental groups from around the country, had become so overwhelming that the problem could no longer be ignored. The break came when a group of Oregon women were able to demonstrate that 2,4,5-T had killed their unborn babies.

During the six year period from 1972 to 1978, over 7,000 acres of the Oregon coastal forests had been sprayed with some 10,000 pounds of 2,4,5-T; along with an unknown amount of 2, 4-D. Several women in the Alsea, Oregon area, who had experienced miscarriages that appeared to follow the chemical applications, began to keep diaries detailing the dates of the spraying and of their miscarriages, and a definite pattern emerged. In six years, 9 women had suffered 13 spontaneous abortions following the herbicide spraying. In June 1978, a letter was sent to EPA outlining this information, which the agency considered the first real data it had received showing human health effects from such exposure.[111]

As a result of this, in 1979, EPA finally took action to restrict major uses of 2,4,5-T. Admitting that between one and two dozen miscarriages by Oregon women living near forests that had been sprayed had been definitely linked to 2,4,5-T, EPA, under pressure from both industry and public interest groups, decided on a compromise. On 28 February 1979, EPA declared 2,4,5-T and Silvex to be imminent threats to human health, and issued an emergency ban, temporarily prohibiting their use on forests and rights-of-way.

This partial suspension came over a decade after it had been conclusively shown that these chemicals caused birth defects in animals and severely adverse health effects in humans. But at the same time EPA suspended some uses of 2,4,5-T and Silvex, the agency announced that the other major uses of these substances—such as for spraying fruit crops, rice fields, and rangeland used for livestock—could continue as before!

In terms of its official position on the safety of 2,4,5-T, EPA had now come full circle. After a decade of denying that 2,4,5-T was a hazard to humans and refusing to ban its use, the agency was now admitting that the chemical was a killer of unborn children. At a 1 March 1979 press conference, Deputy Administrator Barbara Blum laid out the case against 2,4,5-T:

> ...dioxin, even at very low levels, causes severe reproductive effects—miscarriages and birth defects—and tumors in laboratory animals... New studies in the Alsea basin area of Oregon... spanning a six year period... show a high miscarriage rate shortly after the spraying of 2, 4, 5-T in the forests... the miscarriage rate appears to be correlated to the amount of 2,4,5-T sprayed. Notice this dramatic peak in June which follows the heaviest use of the herbicide by only 2 months. This peak occurred consistently in each year examined... There is a remarkably high miscarriage rate in June in the Alsea study not seen in the control area.[112]

EPA concluded from the studies that "there is a statistically significant relationship between the spray season and the high miscarriage peak which follows application of 2, 4,5-T by 2 to 3 months." "It's a remarkable correlation," said Blum.[113]

EPA estimated that its emergency order prevented exposure of some "4 million people who may be unknowingly and involuntarily exposed as a result of those uses."[114]

It should be emphasized, however, that EPA's suspension, unlike a cancellation action, was a temporary measure, and can be lifted by the agency at any time.

EPA's partial suspension had the effect of preventing the spraying of 6.9 million pounds of 2,4,5-T on 2,843,000 acres, while allowing the use of 2.3 million pounds on 2 million acres to continue just on rangeland and rice,[115] in addition to an unknown amount for other uses. Uses of Silvex allowed to continue include its annual application to: 50,000 acres of apples, 8,300 acres of prunes and 115,000 acres of sugarcane, and to an unknown extent on pears and for such non-crop uses as fencerows, storage areas, and parking lots.[116]

EPA estimated that the economic impact of cancellation of these uses of 2, 4,5-T and Silvex would, if undertaken, for the most part be insignificant and would not affect the retail price of apples, pears, and sugar. The price of prunes would go up somewhat, rice could increase five percent in three years, and beef might rise by less than one percent.[117]

But having made the case against 2,4,5-T and Silvex as imminent threats to human health whose use cannot be justified on economic grounds, EPA still could not bring itself to suspend the chemicals entirely. Thus, their widespread use, and the fight to ban them, continue.

The Fight Goes On

Because of EPA's suspension, 2,4,5-T *theoretically* cannot be used in forestry, rights-of-way, and pasture applications; lakes, ponds, and ditch banks; around the house and recreation areas and on food crops. But it is still readily available for other uses, and some spraying is doubtlessly taking place in these prohibited areas. Indeed, how is a person supposed to distinguish between "pastureland," where it is prohibited, and "rangeland," where it is allowed?

Allowed uses specified by EPA include rice fields, rangeland, (non-pasture grazing land for livestock), and non-crop applications such as at airports, lumberyards, refineries, storage areas, vacant lots, tank farms, industrial sites and areas, and on non-food crop areas, fences, hedgerows, and "wasteland."[118]

It is unclear why EPA is still allowing the widespread spraying, in inhabited and food producing areas, of a substance the agency knows causes cancer, birth defects and other disorders, especially in light of evidence that cattle which graze on land sprayed with this herbicide accumulate it in their fat.

EPA's official position document on these chemicals admits that their continued use carries grave risks:

> EPA believes that human exposure from the use of 2,4,5-T and Silvex on rice may be broad and substantial due to herbicide drift during and after application, and that more diffuse exposure is possible through the water environment and through crayfish, catfish, and other food sources...Beef monitoring shows low levels of dioxin in a limited number of samples from beef that grazed on 2, 4,

5-T treated range ... Little is known about the potential for
dietary exposure to Silvex and/or TCDD from the uses of
Silvex on food crops, except for apples on which Silvex
residues have been detected.[119]

In fact, beef fat from cows which grazed on rangeland sprayed
with 2,4,5-T has been found to contain dioxin at levels as high
as 60 parts per trillion. This is a much higher level than that
which has caused serious health effects in laboratory animals.
Dioxin has also turned up in mothers' milk from the western
U.S.[120]

As Bill Butler, then of the Environmental Defense Fund
(EDF) and now with the National Audubon Society,
characterized the situation: "The American population is being
used as Dow's guinea pigs while these substances are on the
market,[121] and the environment is their laboratory." And
Harvard biologist Dr. Matthew Meselson stated on the NOVA
television special, "A Plague on Our Children," considering the
amount of dioxin getting into the American diet, it is possible
that "thousands of Americans die every year from tumors
which are induced because of their exposure to dioxins."[122]

Nor has the chemical industry taken EPA's decision lying
down. Eleven chemical companies and industry associations, led
by Dow, have filed a legal appeal against the partial ban.[123] Dow
still insists that "2,4,5-T is about as toxic as aspirin."[124]

People who wonder why the public is not being protected
from lethal poisons should consider the fact that one group
which protested EPA's partial ban on 2,4,5-T was an organization composed of public officials charged with this responsibility: the Association of American Pesticide Control Officials.
This association of state personnel, who supposedly enforce
pesticide laws and regulations, drafted and sent to EPA a
resolution in March 1979, urging the Agency to reconsider its
action.

Thus, in its compromise decision, EPA satisfied neither side.
So the fight over 2,4,5-T continues, with the chemical industry
trying to have banned uses reinstated, and environmentalists
working to ban those uses still allowed.

By April, 1981, EPA had stopped its attempt to ban 2,4,5-T,
and was negotiating a settlement with Dow based on more
restrictive labeling requirements that would permit its
continued use.[125]

Waiting For More Dead Bodies

Scientists had known for decades that 2,4,5-T and dioxin caused tumors, birth defects, and other serious health damage to humans and laboratory animals exposed to it; but EPA refused to act on this data until it had become apparent and undeniable that human corpses had begun to accumulate. As EPA Assistant Administrator for Toxic Substances Steven Jellinek stated shortly after the partial suspension, "Now we have human evidence. We have dead fetuses."[126]

For many Americans— including those who have lost their health, their children, or their lives—EPA's action was too little, too late. A decade ago, when EDF first petitioned EPA to suspend *all* uses of 2,4,5-T, the group based its request on the well-known, well-documented information that had been available, in some cases for 20 years, on dioxin's toxicity.

If one thing is clear about the present situation it is that the Alsea, Oregon deaths—which EPA now says were apparently caused by exposure to the herbicide 2,4,5-T—were completely predictable and thus preventable. So were the many other deaths that have occurred and will occur among the untold millions of Americans who have been exposed to dioxin and carry the residues in their bodies.

Nor does EPA appear to have learned any lessons from its experience with 2,4,5-T and the deaths it now admits were caused by it. On 29 April 1980, the agency announced that it had decided not to ban or even restrict the use of the other closely-related phenoxy herbicide 2,4-D, the nation's most widely used weed killer. On 10 September 1980, Assistant Administrator Steven Jellinek testified before the Senate Veterans Affairs Committee that "...our review of studies on 2,4-D led us to conclude that evidence of adverse effects was not sufficient to justify short-range regulatory action." But he also admitted that "there is a basis for concern about reproductive and mutagenic effects."[127] Apparently, as with 2,4,5-T, before deciding to act, EPA is awaiting some more dead bodies.

Agent Orange and Vietnam:
A Preview Of What's Happening Here?

Another group of Americans who have been exposed to dioxin-laden 2,4,5-T and to 2,4-D—and have suffered adverse health effects as a result—are veterans of the war in Vietnam. Yet, despite the overwhelming evidence of the harm done by these chemicals, the government has adopted an official position that no damages can be proven. Nevertheless, an examination of the effects of these chemicals on humans and the environment in Vietnam may show us the harvest we can expect to reap from their use in this country.

During the Vietnam war, Agent Orange—a 50-50 mixture of 2,4,5-T and 2,4-D—was sprayed by the U.S. over South Vietnamese jungle and farmlands to deprive enemy soldiers of cover and to destroy the crops on which they relied for food. One Air Force C-123, spraying Agent Orange at a rate of 4 gallons per second, could treat a strip of land 80 yards wide and 10 miles long in four minutes, defoliating the area and turning it into a wasteland in a few days.[128] "Operation Ranchhand" pilots had a motto: "Only we can prevent forests."

Over 11 million gallons of Agent Orange, containing some 50 million pounds of 2,4,5-T and 350 pounds of dioxin, were used until the spraying program was halted in 1970 after the National Academy of Sciences concluded that 2,4,5-T was causing birth defects among Vietnamese villagers.[129] This followed numerous published reports in South Vietnamese newspapers of women exposed to herbicides giving birth to babies with such defects as cleft palates, twisted limbs, and incomplete spines and faces. According to later reports from Vietnam, the known incidence of a rare form of cancer of the liver increased five-fold during the time of U.S. defoliation, and this has been attributed to the carcinogenic effect of dioxin.[130] (Ironically, around the same time that the use of the 2,4,5-T/ 2,4-D combination was suspended in Vietnam, their use dramatically increased in U.S. forests.)

As of February 1982, 12,250 Vietnam veterans had filed medical disability claims and asked for compensation based on exposure to Agent Orange, blaming it for birth defects in offspring, cancer, nervous disorders, dizziness, personality changes, chronic coughing, impotence, liver and kidney disease, muscular weakness, loss of sex drive and appetite, insomnia, blurred vision, ringing in the ears, skin rashes, and other ailments. As of June 1980, some 2,000 claims had been filed, of which the Veterans Administration (VA) had allowed only three! These were 10 percent disabilities, which allow a monthly payment of $48 for chloracne scars. The other alleged damages are still not officially recognized by the VA as having been caused by Agent Orange.[131] The Agency refuses to acknowledge any illnesses were caused by it, but did allow 852 of the 12,250 claims for other reasons.

In large part because of the VA's refusal to act, Congress in 1981 passed legislation giving priority medical care to Vietnam veterans who were exposed to defoliants.

The VA has estimated that several hundred thousand Vietnam veterans may have been exposed to the herbicide,[132] but recent tests conducted for the agency and released in July 1980 indicate that Agent Orange contamination is much more widespread among veterans than previously believed. When Dr. Michael Gross, director of the University of Nebraska's Spectrometry Laboratory, took tissue samples from a group of men who served in Vietnam—a third of whom did not believe they had been exposed to the chemical—traces of it were found in the entire test group![133] Congressman David Bonior (D-Mich.), Chairman of the Vietnam Veterans in Congress, believes that "all 2.5 million Vietnam veterans may have been exposed to the herbicide because of its entry into the food chain and water system."[134] In many cases, the dioxin could accumulate in the fat tissue and not cause symptoms until years later, when the person exposed loses weight and breaks down the fat, thus releasing the dioxin.

Despite Dow Chemical's knowledge of the chemical's hazards, U.S. military personnel were assured of the safety of Agent Orange, and it was often handled carelessly. Marines would engage in playful spray fights, dousing each other from head to foot with the chemical. Career Air Force officer Richard King, who flew spray missions in Vietnam, has told of seeing pilots at demonstrations for the press "stick their fingers into cannisters of the stuff and then lick them to show how safe it was." Eventually, King came to suffer from numerous ailments, including ringing in his ears, liver problems, cancer of the prostate, and dry eyes and skin. He also became impotent and unable to control his bowels, and had to wear diapers.[135]

Paul Reutershan, an organizer of Agent Orange Victims International, who had been a helicopter pilot in Vietnam, upon learning that he had incurable abdominal cancer, expressed the feeling of many veterans: "I got killed in Vietnam, I just didn't know it at the time."[136] Reutershan died in 1978.

The statistical data and projections on Agent Orange's effects on human health are complex and confusing, and it is easy to lose sight of the fact that each one of these "samples" is a human being. Terming the government's failure to adequately investigate dioxin's effects on Vietnam veterans "a national disgrace," Congressman Bob Eckhardt (D-Tex.) held hearings in June 1979 on this issue. One of the emotional highpoints of the hearing was the appearance of Michael Ryan, a Long Island policeman who was exposed to Agent Orange in Vietnam in 1966 and has since suffered extreme weight loss, migraine headaches, and other illnesses. He brought with him to the witness table his daughter Kerrie, who was born with severe deformities although neither he nor his wife had any family history of birth defects. As *Washington Post* reporter Margot Hornblower described the scene:

> During the emotion-laden hearing, Kerrie, a frail child with short brown hair, sat in her wheelchair gazing wide-eyed at the television cameras, the Congressmen high on the wood-paneled dais and the roomful of lobbyists and reporters.
> "She's a dynamite little kid," her mother told the committee.
> Kerrie was born eight years ago with 18 birth defects: missing bones, twisted limbs, a hole in her heart, deformed intestines, a partial spine, shrunken fingers, no rectum. During surgery a blood clot developed and she suffered brain damage. Doctors say she will never walk.[137]

Unfortunately, the U.S. government has no help it can give to Kerrie because of the damage Agent Orange may have done to her. She should try to understand that the dangers of the chemical remain officially unproven.

In addition to being used in Agent Orange, 2,4,5-T was also an ingredient of other herbicide formulations used in Vietnam such as agents purple, pink, and green, which had a tenfold higher dioxin content than Agent Orange. All told, from 1962 to 1971, an estimated 107 million pounds of herbicides were sprayed on some 6 million acres of South Vietnam.[138]

Following the ban on Agent Orange use, the Air Force, after initially planning on selling its surplus stocks in South America, eventually had two million gallons of it burned in a ship at sea in a high-temperature incinerator, reportedly resulting in the complete combustion and destruction of the chemical.[139] Afterward, the elaborate disposal plans called for the drums in which the chemical had been stored to be thoroughly washed, crushed, and then melted down at 2,900 degrees Fahrenheit, a temperature which would guarantee total destruction of any remaining dioxin.[140] Such careful and well planned precautions contrast with the fact that every year millions of pounds of the same chemicals used to make Agent Orange—2,4,5-T and 2,4-D—are still being sprayed over rural and urban areas throughout America. The results—the inevitable human health effects and contamination of the food chain—have been remarkably similar to what ensued in Vietnam.

The post-mortems conducted on Vietnam show us what we may face in America in coming years. A study by the National Academy of Sciences released in 1974 concluded that the ecological damage done by herbicides to South Vietnam—our ally, lest we forget—may take at least a hundred years to heal. The report noted that "serious and extensive damage" had been done to the inland tropical forests and that over one-third of the coastal mangrove swamp forests—a major breeding ground for fish and shellfish—had been destroyed. The Academy also gave credence to reports that many children of the Montagnard hill tribes had been killed by the defoliants. Interviews with the Montagnards indicated that after a plane flew overhead "spraying smoke," virtually all of the people in a village experienced intense coughing, abdominal cramps, and massive skin rashes. Often "lots of children died"—38 in one village—including babies carried on the backs of their mothers into fields that had been sprayed.[141]

The persistent reports of dead and deformed babies from areas in the U.S. sprayed with 2,4,5-T and 2,4-D show that we have brought home to America one particularly hideous aspect of the war in Vietnam.

The Long History Of Dioxin's Dangers

The dangers of dioxin and 2,4,5-T have long been known. Had this information not been suppressed, denied and ignored by the U.S. government and the chemical industry, the damage caused in recent years by these compounds could have been avoided.

In 1965 and 1966, the federal government commissioned a study using laboratory animals that showed 2,4,5-T's ability to cause serious birth defects, such as kidney abnormalities and cleft palates. (The chemical had been registered since 1948 for use against weeds.) For four years, these results were suppressed, and were finally released only through the efforts of Dr. Matthew Meselson.[142] In later tests, even when dioxin was present at the lowest levels—less than one part per million—2, 4, 5-T still caused birth defects.[143] But, still, the government refused to ban the chemical.

Evidence of 2,4,5-T's and dioxin's toxicity is not confined to test animals; they also have a long history, going back at least three decades, of causing harm to humans exposed to them.

The first known instance of the manufacture of 2,4,5-T causing human illness occurred as far back as 1949, when a Monsanto plant in Nitro, West Virginia exploded, and 228 workers contracted chloracne. Besides the disfiguring, pus-filled skin eruptions, the symptoms that appeared included liver damage, loss of sex drive, fatigue, irritability, insomnia, nervousness, difficulty in breathing, and nerve damage. Between 1949 and 1976, there were a total of 16 different incidents reported at chemical plants in the U.S. and Europe involving human overexposure to dioxin.

In 1952 and 1953, workers exposed to dioxin at two German manufacturing plants developed chloracne and severe liver damage. At one of the plants, not only did all of the male workers develop chloracne, but many of their wives, children, and even their pets did so as well, apparently as a result of exposure to the chemical brought home on the workers' clothes. By 1957, dioxin had been publicly identified as the toxic chemical responsible for the illness of these workers.[144]

In 1963, after a Dutch plant had an explosion, 50 workers also contracted chloracne, as well as damage to their internal organs and psychological disorders. The plant was shut down for ten years, but still could not be decontaminated. According to Dr. Marion Moses, of the Mount Sinai School of Medicine's Environmental Services Laboratory, this accident demonstrated the extreme persistence and stability of the chemical:

> The entire operation was shut down and they were unable to decontaminate the building. The way they solved the problem was taking it apart, embedding it in concrete, and dumping it into the Atlantic Ocean. There is a report of an excess of cardiovascular deaths in a small group of workers involved in the cleanup after the explosion...

Nevertheless, despite all the earlier evidence of 2,4,5-T's hazards, and numerous other accidents involving death and disability of workers, Dow Chemical in 1968 issued a press release claiming that the chemical was "absolutely non-toxic to humans or animals." [145]

In 1969, a National Cancer Institute study using laboratory animals found a connection between dioxin/2,4,5-T and birth defects. Shortly thereafter, in October 1969, the Science Adviser to the President, Dr. Lee DuBridge, announced that because 2,4,5-T and dioxin had been linked to birth defects in laboratory animals, their use would be restricted.

And in December 1969, the Advisory Panel on Teratogenicity of Pesticides of the HEW Secretary's Commission on Pesticides and their Relationship to Environmental Health (the so-called Mrak Commission) concluded in its report that 2,4,5-T, some forms of 2,4-D, and several other "currently registered pesticides to which humans are exposed and which are found to be teratogenic ... should be immediately restricted to prevent risk of human exposure." As a result, the U.S. stopped its use in Vietnam of Agent Orange, and the Agriculture Department in May 1970 took action to cancel registration for 2,4,5-T products used on food crops, in gardens and recreational areas, where it could contaminate water supplies, and around the home. But these restrictions only affected a small amount—less than 20 percent—of the 2,4,5-T being used in the United States! [146]

Nevertheless, in May 1970, the major purveyors of 2,4,5-T—Dow, Hercules, Inc., and Amchem Products Inc.—appealed the ban on the chemical's use on rice crops. [147] Under the rules governing "cancellation" of a chemical's use, the appeal meant that it could continue to be used until the resolution of the process years later. (A "suspension," on the other hand, takes effect immediately.) Mainly through legal action, Dow was able to block further efforts by EPA to apply restrictions on 2,4,5-T until two or three years later. (In 1970, the newly-created EPA was given the pesticide regulation responsibilities formerly administered by USDA.)

By this time, the dangers of the chemical were widely recognized by many scientists and environmentalists. Thomas Whiteside was writing lengthy articles in the *New Yorker* on the hazards of 2,4,5-T; only the chemical industry and the U.S. government were still unwilling to admit to its hazards.

"The Most Horrible Tragedy Ever Known..."

One important factor in the government's decision to suspend some relatively minor uses of this chemical were the Congressional hearings held in April 1970, by the late Senator Philip Hart (D-Mich.) on dioxin, Agent Orange, and 2,4,5-T, in which almost all of their hazards that are today generally accepted were discussed and described. At the time, Senator Hart made a statement, based on the evidence then available, which, if heeded, could have saved countless lives and prevented enormous contamination of the environment:

> The questions which have been raised recently concerning the hazards of 2,4,5-T and related chemicals may in the end appear to be much ado about very little indeed. On the other hand, they may ultimately be regarded as portending the most horrible tragedy ever known to mankind. What does emerge clearly from this uncertainty is that we must take steps to eliminate it. In view of the potential disaster that could befall us—or conceivably has insidiously already befallen us—absolutely no delay is tolerable in the search for answers to the questions posed.[148]

In his testimony, Surgeon General Jesse Steinfeld presented a chronology of research that had been done on 2,4,5-T, showing that experiments conducted as far back as June and November of 1966, and in January of 1968, demonstrated the chemical to be teratogenic (fetus deforming).[149] He also announced the immediate suspension of registration for most forms of 2,4,5-T used around the home, on lakes, ponds, and ditch banks, and on food crops stating:

> New information reported to HEW on April 13, 1970, indicates that 2,4,5-T as well as its contaminant dioxins, may produce abnormal development in unborn animals ... These actions do not eliminate registered use of 2,4,5-T for control of weeds and brush on range, pasture, forest, rights-of-way and other non-agricultural land.

Dr. Steinfeld pointed out as well that dioxin produced birth defects in laboratory animals "in 10,000 to 30,000 times smaller a dosage than 2,4,5-T."[150]

But conservationist objected to the non-suspension of other, similar chemicals, such as 2,4-D, PCP, and Silvex, a more common ingredient in garden products than its almost identical relative 2,4,5-T. They also criticized the failure of USDA immediately to issue a total, publicly-announced recall of 2,4,5-T and related compounds from retail stores and homes, to be reimbursed by manufacturers.[151] The suspension of most additional uses of 2,4,5-T and Silvex almost a decade later proved the validity of these charges.

And since pesticide drops and dust particles are carried by the wind into other yards, screen porches, and open windows, one person in a residential area who sprays his yard to get rid of weeds can contaminate many of the neighbors' homes.

Harrison Wellford, of the Center for the Study of Responsive Law, also pointed out that, as the Mrak Commission makes clear,[152] if 2,4,5-T were causing a rash of excess birth defects, these would probably not be detected and attributed to the chemical:

> In the case of 2,4,5-T, the most common defects produced in test animals are kidney abnormalities and cleft palates, neither of which is unusual in humans. Had thalidomide produced such ordinary malformations instead of bizarre and unusual ones, it probably would never have been discovered. Thus, any birth defects produced by human exposure to 2,4,5-T are unlikely to be traced to the weedkiller because they are already common in the population.[153]

The Coverup By Dow And USDA

During these hearings, despite what was known at the time, the witnesses from the Dow Chemical Company assured the committee of the safety of 2,4,5-T even with its inevitable dioxin contaminant. But even so, Dow's Vice President and Director of Research, Dr. Julius Johnson, admitted that Dow had been aware of dioxin's presence in 2,4,5-T and the effects of these chemicals on human health since 1950. He also acknowledged that the company had failed to notify the FDA and the USDA of health problems caused by these substances in Dow workers in 1964.[154] This was a particularly significant and damning admission, since USDA, rather than conducting its own research or studying published data, relied almost entirely on the manufacturers of pesticides to provide toxicological information on their products.

Dow's testimony on 2,4,5-T's safety was buttressed by that given earlier by the Agriculture Department, which insisted that there was no evidence that registered uses of 2,4,5-T were hazardous. Testifying on behalf of USDA, Dr. Ned Bayley, Director of Science and Education for Agriculture, assured the committee that 2,4,5-T was relatively harmless. Even when Senator Hart pressed him and cited recent studies showing the chemical to cause birth defects, Dr. Bayley refused to admit to any second thoughts about its safety:

> *Senator Hart:* Are you aware that the preliminary results of tests conducted by FDA, Dow, and by the National Institute of Dental Research and by the National Institute of Environmental Health Sciences all indicate that 2,4,5-T contaminated with no more dioxin than is found in the currently produced 2,4,5-T is teratogenic?
>
> *Dr. Bayley:* We are fully aware of this ... we do not believe this in any way changes the hypothesis that the low level of dioxin is safe.[155]

This led to the following exchange between an incredulous Senator Hart and Ned Bayley:

> *Senator Hart:* And yet this morning we have heard testimony that preliminary tests suggest that 2,4,5-T, when contaminated by dioxin comparable to that found in

currently produced 2,4,5-T, is teratogenic in three species; that the Mrak Commission or a panel advisory to it said that the teratogenic effects in one or more species should be grounds for immediate restriction of pesticide use; that residues of 2,4,5-T are now found on approximately one out of every 200 food samples analyzed by FDA; that we can't be sure of the amounts of tetradioxin in 2,4,5-T now being sold...that no evidence suggests that these dioxins are not persistent or cumulative in human tissue, and that some evidence would indicate perhaps they are.

If you accept that as a premise, in view of all this, would you say that you are sure that registration of 2,4,5-T for use directly on food crops does not constitute a hazard to man?

Dr. Bayley: I would say that the information we have does not give us indication that it is a hazard to man in accordance with the registered uses.

Senator Hart: Your position is that they do not constitute a hazard?

Dr. Bayley: Yes, sir...[156]

(Dr. Bayley continued to rise through the ranks at USDA, parent agency of the U.S. Forest Service. In August 1980, Bayley was named Acting Assistant Secretary for Natural Resources and the Environment, a post giving him jurisdiction over all of the Department's environmental dealings, including pesticides. He retired in 1981.)

Much of the rest of the testimony presented at these hearings, including some by government officials, refuted Dow's and the USDA's assurances of 2,4,5-T's safety. Dr. Jacqueline Verrett, of the Food and Drug Administration (FDA) discussed the toxic and cumulative effects of dioxin, and how its presence in commercial poultry feed was implicated in causing the death of millions of chickens since 1957 in epidemics of "chick edema."[157]

Dr. Samuel Epstein, then of Harvard Medical School, discussed various experiments conducted by Dow, the National Cancer Institute, FDA, and the National Institute of Environmental Health Sciences, all of which showed 2,4,5-T to cause such birth defects as missing limbs, cleft palates, and other abnormalities in various species of laboratory animals. Observing that "we have had information on dioxin for 20 years," Dr.

Epstein concluded that, "phenoxy herbicides should, under no circumstances, be used on crops. I would submit that there is a strong presumption for suspension of their use under any circumstances in the environment."[158]

Playing Russian Roulette With Herbicides

In 1974, Senator John Tunney (D-Calif.) held hearings on the matter, in which a top EPA scientist testified that the chemical should be banned for all uses. Dr. Diane Courtney, head of the toxic effects branch of EPA's National Environmental Research Center, told the Senate Commerce Committee's Subcommittee on the Environment on 9 August 1974: "I think it should not be used in any way at all. We are using a chemical that we can barely handle." Citing tests on animals that resulted in cleft palates and other birth defects, she said her main concern was dioxin, "by far the most toxic chemical known to mankind." She also testified that dioxin caused birth defects in doses so low they could not be measured, and that it was present in some beef and dairy products from cattle that had grazed on pasture treated with 2,4,5-T.

At hearings held by the subcommittee a week later, EPA officials, including Edwin Johnson, now head of the pesticide office, testified that they were not aware of any substantial safety question question which would even require a public hearing, and that they favored continued use of the herbicide. Johnson even informed Senator Tunney that he was so confident of 2,4,5-T's safety that he would be willing to eat ice that had been treated with it![159]

Realizing that the agency's bureaucracy had no intention of acting on 2,4,5-T, Senator Tunney predicted accurately that EPA would allow the herbicide to continue to be used "until there is a substantial adverse impact upon the health and welfare of the population."

> So in effect we are playing Russian Roulette...we know it is harmful, and it is just very difficult at the moment to tell how harmful it is, and therefore we will just put our heads in the sand and pretend it doesn't exist.[160]

EPA thus acted precisely as predicted: it waited until it could prove human deaths had occured before restricting 2,4,5-T.

The agency is continuing to use the U.S. population as its test group while allowing other major uses of the herbicide to continue.

Seveso, Italy: Destroyed By Dioxin

As far as is known, there has not *yet* been a major accident in the U.S. involving the large-scale release of dioxin in an urban area. But if and when such a situation occurs, we at least know some of what to expect from the experience of an Italian town that was virtually destroyed—rendered uninhabitable—by dioxin in 1976.

The worst known accident involving dioxin occurred near Seveso, Italy on 10 July 1976, when a chemical plant manufacturing 2, 4, 5-trichlorophenol (similar to 2, 4, 5-T and containing dioxin) exploded, releasing a vaporous cloud of this substance over an extensive inhabited area. (This incident is described in detail in a lengthy article in the 4 September 1978 edition of the *New Yorker* by Thomas Whiteside, who has researched and written about dioxin for over a decade.) The plant, belonging to the ICMESA company, of the Swiss-based Givaudan Corporation (itself owned by the multinational drug company Hoffmann-La Roche), was producing 2, 4, 5-trichlorophenol for use in making the widely-used antibacterial chemical hexachlorophene.

Before long, cats, rabbits, chickens, and other pets in the area began to sicken and die by the hundreds; birds were described as "falling from the sky," and soon hardly any seemed to be left alive around Seveso. Children became sickly, some developing burn-like skin lesions. Autopsies performed on the dead animals showed extensive liver damage.[161]

When the extent of the contamination became apparent, one part of the town, containing 739 people, was evacuated and sealed off by a fence nine feet high and four miles long that took seven months to erect. A larger area containing 5,000 people was sealed off to nonresidents, and 1,800 children under fifteen were taken during the day to schools or nurseries outside of the town to minimize their exposure to dioxin.

After a year or so, the short-term toxic effects of the dioxin, which penetrated at least a foot into the soil, had made their mark. Some 81,000 domestic animals had either died of dioxin poisoning or had been destroyed as a sanitary measure. In one area, thirteen pregnant cows suffered ten spontaneous abortions, and two more calves died shortly after birth; only one pregnancy of the thirteen resulted in a healthy calf.[162] There were 51 spontaneous abortions among the pregnant women of the village, a rate of almost one in four.

Prior to the opening of the 1977 school year, tests found that some 124 schools in nearby municipalities were so contaminated with dioxin that they had to remain closed pending cleanup procedures and additional testing. Some 280 children in the area came down with what appeared to be chloracne, an affliction common among workers involved in manufacturing substances containing dioxin, which is characterized by pus-filled skin eruptions over extensive portions of the body that can last for years, blurred vision, liver damage, depression, and loss of sex drive.

The long-term effects on residents of the area may not be felt for decades or generations. Between sixty and one hundred pregnant women had abortions (some legally, others illegally) for fear that their fetuses would be deformed, after the Catholic Church sanctioned such operations for many of those exposed to dioxin.[163]

Some of the women who underwent abortions were found to have incurred damage both to their chromosomes and those of their fetuses.[164] In December 1980, Givaudan announced that it would pay $109 million in compensation for damages caused by the chemical explosion.[165]

The evacuated area of Seveso, it is estimated, may not be inhabitable for another half century or more. It will, however, serve a purpose if this abandoned area, empty and lifeless, helps to remind us of our folly in permitting the chemical industry to inflict its poisons on us to make its yearly profits.

Other Dioxin Tragedies

Since dioxin, 2,4,5-T, and 2,4-D have remained in widespread use, the predictions repeatedly made about the dangers of such chemicals have come to pass in numerous incidents. Again and again, it has been demonstrated that when people or animals are exposed to dioxin, even in minute concentrations, the short term effects can be devastating.

One particular incident shows how hazardous dioxin can be. On 26 May 1971, an oil salvage disposer sprayed oil contaminated with dioxin on a horse farm near St. Louis, at Moscow Mills, Missouri, as a dust suppressant. Within four days, hundreds of sparrows and other birds that roosted in the barn were found dead on the ground. Within the next few weeks, hundreds of more birds died, as well as cats, rodents and horses. Even after the dioxin-laden soil to a depth of 6 to 8 inches had been removed in October, 1971 and replaced with uncontaminated river sand, the horses continued to sicken, lose their hair and die. In the spring of 1972, an additional six inches of soil were removed, but to no avail.[167]

Eventually, a total of 45 horses, all eleven cats, four out of the five dogs and a large but unknown number of wild birds and insects were killed by the spraying. Twenty-three other horses became ill and had to be sold at a fraction of their normal value. Symptoms shown by the affected animals included loss of hair, weight, and appetite; a staggering gait; nosebleeds; kidney and bladder troubles; difficulty in urinating; ulcers; colic; foaming at the mouth; fever and skin lesions. There were 26 known miscarriages among pregnant mares, as well as six deformed foals born and several stillbirths.

At least seven humans on the farm were also poisoned by the spraying, and experienced headaches, nausea, diarrhea, extreme fatigue, abdominal pains, blood in the urine, and pain on urination. Besides an increased risk of cancer and birth defects, a Senate report noted, those exposed could later incur such latent effects as "liver disease and central nervous system disorders, such as muscle tremors and psychological/psychiatric damage."

A six year old girl, the daughter of one of the stable owners, had to be rushed to the hospital after she suddenly came down with these symptoms as well as kidney and bladder problems. Doctors diagnosed her illness as being caused by a toxic agent, probably one present in the farm's soil. It was not until a year later that the federal Center for Disease Control (CDC) determined that the oil contained dioxin. In fact, the soil was found to contain dioxin at levels of 31 to 33 parts per million.

Another daughter, then ten years old, displayed similar symptoms. When she reached eighteen, she complained of having suffered for the last eight years from skin rashes, irritability, neurological damage, gastrointestinal disorders, severe headaches, nosebleeds, nausea, bodily pain and weakness, and increasing physical disability, all of these requiring numerous periods of hospitalization. The two co-owners of the stable estimate their out-of-pocket losses due to the spraying at $213,000.[168]

Dioxin In Consumer Products

Consumer products contaminated with dioxin have also demonstrated the deadly nature of these chemicals, even in the most minute doses. One example is PCP (pentachlorophenol), which is a by-product of the manufacture of 2,4,5-T and is used extensively as a fungicide and wood preservative. It has also been used in shampoos, paint and laundry starch. (The chemical is not to be confused with the "street" drug called PCP or "angel dust".) This chemical has become widely dispersed throughout society and is a significant source of exposure to dioxin, in some cases causing the death of people and livestock.

PCP contamination of the environment is pervasive. It has turned up in 85 to 100 percent of urine samples tested, as well as in virtually all samples of seminal fluid. The source for this contamination in residents of urban areas is probably food that has been sprayed with herbicides or stored in containers treated with PCP. The chemical has been found in bread, rice, candy, cereal, soft drinks, powdered milk, noodles, sugar, wheat, and water.[169]

PCP may also be a factor in the increasing incidence of sterility and infertility among Americans. An EPA-sponsored study found that PCP bioaccumulates in the sperm cells at levels 20 to 40 times higher than in the semen, thus depressing sperm counts in males.[170]

One of the earliest studies on the extreme toxicity of PCP was published by the National Academy of Sciences in 1957. The study described how a tank truck driver was killed while mixing diesel oil with PCP to make a cotton defoliant solution. After his hand came in contact with the chemical, he washed it immediately, but quickly became ill and was dead by the following day.[171]

Another early report indicating the toxic potential of PCP appeared in the August 1969 *Journal of Pediatrics*, describing how nine newborn infants, born at the same St. Louis hospital, were apparently poisoned there by dioxin. Two of the children died, and the others survived only after undergoing intensive treatment and blood transfusions. The source of these severe illnesses was traced to an antibacterial laundry product which left PCP on sheets, diapers and other clothing. The PCP then penetrated the skin of the children and entered their bloodstreams and bodies. Even after use of the product was discontinued, PCP was still detected in mothers about to give birth and even in newborn infants, the chemical having been absorbed by the mothers from sheets earlier washed with the solution, after which it crossed the placenta and entered the fetus.[172]

In Michigan, in 1976 and 1977, over 400 head of beef and dairy cattle at several farms came down with a variety of illnesses after apparently licking, inhaling the fumes from, or just rubbing against wood treated with a commonly used lumber preservative containing PCP. In one of the affected dairy herds, 61 out of 67 calves were stillborn, some being hideously deformed as well.[173]

In Stark County, Ohio, a herd of dairy cows housed in a new barn died within a few weeks. In South Dakota, a farmer lost ten cows under similar circumstances. There are several hundred thousand log homes in the U.S., many of which pose a threat from vaporization of PCP to anyone who lives, or spends time, in such structures. A 1980 Communicable Disease Center study of log home residents found ten times higher than normal PCP levels in their blood and urine.[174]

PCP's toxicity is well known in Guatemala, where farmers and ranchers have used it on wooden fenceposts. When youngsters crawl over fences to steal fruit or other goods, the dioxin burn they incur marks the intruders and makes it easier for the police to find and arrest them.[175]

Writer Thomas Whiteside, author of *The Pendulum and the Toxic Cloud*, which describes the dangers of dioxin, estimates that about 60 million pounds of PCP are produced each year in the U.S.[176] Much of this eventually ends up in the air after treated wood is burned or incinerated, or it can enter the environment and the food chain in other ways. On 22 July 1980, a cargo ship collision near New Orleans dumped overboard a 12½ ton container full of PCP into the Mississippi River causing the closing of some 400 square miles of prime commercial fishing waters in marshes, bayous, and lakes, and the shutting down of shrimping, crabbing and oyster-catching operations because of the spreading contamination.[177] Such an accident has the potential for spreading dioxin contamination throughout the food chain, including fish, shellfish, shrimp, and birds, and ultimately to humans.

In February 1981, EPA announced that because "the risks posed by the wood preservative pesticidal chemicals are substantial," it was restricting *some* uses of PCP, while allowing its general sale and use to continue.[178]

Another once-popular consumer product that contains dioxin is hexachlorophene. (The sole producer of hexachlorophene is the Givaudan company (see pp. 69-70).

Hexachlorophene is an antiseptic solution that has been extensively used in hospitals, clinics and doctors' offices. Nurses wash their hands in the solution, and newborn infants used to be routinely bathed in it. It was, indeed, an effective anti-bacterial compound, killing not only the germs, but sometimes the babies as well.

Introduced in 1941, it became a base for a variety of cosmetic products, such as baby creams, oils, and powders; shaving cream, and the first hexachlorophene soap, Dial. By the 1950's and '60's, diapers were being impregnated with it, and the compound was being used as a mouthwash; in drugs to treat acne and psoriasis, and in first aid kits, anti-perspirant deodorants,[179] and hundreds of other products.

In 1972, after 35 healthy newborn infants in Paris, France died following a dusting with talcum powder containing a high level of hexachlorophene, the sale of such products was banned. However, the chemical remained in use as a surgical scrub.[180]

Since then, various studies have demonstrated the toxicity of hexachlorophene. It has been shown to cause central nervous system disorders in test animals, and a large number of spinal and brain lesions have been found in premature infants who have been bathed in the solution.[181] In June 1978, a Swedish physician released the results of a six-year study showing that nurses at six hospitals in Sweden who regularly washed their hands in a hexachlorophene solution had given births to an extraordinarily large number of deformed children.[182]

Yet, hexachlorophene and other dioxin-contaminated products are still in widespread use in such products as disinfectants, hospital scrub solutions and cosmetics, and will presumably remain so until the inevitable heath disaster forces the government to take action.

The Inevitability Of Disaster

There is no way to know how many other incidents of the sort described have occurred in which the cause has remained unknown. What can be said for certain is that with the millions of pounds of dioxin-contaminated wastes spread throughout the country, and with more being produced each day, future dioxin tragedies are inevitable.

Moreover, as tragic and destructive as have been the accidents involving dioxin, they may one day pale in comparison to the damage that might intentionally be done by a madman or a terrorist armed with dioxin. Having dioxin and products containing it so readily available to the public also puts these substances within easy reach of potential terrorists, extortionists, or even juvenile delinquents who could poison reservoirs, wells and other water supplies, or threaten to do so. A drop or two of dioxin put in a punch bowl, a water fountain, or even a swimming pool could have disastrous consequences. It is estimated, for example, that if a few ounces of pure dioxin, which could fit in a small flask, were placed in New York's water supply, it could kill the city's entire population.

In December 1980, chlordane or heptachlor, pesticides so toxic that they are being phased out, were deliberately injected into water mains in Pittsburgh, Pennsylvania serving 20,000 people. The contamination was discovered after many people complained of nausea, diarrhea, chest pains, and neck and muscle aches; the water supply for thousands of residents had to be cut off. Had the polluters used chemicals containing dioxin, the result could have been massive loss of life.[183]

Moreover, even if all products containing dioxin were banned, the chemical would remain a problem for future generations. Huge concentrations of dioxin-contaminated chemicals are found at many of the 30 to 50,000 hazardous waste sites scattered across the country. At several Hooker Chemical company sites in Niagara Falls, New York, studies have revealed enormous amounts of dioxin wastes. At the most publicized of these—the infamous Love Canal—Hooker dumped an estimated 460,000 pounds of TCP (trichlorophenol), a chemical often containing dioxin that is used to make 2, 4, 5 -T .[184] In *The Pendulum and the Toxic Cloud*, author Thomas Whiteside calculates that this much TCP could contain some 150 pounds of dioxin: "many times more than the total amount of dioxin reckoned to have been released as a result of the Seveso explosion, and an amount that may equal the estimated total dioxin content of the thousands of tons of Agent Orange sprayed over Vietnam..."[185] (Later estimates put the Vietnam figure at 350 pounds of dioxin.)[186]

Hooker's nearby Hyde Park site contains almost 15 times as much TCP as Love Canal—some 6.6 million pounds. This means that the Hyde Park waste site, by itself, may contain several *thousand* pounds of dioxin—enough to kill every human on earth![187]

From 1954 to 1977, the Hooker Chemicals & Plastics Corporation dumped some 400,000 cubic yards of toxic chemical wastes, much of it laden with dioxin, in the vicinity of its Montague, Michigan plant. The result has been contamination of local wells and the underground aquifer containing some 2 billion gallons of groundwater. These chemicals have been flowing into White Lake, a major recreation area, at a rate of 875 pounds a day. Dioxin has been found on the site at levels of 240 parts per trillion. These wastes were also dumped at sites elsewhere in Michigan and Ohio.[188]

Before exploding and going up in flames, the Chemical Control Corporation's Elizabeth, New Jersey dump site was reported by its owner to have contained some 1,000 55-gallon drums of wastes laden with dioxin (see pages 147-50).

In late 1981, it was announced that research was underway to test and further develop bacterial strains that seem to be capable of consuming and degrading 2,4,5-T in the soil, and field tests were expected to be conducted in the spring of 1982. The potential applications and feasibility of using such techniques to de-toxify contaminated soil and waste dumps remain unclear.[189]

The prevalence of dioxin at dump sites raises several questions for which there are not yet clear answers: As dioxin seeps into the groundwater and contaminates drinking water, how will this affect people? If an attempt is made to clean up the dumps, what will be done with the dioxin? Where will it be put, and how will it be transported? Considering the enormous costs and dangers involved, who will do the cleaning up, and who will pay for it?

While we are pondering these matters, massive amounts of dioxin wastes are still being generated, and are going somewhere, though it is doubtful that much of it is being properly disposed of. What is certain is that as long as we permit the manufacture and use of herbicides and other products containing dioxin, we will continue to court disaster. While we contemplate the "documented" dioxin tragedies of Vietnam, Seveso, and the dead children of Alsea and elsewhere, we can only wonder what future disasters await the 230 million American hostages of the chemical industry.

What You Can Do

As has been noted, the spraying of herbicides over forests to increase timber production appears to have the opposite effect. Hiring people to remove weeds, undesirable trees, and other unwanted vegetation by hand provides employment and also is cost competitive with the spraying of herbicides. This method does not cause any pollution either.

It should be remembered, however, that in addition to providing natural fertilizer and nitrogen to the soil, lush vegetation gives shelter to creatures that are important to a healthy ecosystem.

Home gardeners should, if possible, remove weeds by hand, frequent tilling, hoeing, and pruning to exhaust their root systems. Mulch can be used between plants to reduce the spread of weeds and retain moisture in the soil.[190] Tight turf is often a good way of controlling crabgrass; and for farmers, according to CEQ, "Crop rotation is very effective in reducing weeds..."[191]

If you insist on using chemical herbicides, apply them sparingly and only on the target plant.

Finally, learn to appreciate the value and beauty of plants. What is a weed to one person may be a lovely flower to another and may be an important part of the plant community. If you let areas of your yard grow naturally, you may be amazed at how lovely they look. And you may be delighted by the birds, chipmunks, and other creatures that are attracted to them. Perhaps the main advantage of learning to live with "weeds" is that, in many cases, there is no alternative to doing so.

Chapter Two

Pesticides: Feeding The Insects, Poisoning The People

Despite the ... dramatic increase (about 10-fold) in chemical pesticides used on U S cropland during the past 30 years, annual crop losses from all pests appear to have remained constant. Losses   caused by insects may have nearly doubled.

President's Council on Environmental Quality, December, 1979[1]

America in the 1980's is a land virtually drowning in pesticides. Today, the amount of these chemical poisons being used staggers the imagination.

While we are now producing and using many times more pesticides than we were in the 1950's, at a far higher cost, we are losing a greater percentage than ever of our crops to insects. One reason for this is that pesticides kill off birds, predatory insects and other beneficial creatures that are part of the natural system that keeps harmful insects under control. Pesticides are actually hurting our efforts to grow food, and what is being grown is contaminated with poisons.

The ultimate victims of our pesticide folly are not the insects, but ourselves. As the President's Council on Environmental Quality (CEQ) points out,

Pesticide residues are common in U.S. food and
water... the effects of long-term, low-level dosages have
not been adequately studied... Pesticides have been
implicated in cancer... heavy pesticide use in southeastern
U.S. cotton and vegetables crops was associated with
human cancer mortality.[2]

Other recent reports by the Surgeon General, CEQ, and a
task force of federal agencies (discussed on pages 247-49) have
concluded unequivocally that exposure to pesticides and other
toxic chemicals has played a major role in bringing about the
current cancer epidemic, the highest rate ever recorded. One
American in four now contracts the disease, and the rate is
climbing at an alarming speed.

Although the Environmental Protection Agency has
identified several hundred pesticide ingredients that cause
cancer, it has effectively restricted only a handful of these.

It is estimated that over 100,000 farm workers are poisoned
seriously enough each year in the U.S. by pesticides to require
medical treatment. But since only a small fraction of the actual
poisonings is ever reported, some authorities believe that figure
should be multiplied several times to approximate the real
number.

Thus, while we do not seem to be very proficient at
eradicating insects, we are quite effectively poisoning the
American population. CEQ summarizes the consequences of
decades of casual overuse of chemicals as follows:

Over the long-term, this treadmill chemical approach
has proved to be self-defeating, only engendering such
serious problems as insecticide resistance, human
poisonings, and environmental pollution.[3]

Some agencies of the U.S. government now stress that the
massive use of chemical pesticides is totally unnecessary and
that biological methods can grow more food at less cost without
poisoning the environment and the human population. Yet the
indiscriminate use of these toxic chemicals continues with few
real restrictions. The outlook is for greatly increased
production and use of these poisons, with impacts on the
human race that could have ominous implications for our
future survival.

The More Used, The Greater The Losses

The use of chemical pesticides on U.S. croplands has increased tenfold in the last thirty years. The production of synthetic organic pesticides alone (carbon-based compounds made from petroleum) in the U.S. has increased even more dramatically— from 464,000 pounds in 1951 to 1.4 billion pounds in 1977.[4] This means that we are producing 3,000 times more pesticides today than 30 years ago (much of which is exported abroad). The result has been widespread contamination of food crops, rivers, lakes, fish, wildlife and humans, usually without achieving the desired effect of insect control.

Despite the dramatic increase in chemical pesticide use on U.S. cropland in the last 30 years, says the President's Council on Environmental Quality (CEQ), "annual crop losses from all pests appear to have remained constant. Losses caused by weeds may have declined slightly, but those caused by insects may have nearly doubled."[5]

People tend to forget that prior to the large-scale development and use of chemical pesticides after World War II, American farmers grew huge amounts of food without poisoning their product. While American farmers have achieved record crop yields in the last 25 years, this success can be attributed to such factors as unusually favorable climatic conditions, the development of fast-growing and disease-resistant types of crops, the increased use of better fertilizers and farming machines, and improved irrigation and cultivation practices. The burgeoning use of toxic insect and weed killers has played a much more complex and often damaging role, and has often been counter-productive in achieving the desired goal.

According to the late University of California entomologist Robert van den Bosch, who studied and "battled the bugs" for over a quarter of a century, "... thirty years ago, at the outset of the synthetic-insecticide era, when the nation used roughly 50 million pounds of insecticides, the insects destroyed about 7 percent of our preharvest crops; today, under a 600 million pound insecticide load, we are losing 13 percent of our preharvest yield to the rampaging insects."[6]

In 1972, in *The Darkening Land*, William Longgood wrote of the futility of trying to poison off insects, noting that we have poisoned ourselves instead. In so doing, we have destroyed and disrupted the natural systems that keep insects in check, such as "birds [which] eat their weight in bugs every day; no insect ever becomes resistant to a bird:"

> Each spraying makes more spraying necessary ... The farmer desperately turns to new and more powerful poisons. More imbalances result. More poison residues are in the crops for people to eat ... Even after all the spraying, losses due to pests are about the same as they were 50 years ago—about ten percent ... We have destroyed the old, and the new does not work. We set out to poison the bugs so we could feed ourselves. We wind up feeding the bugs and poisoning ourselves.[7]

Sometimes, the application of pesticides to crops poisons the soil for decades and causes the toxic chemicals to continue to be absorbed by unsprayed crops for many years thereafter. Carrots, for instance, appear to accumulate greater quantities of pesticides than other foods, and in some cases can even absorb and concentrate higher levels than the soil contains.[8]

Moreover, only a very small amount of the pesticides used ever lands on target. According to a recent study on pesticides by CEQ, "only 1 percent or less of the ingredients of some insecticide sprays applied by aircraft intercepts the target insects. No more than 25 to 50 percent may even land in the target area (e.g., crop field). The remaining 50 to 75 percent may be lost through volatilization and drift and may be carried many miles away."[9] In California alone, it is estimated that some 17 million pounds of aerial-applied pesticides landed on non-targeted property in 1977.[10]

Even the accurate targeting of pesticides often achieves the opposite of the effect intended, for insects frequently evolve immunity to these chemicals.

By 1975, some 305 species of insects were known to have developed genetic resistance to at least one chemical pesticide, and 59 other species were suspected of having developed resistant strains. In the major agricultural state of California, 75 percent of the most serious insect infestors of crops had become immune to at least one pesticide.[11]

The use of chemical agents can also destroy the natural enemies of insect "pests" which help keep their numbers in check, such as parasites, disease-causing organisms and beneficial predatory insects that feed on the target species, such as ladybugs, praying mantises, spiders, centipedes, lacewings, wasps and beetles, as well as frogs and birds. Thus, spraying can, and often does, cause the resurgence and rapid increase of the target "pest" population to a higher level than that existing before treatment. This ecological disruption can also induce the outbreak to "pest" status of other organisms whose natural enemies are destroyed. In order to control and suppress these outbreaks, even higher dosages of chemicals are used by farmers, creating a vicious circle.

In his book *The Pesticide Conspiracy*, entomologist Robert van den Bosch, describes how spider mites, once a minor crop pest, have now become a major one because pesticides have killed off many of their natural enemies. According to him, "Today in California, the once relatively unimportant spider mites cost the Agri-economy more than $116 million annually, double the losses caused by any other 'insect' pest group and five times what they cost the economy fifteen years ago."[12]

Since predatory insects are much less numerous than their prey, they have much more difficulty recovering from pesticide application, and are often unable to do so for the remainder of the season. This is especially true if the absence of predators causes more severe pest outbreaks and heavier spraying of insecticides. CEQ and the State Department thus reported in their *Global 2000* study, "Insecticides are probably largely responsible for new insect pests in Agriculture:"

> ... in such predator-free environments, species that
> have previously been benign become major pests. Mites,
> which are now major pests in fruit culture, became pests
> only after the start of heavy use of pesticides in fruit
> cultivation. Similar histories can be found for many insect
> species, including Hessian flies, and several major pest
> species of the cotton plant.[13]

Also, fumigating the soil with nematicides (chemicals that kill root-bearing creatures) can destroy organisms necessary for developing vine and tree crops and thus reduce plant growth.[14] The result in all these cases can be far more destructive to crops than normal insect infestation.

Chemical farming also tends to "burn up" the soil, using up its nutrients, killing off the microorganisms and compacting it so that water is less easily absorbed and runs off more rapidly, carrying the rich topsoil with it.

Another serious problem caused by pesticides is the destruction of honey bees, which each year produce $50 million worth of honey and beeswax. Honey bees are responsible for pollinating crops that provide one-third of our diet, including some 80 percent of America's deciduous vegetable, fruit, legume and oil seed crops; the annual value of crops pollinated by bees exceeds $1 billion. In just the state of California, in 1970, 89,000 honey bees colonies were lost to pesticide poisonings. Nationwide, it is conservatively estimated that well over 400,000 bee colonies are destroyed or severely damaged by pesticides each year.[15]

Poisoning Our Food

There are some 400 pesticides used on the food we eat, and three separate agencies are charged with the responsibility to limit the contamination of food with residues of these pesticides: EPA, USDA, and the Food and Drug Administration (FDA). Theoretically, these agencies monitor foodstuffs to ensure that traces of pesticides do not exceed the "tolerance levels" that have been established. In reality, no such system exists to effectively limit the public's exposure to toxic chemicals.

Tolerance levels are usually set arbitrarily and are based not on health safety considerations but on the amount of pesticides commonly found on agricultural products. About 6,000 of these tolerances, or exemptions to them, have been established by EPA, but these have, in fact, had little effect other than to recognize the existence of harmful and in some cases lethal substances in our food. A 1975 report by Congress' General Accounting Office, on the FDA's program to regulate tolerance levels for 233 chemicals, found that the agency could not detect 87.5 percent of the herbicides, 85 percent of the fungicides, 62 percent of the insecticides and 86 percent of the rest of the pesticides that contaminate food products.[16]

Because of the lack of evidence to justify most tolerance levels and FDA's inability even to identify the presence of most pesticides, the GAO concluded that the safety of America's food supply could not be assumed, and "EPA has little assurance that human health and the environment are being adequately protected from possible pesticide hazards."[17]

In the case of the several hundred pesticide chemicals thought to be carcinogenic, EPA cannot generally set tolerance levels, since there is not believed to be any safe level of exposure to cancer-causing substances. But to get around this awkward barrier to the free use of carcinogens, EPA has established what it calls "action levels" of pesticides in food, which are the amount at and below which traces of carcinogens will be permitted. As with "tolerance levels," "action levels"—which have been set for such deadly and restricted chemicals as aldrin and dieldrin, chlordane, DDT and heptachlor—are based on normal levels that usually show up in food, and not on public safety considerations. A June 1979 GAO study on pesticide residues in food concluded that "The safety of some existing action levels is clearly not supported by scientific evidence."[18]

GAO's June 1979 report indicated that the situation had hardly improved in the four years since its 1975 study. This recent report showed that dangerous levels of cancer-causing pesticides, many of which have been largely banned for use in the U.S., are found on food being imported from abroad. Much of this pesticide contamination results from the use of chemicals that were manufactured in the U.S. (even if they were banned for use here) and exported abroad, sometimes by the U.S. Agency for International Development. FDA analyzes less than one percent of imported food shipments, and when it does check, there are hundreds of pesticides whose presence the agency cannot detect (see pages 218-21).

Ironically, FDA not only fails to protect the public from carcinogens in food, but the agency's regulations encourage and even require the large-scale use of pesticides in some cases where this would otherwise be unnecessary. FDA has, since the 1930's, lowered the permitted levels of insects and insect parts found in food, even though foodstuffs can never be made insect-free. As the FDA itself admitted in a 1974 publication, such insects and their parts are generally harmless and present no health hazard.[19] (Some would even argue that they enrich the food by providing a good source of protein.)

The main result of this FDA policy is less, not more, protection for the public, since it encourages a 10-20 percent increase in the use of pesticides on fruit and vegetable crops. Indeed, the National Academy of Sciences and the President's Council on Environmental Quality have pointed out that in growing vegetable crops for canners or frozen food processors, these government regulations are the "primary determinant of pest control practices." Some growers of Brussels sprouts, for example, use pesticides solely on account of these FDA standards. These requirements also cause large amounts of food crops to be unnecessarily wasted, as they are deemed unsuitable for commercial sale. Similar problems are created by the commercial emphasis on the appearance of fruits and vegetables and the rejection of blemished products. In its 1980 publication *Integrated Pest Management*, CEQ called for changes in this FDA policy which, CEQ wrote:

> ...favors pesticides at unnecessary costs to the consumer...there is no apparent health hazard from ingesting small plant-feeding insects...It may be that the regulations here are more stringent than necessary.[20]

Indeed, a review of the government's policy in setting and enforcing tolerance levels for toxic pesticides leads to the inescapable conclusion that the program exists primarily to reassure the public that it is being protected from harmful chemical residues. In fact, the program, as currently administered, does little to minimize or even monitor the amount of poisons in our food, and serves the interests of the users and producers of pesticides rather than those of the public.

Poisoning Farm Workers

The most direct victims of the government's lenient and contradictory regulations on the use of poisonous pesticides are the more than 5 million farmworkers and their families. Often uneducated, illiterate, poverty-stricken and unable to find other jobs, they are poisoned each year by the tens of thousands, sometimes fatally.

No precise figures on this problem of farmworker poisoning are available, since the government does not keep them. But official estimates of the number of serious pesticide poisonings in the U.S. each year range up to 100,000 and over. Worldwide, the figure is put at 500,000 annually, with one percent—5,000—being fatal.[21] However, there is strong evidence that only a small fraction of actual pesticide poisonings—perhaps as few as one percent—are reported, which makes these estimates a tiny fraction of the actual cases.

Congressional hearings held in June 1979 by Congressman Bob Eckhardt (D-Tex.), then Chairman of the House Sub-committee on Oversight and Investigations, provide some idea of the alarming proportions of the problem. In describing how EPA "has basically abdicated the responsibility for worker safety regulations," Ralph Lightstone, an attorney for the California Rural Legal Assistance, gave a shocking account of how farmworkers in that state are being poisoned by pesticides. In California, the only state that tries to keep track of such poisonings, some 1,400 cases a year of people poisoned seriously enough to require medical help are reported, and the disease rate for agricultural workers is almost three times higher than for industrial workes.[22]

However, these official state figures seriously understate the real incidence of poisoning, since the vast majority of cases are either not reported or not properly diagnosed. For example, when 118 workers in Madera, California were poisoned in 1976, the state counted only the six that were officially reported. A study of four separate incidents of pesticide poisoning in California found that of 175 workers involved, only 64 received medical treatment, and only 45 of these cases were reported.[23]

In September 1978, because of drifts from chemicals being sprayed on cotton fields, Fresno County schools had to be closed for a week after children complained of respiratory problems, but these children were not included in the official list of poisonings.[24]

Dr. Ephraim Kahn, Chief of the California State Department of Health's Epidemiological Laboratory, has estimated that no more than one percent of the actual number of pesticide poisonings in California are reported. If this is the case, then over a hundred thousand people are poisoned each year just in that state alone! Nationwide, the figures may be similarly underestimated. An HEW official estimated in 1970 that 800 people are killed and 80,000 injured nationwide each year because of agricultural pesticides. In 1980, CEQ put the number of serious poisonings at 100,000 annually.[25] Based on calculations by Dr. Kahn and other authorities, the number of these poisonings that occur each year across the country may be in the millions. Indeed, it is virtually impossible for the average farm worker to avoid dangerous and illegal exposure to toxic pesticides.

Even if all obvious cases of short-term pesticide poisoning were tabulated, there would be no way to know the extent of harm done to workers in terms of shortened lifespans, genetic damage resulting in future birth defects, an increased risk of cancer, and other subtle damage that usually does not show up until years later. Studies by EPA have found extremely low sperm levels among a third of the farmers, farm workers and pesticide applicators examined,[26] indicating a risk of sterility among people who work with pesticides.

The pesticide industry and government regulators repeatedly claim that such chemicals, if "properly used" according to label instructions, are safe and effective. But many if not most farm workers are either illiterate or do not read and understand English; those who do rarely pay any attention to labels.

It is, of course, no secret that even simple labels are widely disregarded. A November 1969 report on pesticides by the House Committee on Government Operations stated that "most pesticide users do not read labels and those who do find the labels difficult to understand."[27] A University of California study found that only 14.5 percent of the farmworkers tested could understand warning labels.[28] (This problem is further discussed on pages 107-111.) And when pesticides are applied by "cropduster" airplanes, it is done at the convenience of the farmer, often with little regard for whether or not workers are in the fields.

Conditions similar to those in the West prevail throughout the various farm areas of the country. But farm workers—especially the migrants—are so intimidated that they seldom voice public complaints, and the dangerous and dismal conditions seldom receive attention unless and until a tragedy occurs.

A case in point is that of Charles Thomas, a nine-year-old boy who died in August 1971 after being sprayed by airplane several times while picking cucumbers in New Jersey. Shortly after being "dusted," the boy began to vomit and his face swelled up, as did his arms, legs and chest. He continued to worsen and finally died, with the cause of death tentatively identified as insecticide poisoning that destroyed the boy's lung tissue. His father, John Thomas, who had been a migrant worker for over 40 years, and three other Thomas children were looking on from an adjacent field where they were picking tomatoes when the incident occurred. But the father was not initially concerned, he said, because "they spray all the time while we're in the fields."[29]

Other workers described conditions in a similar manner to *New York Times* reporter Ronald Sullivan. Lilly McKay, a 52 year old migrant worker, said that the laborers were usually not informed when a field in which they were working was to be sprayed. The only warning was the sound of the approaching spray plane. "By the time it's over you, it's too late or you're too tired to run," she said. "We sometimes have to lay flat down on the ground to keep from getting hit, these planes come by so close."[30]

Such conditions are rarely reported to the authorities for fear of reprisals being taken by the farmers. According to the head of the Camden, New Jersey Legal Services Agency, he often received information on workers being sprayed and becoming ill, but they were afraid to report such incidents out of fear of being ejected from the camps in which they lived.

The owner of the farm where Charles Thomas was fatally sprayed, Joseph A. Pizzo, denied any responsibility for the death, but refused to comment any further on the conditions on his farm or even to say what types of pesticides were being sprayed. Indeed, the most remarkable thing about the not unusual incident is that it rated mention in a newspaper. Thousands of similar cases are documented each year, and sometimes reporters do exposés on the situation. Occasionally, a politician here or there waxes indignant and promises to "do something" about the problem. But nothing really ever changes in the poisoned world of the farm field worker.

The Government Looks The Other Way

A major reason that pesticide poisonings are so widespread is the laxity or lack of regulations established and enforced by EPA and state authorities. For example, EPA allows workers to reenter a field that has been sprayed by such toxic pesticides as endrin and ethyl parathion after 24 to 48 hours. But California prohibits reentry from 30 to 60 *days* afterward. Moreover, as Ralph Lightstone points out, the laws are rarely enforced and violators of the few regulations that do exist are not generally punished in any meaningful way. As he testified at the 1979 hearings:

> ... The largest penalty on record for the poisoning of workers in California was for the poisoning of 118 workers in Madera, California in 1976. The penalty was $1,750, or $14.83 per worker. That is hardly a deterrent. The grower in that case, who was found guilty of numerous violations of pesticide safety regulations, got his wine grape crop harvested and presumably made a handsome profit ... $1,750 is no deterrent at all.[31]

None of this is surprising when one realizes that EPA, in most states, has turned over the responsibility for enforcement to state agricultural departments. As Ralph Lightstone points out,

> the people responsible for enforcing pesticide protection are primarily concerned with promoting the agricultural industry ... The agriculture departments see themselves as promoting agriculture, and have a difficult time finding the interest and will to enforce worker protection laws.[32]

Congressman Eckhardt's hearings produced invaluable information and received widespread publicity. But Ralph Lightstone, for one, was pessimistic that this industry-dominated situation would be changed for the benefit of farmworkers. In 1969, he pointed out, then-Senator Walter Mondale, Chairman of the Subcommittee on Migratory Labor, held a series of hearings on the same subject:

> Each issue which has been raised here today, the incidence of death, acute poisoning from pesticides, lack of

information about the chronic effects, drift of pesticides onto workers and others, failure of the health care community to diagnose and treat pesticide poisoning cases, lack of reasonable standards, and enforcement of these standards by responsible government agencies; each of those issues was aired at those hearings one decade ago.[33]

The primary difference between 1969 and 1979, Lightstone observed, is that the problem has gotten worse: the overall use of pesticides has more than doubled since then; a multitude of new and untested pesticides has come into use and the few that have been largely banned, such as DDT, have often been replaced by even more toxic chemicals.[34] It should also be noted that when Mondale served as Vice President from 1977 through 1980 and was in a position to do something about the problem, he showed no real interest in it, and the situation continued to worsen.

Poisoning The Children

One group of field workers especially at risk are the 10 and 11 year old children used to pick berries that have been sprayed with toxic chemicals. Children are much more susceptible to the effects of pesticides than are adults. Although the 1974 amendments to the Fair Labor Standards Act ban the agricultural employment of children under 12, except on family farms, the 1977 amendments allow the Labor Department to grant waivers for the hand picking of short season crops if the "level and type of pesticides and other chemicals used would not have an adverse effect on the health or well being" of the minors.[35] But despite these supposed safeguards built into the law, 10 and 11 year olds are regularly exposed to dangerous chemicals by absorbing them through their skin, breathing in their vapors and eating unwashed berries as they pick them.

For example, two fungicides commonly used on straw-berries—captan and benomyl—cause serious health problems. The regulatory agencies that have reviewed their effects, including the Occupational Safety and Health Administration (OSHA) and EPA, have concluded that these chemicals and others used on berries pose "serious and unacceptable risks" to the children who pick them.[36]

Benomyl is considered to be mutagenic (gene altering) and teratogenic (causing birth defects), and causes such reproductive disorders as testicular damage and reduced sperm activity and production. Its breakdown product MBC has been shown to be a cancer promoter.[37] The fact that it and captan remain in widespread use despite their known hazards, even on crops harvested by children, provides an instructive study in how the government fails to protect the public as the law requires.

Captan, 10 to 17 million pounds of which are used on crops each year in the U.S., has been shown to cause cancer, birth defects and genetic damage in laboratory animals.[38] That captan causes birth defects is not surprising in light of the fact that its chemical structure is similar to that of thalidomide,[39] the drug taken by pregnant women in the 1960's which caused severe deformities in many children.

The dangers of captan have been known for well over a decade. The 1969 HEW advisory panel on pesticides recommended unanimously that human exposure to captan be considered a potential health hazard and listed it as one of the teratogenic chemicals which "should be immediately restricted to prevent risk of human exposure." Eleven years later, in August 1980, EPA finally got around to starting the public review process that could, in two years or more, lead to captan being banned or restricted. EPA stated that the review was being undertaken "due to evidence showing that captan induced mutagenic and oncogenic [tumor-producing] effects in test animals."[40]

In addition, said EPA, it "had evidence suggesting that it [captan] caused not only cancer [and] mutagenic effects...[41] Captan may produce chronic or delayed toxic effects on organs and functions of the body including the lungs, central nervous system, metabolism, kidneys, reproductive system and others. Currently available data strongly suggest that captan may have teratogenic, fetotoxic and hypersensitive effects." The agency also observed that "the diet of the general population is exposed to captan because it is used on a wide variety of fruits and vegetables" as well as cotton.[42]

Incredibly, a dozen studies used by the manufacturers to prove captan's safety were later found to have contained false information. EPA has pointed out that in 1979, when reviews were conducted on earlier studies on captan that were used to demonstrate its safety, they were found to be invalid: "The audits disclosed a large number of Industrial Bio-Test Laboratories testing deficiencies and resulted in the Agency's determination that none of the 12 studies were [sic] valid."[43] This same laboratory, in Northbrook, Illinois, has been involved in providing false safety data on other toxic chemicals as well, such as leptophos (see pages 235-36). A subsidiary of Nalco Industries, the Laboratory has, in fact, tested and submitted safety data on hundreds of chemicals which the government has used as the basis for allowing such products to be produced and sold. The validity of much of that data is now being questioned. (See pages 121-23).

Despite its known hazards, captan is still used throughout the U.S. and the world on a variety of crops, including cherries, apples, raspberries, cantaloupes, currants, pears, eggplant, lettuce, endive, cucumbers, potatoes, tomatoes, onions, beans, blueberries, lemons, beets, carrots and cauliflower.

There are dozens of other equally or more toxic pesticides used on crops that are harvested by young and relatively susceptible children, in clear violation of the letter and spirit of the law. If the government refuses to take action to protect children from such chemicals that it has unequivocally determined to be hazardous, and continues to allow such an intolerable and illegal situation to persist, can there by any reason to hope that we will ever remedy our other major pesticide problems?

Poisoning Suburbia

Rural areas are not the only regions subjected to this flood of poison. A surprisingly large amount of pesticides are used not only by farmers but by urban homeowners on lawns, trees, gardens and to control insects in the home. These and other non-agricultural uses account for a third or more of the total quantities of pesticides sold in the U.S. A 1972 EPA survey showed that over 92 percent of homeowners used pesticides, and did so at a higher amount per acre than farmers. In addition, pest control firms do over one billion dollars a year in business. The result of all this is pollution of urban air, water, dust, grass and the general environment with hazardous chemicals representing a considerable threat to children, pets and indeed any and all residents of such metropolitan neighborhoods.[44]

Although farm workers are probably the group most heavily exposed to pesticides, residents of rural and suburban areas who live near cotton fields or other agricultural lands are also subjected to unavoidably heavy exposure to some of the most toxic of these poisons.

For example, Suzanne Prosnier, a Scottsdale, Arizona housewife, described to the Eckhardt subcommittee how her family "almost immediately" lost their former good health in 1973 after moving into their suburban home across a highway from some cotton, lettuce, barley and sugarbeet fields which were regularly sprayed with herbicides and pesticides.[45] Her 10-month old baby, who had never previously been ill, suffered repeated respiratory infections, "giant hives," headaches and cramps, and would wake up in the middle of the night thrashing about his crib and writhing in pain. During the day, he was listless, sick and lost his appetite. The ten year old son complained of stomach aches, sore throats, fevers, severe chest pain and visual disturbances during which he felt as if his room was spinning and the walls closing in on him, leaving him drained and weak the following day.

Mrs. Prosnier and the other members of her family also experienced a variety of similar illnesses, including frequent headaches, stomach aches, sore throats, skin rashes, dizziness, disorientation, high blood pressure, bladder problems, blotches under the eyes, chest pains, bad coughs and rectal bleeding. One daughter stopped menstruating and the older son developed an ulcerated throat and "experienced several episodes of profuse salivation...he had to sit with a bucket under his mouth because the saliva literally drooled into the bucket." Mrs. Prosnier described her husband as "one of those very unusually healthy people; he never had a headache or a cold in 25 years of marriage, he has never been ill in that time." But he also came down with gastrointestinal disorders, frequent chest pains and headaches.[46]

In March 1975, Mrs. Prosnier, pregnant with her sixth child, began to hemorrhage, and in April gave birth to a small and premature baby with a birth defect requiring surgery—her only child with any abnormality. In May, the family's dog became ill, its eyes glazed over and its legs were paralyzed. The veterinarian diagnosed the problem as poisoning.

Guests in their home experienced headaches, skin rashes, sore throats and vomiting. Mrs. Prosnier testified that many other women in the neighborhood had miscarriages, still-born, abnormal and premature babies, and that she personally knew of 10 neighbors who died of leukemia in recent years, and several others who had leukemia and other forms of cancer. A house-to-house survey along a 7 mile strip of streets revealed that over half of the families had suffered from serious symptoms of illness. Physical education teachers in nearby schools were advising against allowing the children to go on the playgrounds because so many were becoming ill.

In 1976, when the three older children went off to college, "their health improved dramatically; they had no more sore throats, no more stomach aches, the skin rashes were gone and they felt 100 percent better." But the rest of the family continued to suffer such symptoms and they became especially severe in 1978, when the spraying was greatly increased because of heavy insect infestations. In the spring, the 5-year old child developed pneumonia, as did others in the area, in addition to chronic bronchitis and respiratory ailments. On the fourth of July weekend, the entire family became violently sick with stomach cramps, vomiting, diarrhea and muscle pains and they remained ill all month long.[47]

When the family travelled to San Diego in early August, they recovered and felt fine, but upon returning home, their symptoms reappeared. On one occasion, the children came running inside crying, saying their noses were burning and they felt sick. When Mrs. Prosnier went outside to investigate, she said, "my nose instantly was on fire. I could feel it choking me, a burning sensation halfway down into my chest." When she learned that morning that the fields near her house were being sprayed, she called state, city and county health and air quality agencies, as well as EPA, but "I got a continual passing of the buck; each agency said call the next agency. I went full circle."

When she finally reached the people with jurisdiction over the matter—the state pesticide control board—she was told by a spokesman there, "Lady, the stuff they are using out there isn't any more harmful than table salt or aspirin." When she checked with a local hospital's emergency division, she was told that the chemicals being sprayed were organophosphates which "are toxic and are known to cause severe asthmatic attacks and all of the illnesses which I described to them."

Later in 1978, on 20 October, the Arizona Department of Health Services issued a "medical update" to doctors, entitled "Organophosphate Toxicity," warning the medical profession about the dangers of the chemicals being sprayed and informing them of how to treat poisoning cases. In so doing, it confirmed Mrs. Prosnier's suspicions about the source of her family's illnesses:

> Between Labor Day and Columbus Day, the Department of Health Services and several other governmental agencies received hundreds of calls from citizens complaining of symptoms they felt were related to exposure to pesticides. All were residents in neighborhoods near cotton farms where the fields had been sprayed with organic phosphates and other pesticides.
>
> This year, spraying has been more frequent than in the past, an attempt to combat the severe pest problem ... The increase in the number of complaints parallels a marked increase in the use of chemicals...[48]

The memo went on to warn of the hazards of "the prototype organophosphate ... parathion [which] may be absorbed through any portal in the body, including intact skin ... There are reports of fatalities for children after only 2 mg. of parathion." It listed the common symptoms of organophosphate poisoning as "headache, nausea, abdominal cramps with diarrhea," as well as visual problems, "profound weakness, mental confusion and disorientation, and excess salivation and respiratory tract secretions which may lead to respiratory distress ... convulsions, incontinence, coma, and ultimately death."

Mrs. Prosnier later learned why the state's pesticide control board had assured her that the chemicals being sprayed were no more toxic than salt or aspirin. Rather than being a true regulatory body, the board is composed for the most part of representatives from the chemical interests, including pesticide salesmen and agricultural interests, such as the citrus and cotton industries. Moreover, Arizona's attorney general ruled that the state health department had no authority over pesticides, as this was the jurisdiction of the pesticide control board. As Congressman Albert Gore (D-Tenn.) observed, the board is

> set up in a way that guarantees, or seems to guarantee, that the industry's interests will be paramount. Out of 13 members on the board, by statute only 2 of them are public members, and the other 11 come from interests which directly benefit from unrestrained use of these chemicals.[49]

This prompted subcommittee Chairman Eckhardt to remark, "[I]t looks to me as though it can't be said that the fox is guarding the chickens. The fox, the weasel, the owl, and a number of other predators are on the board."

Two-and-a-half years later, Mrs. Prosnier says that the spraying in her neighborhood has "eased up considerably," as have the health problems of her family. But heavy spraying was carried out twice in 1980 and once in 1981, each time resulting in members of her family and "lots of other people" becoming sick with bloody noses, deep chest coughs, bloody urine and other familiar symptoms. "There's just no question or doubt it's related to the spraying," she insists.

After one such episode, her mother, who lived nearby, became ill and had to be hospitalized. But she continued to deteriorate, and died a few months later, with "cancer all over her body." Although the mother's parents lived to their 80's and 90's, she died in her early 60's.

Many of the families who had originally complained about the pesticide spraying have moved away to escape the contamination, Mrs. Prosnier says, but she knows an unusually large number who have died of leukemia, undergone surgery, or had their health destroyed: "so many of them...I know many who are miserable; they'll never be the same."

She closely follows the obituary notices, and believes that there is a clear pattern of seasonal peaks correlating to the spraying schedule. Mrs. Prosnier herself has developed skin cancer.[50]

Fighting—And Spreading—The Fire Ant

The two-decades long chemical war waged against fire ants in the southeastern United States provides a good example of the dangers and futility of relying on toxic chemicals to eradicate or control an insect "pest." In this program, the people and the environment of the South ended up being poisoned—but the fire ants prospered.

The principal weapon in the war against the ants has been mirex, a cancer-causing pesticide that contains and degrades into kepone, the chemical that has contaminated and caused the closing of major fishing areas in Virginia (see pages 231-22). Mirex has been so widely used throughout the southern U.S. against the fire ants that up to one-third of the people there (and almost one-half in Mississippi) carry residues of it in their bodies. The dangers it poses to the environment and to human health, including its ability to cause cancer and birth defects, have been well known for over a decade. It is now found in mothers' milk and can be transmitted to unborn children through the placenta.[51]

Mirex was preceded by other toxic chemicals, such as dieldrin and heptachlor, which caused such devastating losses among wildlife, pets and livestock that their use had to be restricted (see pages 354, 364-65). And when mirex itself was finally banned in 1978, it was replaced by ferriamicide—a chemical made with mirex and which breaks down into a compound called photomirex that is up to 100 times more toxic than mirex and can cause cancer, liver damage and reproductive problems.[52] Soon, ferriamicide was banned as well. In fact, as Bill Butler of the Environmental Defense Fund pointed out at the time, "Every major pesticide used in the 20 year federal state fire ant control program has been applied before being adequately tested, and subsequently cancelled because of the environmental damage and human health risk."[53]

Ironically, the massive use of mirex for two decades may have actually helped spread the fire ants. A 1977 U.S. Department of Agriculture study found that fire ant populations greatly increased after the chemical was applied, in part because it killed off indigenous ants that competed for food with the fire ants. Indeed, after the expenditure by the federal government of some $100 million, and by the states of $70 million, on mirex and other pesticides, the ants now infest a larger area than ever: they have expanded from 200,000 acres in 1932, to 31 million acres in 1963, to over 126 million acres in 1971, to over 230 million acres in 1981.[54]

A final irony is that the dangers posed by the ants, which are clearly a nuisance in some areas, have been greatly exaggerated. When journalist Robert S. Strother decided to write an article for the June 1959 issue of *Reader's Digest*, he became intrigued with the propaganda circulated by the U.S. Department of Agriculture (USDA) on the alleged menace of the ants, which, USDA contended,

> had captured much of the South's best farm land and were eating their way north and west, sucking plant juices, killing young wildlife, and swarming in vicious assaults on men in the fields. Their onslaught, if unchecked, might not stop short of California and Canada.

After visiting the war zone and talking to farmers and agricultural experts, Strother concluded that "the fire ant is not a serious crop pest, it may not be a crop pest at all." One authority told him:

> Damage to crops by the imported fire ant in Alabama is practically nil. This department has not received a single report of such damage in the past five years. No damage to livestock has been observed. The ants eat other insects, including the cotton boll weevil. It is a major nuisance, but no more.

Indeed, the "facts" propagated by USDA seemed to be just the opposite of the truth:

> Though USDA circular no. 350 asserted that the imported fire ants often attack newborn calves and pigs, are fond of quail and chase brooding hens off their nests to eat their chicks, researchers . . . could rarely induce fire ants, even starving ones, either to eat plants or attack young animals. Instead, the insects became cannibalistic and ate one another.[55]

In fact, a study by the National Academy of Sciences issued in 1972 concluded: "Information accumulated during the last 15 to 20 years shows that the imported fire ant is of virtually no importance as a direct pest to crops, livestock, or wildlife." The study went on to report that fire ants are in some ways beneficial, since most of their diet consists of a variety of insects, including termites, boll weevils, ticks, borers and other potentially harmful creatures.[56]

The main reason this ill-advised program continued for so long is because of pressure from politicians throughout the South. And behind the politics was a consideration so often found at the root of these problems: money. A major factor at stake in this fight was the annual $20 million in matching state and federal funds that go for fire ant control to state agriculture commissioners in the South. As *Washington Post* reporter Bill Richards wrote,

> Without the federal pesticide approval, the money might be lost and with it, a powerful political weapon for the southern commissioners, many of whom are elected and all of whom hold unusual political power in the rural South. "Forget the ants," one federal official said. "The key is the money. It lets those guys make jobs, launch fleets of spray planes, and get up at the Fourth of July picnics and tell everyone that without us to get those ants, you would not be here."[57]

The Two Major Families Of Poisons

The great majority of pesticides in use today fall into two chemical families, the members of which are similar in their characteristics and effects on humans.

One group, known as chlorinated hydrocarbons, or organo-chlorines, include such deadly chemicals as DDT, aldrin, kepone, dieldrin, chlordane, heptachlor, endrin, mirex, PCB's, benzene hexachloride (BHC), toxaphene and lindane. All of these have been banned or drastically restricted, or are being reviewed for such action, because of their ability to cause cancer, birth defects, neurological disorders, and severe harm to wildlife and the environment. They are extremely stable, persist in the environment, are passed up the food chain (leaf-moth-fish-human), and, being fat soluble, bioaccumulate in humans.

Two of these chemicals undergoing cancellation review are the most widely used insecticides in the country. Utilized at a rate of over 100 million pounds a year, toxaphene is the most heavily applied outdoor insecticide. And lindane is the most popular for indoor use, exposing an estimated 126 million Americans a year to this cancer-causing compound, according to one published EPA estimate.[58]

These and the multitude of other pesticides of this group still in widespread use, can be expected to exhibit similar, if not as readily apparent, characteristics as those that have been largely banned.

Chlorinated hydrocarbons have long been recognized for their extremely toxic and persistent nature. Since the members of this group of chemicals exhibit similar characteristics and effects, environmentalists have urged that this entire pesticide family be banned or greatly restricted instead of the government taking years and spending inordinate amounts of money, time and manpower to examine and act on each one individually.

Unfortunately, pressure and delaying action by the chemical industry has prevented more than just a handful of these pesticides from being strictly regulated.

But even the ban on major uses of the most obviously toxic of the chlorinated hydrocarbons came too late to prevent the contamination of virtually the entire American population. By the time EPA banned major pesticide uses of dieldrin, BHC, heptachlor, DDT and PCB's because of their link to cancer and other disorders, these chemicals were being found in 97 percent or more of the human tissue samples tested, as well as in most fish, meat, dairy products and mothers' milk (see page 246).

The other major family of insecticides are known as organic phosphates, and consist of such chemicals as malathion, parathion, leptophos and the carcinogenic flame-retardant tris, which, it is feared, may cause cancer in perhaps hundreds of thousands of children who wore sleeping garments treated with it (see pp. 282-92). Although these chemicals are less persistent in the environment and break down more rapidly than chlorinated hydrocarbons, the immediate (acute) effects of organophosphates can be severe and include paralysis, tremors, convulsions, coma and death. Less serious symptoms include drowsiness, confusion, cramps, diarrhea, headaches, anxiety, increased sweating and difficulty in breathing. Since such disorders are common, they are often not attributed to organo-phosphate exposure, and poisoning victims are frequently misdiagnosed.

A study of the long-term effects of such common organophosphates as malathion, parathion, and mipofax upon human electro-encephalograms shows that just one exposure can alter the electrical activity of an infant's brain for years and possibly cause abnormal behavior and learning patters.[59] Other studies at Harvard Medical School on the tendency of organo-phosphates to cause insomnia, irritability, memory loss, difficulty in concentrating and decreased libido, have concluded that "there is a dangerous possibility that organophosphate pesticides have the potential for causing long-term brain damage."[60]

Another group of insecticides, called carbamates, are widely used not only as insecticides, herbicides, and fungicides, but also in medicines, insulating material, clothing and plasticizers. Some of these pesticides, such as carbaryl ("sevin") are extremely toxic to honey bees and other pollinators.

In December, 1980, EPA released its "Decision Document" on carbaryl, produced by Union Carbide, describing several animal studies and other tests showing that the chemical is teratogenic (causes birth defects), mutagenic (alters genes), and that an extraordinary number of workers producing it had dam-aged sperm, depressed sperm counts and were functionally sterile. At the same time, EPA announced that it had decided not to initiate action to restrict carbaryl, 53 million pounds of which were produced in 1972 in the U.S., with half being exported.[61]

There is, in sum, no such thing as a totally safe chemical insecticide. Such compounds are biocides, designed to kill life forms, and even the least dangerous of these substances, if used at all, should be handled with extreme care.

Waging Chemical Warfare On Americans

It is quite natural that many of the pesticides in use today are toxic to man. Indeed, they were designed precisely for the purpose of killing off living creatures, in some cases, humans. Some forms of these chemicals that we employ so casually around our homes and gardens had such lethal and horrid effects when used during World War I that chemical agents generally have been outlawed for use as weapons in war by the Geneva Convention on chemical and biological warfare. (Chemical weapons caused 1.3 million casualties in World War I, 91,000 of which were fatalities. On 22 April 1915, during the Battle of Ypres, a German gas warfare attack quickly killed 5,000 soldiers and disabled 10,000.)[62] However, there is no law to prevent such chemicals from being used on our food and in consumer products.

For example, phosgene, which caused over 80 percent of the deaths due to gas in World War I, is used to produce herbicides and insecticides. It has several other industrial applications and sometimes causes phosgene poisoning, such as when people have used chemical paint removers in poorly-ventilated rooms.[63] Chemical dump sites can also present a potential danger of phosgene poisoning. Following the investigation of an explosion of a farm near Providence, Rhode Island where a mixture of toxic wastes was being stored, a chemist advised local authorities that if some of the chemicals were exposed to heat, they could produce phosgene.[64]

Another insecticide used during World War II may have caused more human deaths in the late stages of the war than did bullets and bombs. The pesticide Zyklon-B, a powder that gives off hydrogen cyanide gas, was used by the Nazis at Auschwitz and other concentration camps to gas to death millions of Jews, gypsies, mental defectives, political prisoners and other victims of the Holocaust.

The ingredients for many other modern day insecticides were first produced to be used as weapons of war against human beings. This is especially true of the group of insecticides called organophosphates, which are among the world's most poisonous substances. The ability of these derivatives of phosphoric acid to kill insects was first discovered by a German chemist, Gerhard Schrader, in the late 1930's. Shortly thereafter, the Nazi government, recognizing the potential for using these chemicals as weapons, classified as "secret" research on and production of them, and began to manufacture both deadly nerve gas and insecticides from the same basic chemicals. Both groups of poisons have a closely related chemical composition and structure, and were designed to affect the nervous systems of their victims and to produce tremors, muscular spasms, convulsions, and death. Both work extremely well at achieving this result, and are absorbed through the skin as well as the lungs.[65]

For example, the deadly nerve gas sarin is quite closely related to malathion, one of the most widely-used ingredients in household insecticides and mosquito spraying programs. Some 20 years ago, Rachel Carson, in *Silent Spring*, described the dangers of malathion, pointing out that it causes severe muscular weakness, among other ailments. Moreover, its effects are multiplied when it is combined with other chemicals:

> A few years ago, a team of Food and Drug Administration scientists discovered that when malathion and certain other organic phosphates are administered simultaneously, a massive poisoning results, up to 50 times as severe as would be predicted on the basis of adding together the toxicities of the two. In other words, 1/100 of the lethal dose of each compound may be fatal when the two are combined.[66]

As Rachel Carson observed, the implications of this inter-action could affect all of us:

> The two need not be given simultaneously. The hazard exists not only for the man who may spray this week with one insecticide and next week with another; it exists also for the consumer of sprayed products. The common salad bowl may easily present a combination of organic phosphate insecticides. Residues well within the legally permissible limits may interact.[67]

Two decades later, malathion and countless other toxic chemicals that the government has determined are dangerous continue to be used in massive quantities, with possibly severe impacts on human health.

Malathion often kills insects in the same way that it and other nerve gases affect humans—by acting on their nervous system. This is what made the chemical attractive to the Nazis when they developed it as a chemical warfare agent during

World War II. Dr. Frank H. Duffy, of Harvard Medical School, reports that his studies show that small amounts of malathion cause long-term alteration in brain waves, resulting in insomnia, irritability and other abnormalities.[68]

One highly publicized use of malathion began in the summer of 1981, when an infestation of Mediterranean fruit flies was discovered in California and Florida. Despite the opposition of environmentalists and initially of California Governor Jerry Brown, huge amounts of the chemical were repeatedly sprayed over large areas of these states, including residential counties adjacent to San Francisco, comprising over 100 square miles and populated by almost 800,000 people. Governor Brown agreed to allow the spraying only after the U.S. Department of Agriculture threatened to quarantine the state's $14 billion fruit crop.

By August, over 1,300 square miles were being sprayed every week. Children, pregnant women, and nursing mothers were warned to leave the target area or at least to stay indoors for a day or two at the minimum after the spraying, and emergency shelters were set up to handle the anticipated flow of "pesticide refugees" (which never materialized in large numbers). Entymologists predicted that the pesticide would have to be sprayed for two years in order to be effective against the "Medflies."[69]

Additional evidence of malathion's toxicity has been ignored or suppressed. In 1981, pathologist Dr. Melvin Reuber was forced to resign from a federally funded research laboratory after he insisted that three National Cancer Institute studies supposedly showing the safety of the pesticide, actually demonstrated that it caused cancer. And after California state pathologist Marc Lappe released an internal study showing that malathion was more dangerous than the state's health services department claimed, he was removed as chief of its hazard evaluation office. By the end of 1981, after some 160,000 gallons of malathion had been sprayed, about $10 million in damage claims had been filed by California residents against the state on the basis of losses to health and property. Acute health complaints included such symptoms as nausea, headaches, and loss of appetite; long-term effects remain to be determined.

In addition, the U.S. Department of Agriculture required that California fumigate much of the fruit and vegetables from infested areas that was sent out of state with ethylene dibromide (EDB), most of the uses of which EPA had in December 1980 sought to ban because of its link to cancer and birth defects. Dr. Adrian Gross, chief scientist for EPA's hazards evaluation division, charged that in approving this use of the chemical, EPA "could mislead the public about the dangers involved," and called EDB "probably the most potent and toxic carcinogenic substance used as a pesticide today." Low level exposure to EDB causes cancer in 80 to 90 percent of the animals tested, and in a very short time period. During nation-wide testing, EPA found the chemical in almost all samples of grain analyzed, so it can be assumed to be present in bread and other food products. Nevertheless, in the fall of 1981, Dr. John Todhunter, assistant administrator of EPA for pesticides and toxic substances, disregarded this data and recommended that the use of EDB be approved.[70]

Since a very small percentage of pesticide poisonings are ever reported in the U.S. or abroad, it is impossible to know the extent of harm that malathion has caused. But in Pakistan in 1976, a poisoning incident occurred which was so massive it could not be covered up: at least 2,900 people were seriously poisoned, and five died when malathion was carelessly used in a village.[71]

Other organic phosphates used throughout the U.S. also exhibit the deadly characteristics of their close relatives in the nerve gas family. One of the most widely used is parathion, the lethal qualities of which were also described long ago by Rachel Carson. She estimated that the amount being used just in California was sufficient to kill 5 to 10 times the world's population. She told of a chemist who swallowed a very small dose, amounting to .00424 ounce, and was paralyzed instantaneously and died before he could take the antidote he had prepared and had at hand. In Florida, two children died after handling an empty bag that once contained parathion. On a Wisconsin farm, two small boys were killed by parathion poisoning on the same night: one got a whiff of fumes from a nearby field of potatoes being sprayed, the other touched the nozzle of the spray equipment. In Riverside, California, eleven out of thirty farmworkers became violently ill while picking oranges from a grove that had been sprayed with parathion two-and-a-half weeks earlier.[72]

In September 1967, parathion-contaminated sugar caused the poisoning of 600 children, 17 of whom died, after they ate some pastries in Tijuana, Mexico. Two months later, another 600 people became ill, and 80 more died, after eating parathion-contaminated bread in Chiquinquira, Colombia.[73] A 1979 report prepared for the Department of Labor calls parathion "extremely toxic," describes how it produces tumors in test animals and notes:

> More than 550 cases of poisoning due to parathion were
> reported in California alone between 1969 and
> 1975...Most of these cases involved workers harvesting
> citrus fruit...In some cases of poisoning, spraying
> occurred more than 21 days before harvest...residues
> have been reported to persist for several years.[74]

The government and the chemical industry are well aware of parathion's dangers, yet permit it to remain in widespread use. Ironically, during the 1972 cancellation hearings on DDT, a major argument cited by that chemical's users and manufacturers was that a ban on it would require the substitution of an even more dangerous compound—parathion—on tobacco, cotton, and other crops. In his decision cancelling most uses of DDT, EPA Administrator William Ruckelshaus pointed out the dangers of parathion and other organophosphates, characterizing methyl parathion as "... a highly toxic chemical," and warning:

> The introduction into use of organophosphates has, in the
> past, caused deaths among users...A survey conducted
> after the organophosphates began to replace chlorinated
> hydrocarbons in Texas suggests a significantly increased
> incidence of poisonings.[75]

This then, is one of the most puzzling and inexcusable facets of the government's pesticide policy: that it permits the massive use of chemicals that it knows, and admits, will cause poisoning and death among those who use them.

The Myth Of The "Banned" Pesticides

As of 1977, there were approximately 33,000 to 35,000 registered pesticide products on the market using some 1,500 active ingredients combined with 2,000 other possibly toxic substances. Over 100 of these pesticides in general use today are thought to cause cancer. Specifically, EPA estimates that about one-third of the active ingredients used in pesticides are toxic, and one-fourth carcinogenic.[76]

In the decade in which it has had jurisdiction over pesticides, EPA has effectively banned the domestic use of only about a dozen of these by cancelling their registrations. (In addition, an unknown number of pesticides have either not been granted registration or were withdrawn voluntarily by the manufacturer).

One often comes across erroneous references to certain highly toxic pesticides, such as DDT, as having been "banned." Ronald Reagan has even complained that "the world is experiencing a resurgence of deadly diseases spread by insects because pesticides like DDT have been prematurely outlawed."[77] However, the sad fact is that after many years of efforts by scientists, conservationists and some government officials, very few restrictions have been placed on pesticides. Despite the overwhelming evidence that many pesticides cause cancer and are extremely damaging to humans and the environment, almost none of these chemicals has ever been "banned" by the government in the true sense of that word. In the very few cases where pesticides have been the subject of suspensions, cancellation proceedings and/or court actions, the results have usually been restrictions or bans placed on some or most uses, while other applications have been allowed to continue.

Even in the handful of cases where all domestic use has been prohibited (such as with kepone), manufacture and export abroad can still be (and often is) legally undertaken. When production for export does continue, it inevitably results in exposure of workers and the public through pollution, dumping of wastes and other accidental and intentional release of the substance, as well as through foodstuffs imported from foreign countries to which the chemical is exported. In the case of DDT, referred to above, not only has it been extensively used in the U.S. since it was supposedly "banned" in 1972, but 44 million pounds are still produced in the U.S. and exported to foreign nations with some of it returning to us on imported food (see pages 348-49).

The way EPA has carried out its policy on "banning" toxic pesticides has been contradictory and nonsensical. In numerous instances, the Agency has proclaimed a chemical hazardous to humans and the environment, and prohibited major uses of the compound. At the same time, it has allowed other uses to continue as before, often depending on label instructions to ensure adherence to the prohibitions. Several examples of such actions being taken to restrict but not ban such cancer-causing chemicals as aldrin, dieldrin, chlordane, heptachlor, DDT, 2, 4, 5-T, mirex, DBCP and others are discussed in detail elsewhere in this book. In these and other instances, the action by EPA has often been presented to the public by the news media as another pesticide being "banned," when in fact no such thing has happened.

In cancelling the registrations for major uses of these most obviously dangerous pesticides, EPA, in many cases, allowed them to be slowly phased out and/or permitted the sale and distribution of existing stocks. This made it possible for large quantities to continue to be sold and used long after they were ostensibly banned, and even for users to stockpile quantities of these poisons for future use.

The implications of this EPA policy should be obvious. Unless each person who purchases such chemicals is forced to undergo a lie detector test, it is impossible to know what use will be made of these products. Indeed, EPA end-use restrictions are considered a joke within the agricultural community, and many farmers use certain chemicals as they always have, regardless of whether or not such applications have been banned.

Moreover, there is no way to determine how and where such carefully restricted and regulated chemicals are being disposed of. Common sense would dictate that if a chemical is so dangerous that it has been declared a threat to humans and the environment, it should simply be banned—its manufacture, sale, transport, use, or possession prohibited to the extent that the law allows.

Pesticides that have been restricted are listed in the EPA publication, "Suspended and Cancelled Pesticides," which gives the conditions, if any, under which these substances may be used. The rules and regulations outlined in this largely incomprehensible document are so complex and unfathomable that their only real use and value are to lawyers and bureaucrats interested in the byzantine world of federal regulations. It is difficult to believe that farmers and agricultural workers, particularly illiterate or Spanish-speaking migrants, could in any meaningful way decipher the wording or understand the meaning of these rules, or the labels that are required to be affixed to pesticide products.

For example, 2,4,5-T cannot be used in "ponds" or on "food crops intended for human consumption," but can be applied to rice, a food crop grown in pond-like areas. It cannot be sprayed on pastureland, where cattle graze, but can be used on rangeland, where livestock also forage. The regulations for endrin prohibit spraying within certain distances from lakes, ponds and steams, varying from 50 feet to ⅛ mile depending on the crop being sprayed. The instructions allow spraying closer to ponds that are owned by the user, but warn against its causing fish-kills. The document also gives detailed instructions on what to do if fish-kills do occur and how to dispose of the fish (burial), and requires the posting of a "No Fishing" sign for half a year or a year after the kill has occurred. Under "aerial application," the regulations warn,

> Do not operate nozzle liquid pressure over 40 psi or with any fan nozzle smaller than 0.4 gpm or fan angle greater than 65 degrees such as type 6504. Do not use any can type nozzles smaller than 0.4 gpm nor whirl plate smaller than #46 such as type D-4-46...[78]

Ground application instructions are equally bizarre. But in the final analysis, these details don't really matter. What happens in real life is that an illiterate migrant worker is given a container of endrin and told to spray it on the crops, and that's that.

Label requirements for other potentially toxic chemicals, even when clearly written, are equally useless. For example, the brochure requires that metaldehyde, used to kill slugs and snails, "must have the following statement on the front panel of the product label: This pesticide may be fatal to children and dogs or other pets if eaten. Keep children and pets out of treated areas."[79] But such a warning in no way prevents this extremely toxic chemical from being widely used throughout the nation in precisely the way the label cautions against. Not even U.S.

government agencies, including those oriented toward the environment, pay attention to the labels. In recent years, the National Park Service has used metaldehyde on the White House lawn and the public grounds around and between the White House and the Capitol. In 1977, 88 pounds were used just in the President's Park, [80] an area adjacent to the White House. Such use in these areas not only potentially endangers the President and his family (Secret Service take note!) but also countless thousands of tourists who flock to the Ellipse, the Mall and other areas around the White House and the Capitol grounds each spring and summer.

As Shirley Briggs, executive director of the Rachel Carson Council, points out, such use cannot conceivably meet the requirements of the EPA directive:

> I see no way anyone is going to keep children and pets out of the Ellipse. Material spread around that much must get into dust and soil and can, in turn, get into the hot dogs, or be transferred from hand to mouth generally. And how can there be that many slugs to kill?[81]

Other extremely toxic pesticides with strict warning labels that have regularly been used around the White House grounds include 2,4-D, Silvex (the use of both of which has recently been restricted there), paraquat, benomyl and malathion.[82] The casual use of these and other deadly chemicals on the White House lawn, despite the potential danger this represents to the first family, Cabinet members, visiting heads of state and other VIP visitors and tourists, makes it difficult to expect that the public will anytime soon be given adequate protection from such hazards.

EPA's labelling and restricted use policies, unintelligible and contradictory as they may seem, do serve several purposes. They keep the public reassured that it is being protected from "banned" and "restricted" pesticides, while allowing the chemical and agribusiness interests to carry on business as usual with many of these products. It is, in fact, an ideal arrangement: it keeps the bureaucrats, the politicians and industry happy, even if the public does get poisoned in the process.

Conservation Through Poisoning

Ironically, millions of dollars are being spent by the federal government to spray pesticides under a conservation program set up for the benefit of wildlife! This is being done by the Interior Department under the U.S. Fish & Wildlife Service's federal aid to the states for fish and wildlife restoration programs.

As part of this program, pesticides are used to manipulate and alter habitat in order to maximize the growth of vegetation preferred by deer, thereby propagating a maximum number of these "game" animals for hunters. This, of course, results in a continued but artificial "surplus" of deer and increases the sale of guns, ammunition and hunting licenses, a prime source of revenue for state and federal fish and game agencies. As the Fish & Wildlife Service's (F&WS) 1978 Draft Environmental Impact Statement (DEIS) put it, "Chemicals are used to alter the composition of the environment to make the habitat more conducive to wildlife. Herbicides may be applied to eliminate undesirable vegetation in order to allow the growth of a wildlife food source."

Under this program, 35,415 acres were treated with pesticides in 1975 at a cost of $331,000. F&WS estimates that 58,722 acres will be under treatment by 1985, at a cost of $983,000. The projected cumulative total for 1975-1985 is 517,754 acres treated at a cost of $7,226,000.

In its chemical "fish control" program, the F&WS, in order to kill off "undesirable" fish and to promote the growth of "more popular game fish," expects to dole out between 1975-1985 $2.6 million for the use of toxicants on 112,585 acres and 2,112 miles of free-flowing streams. In addition, the cost of chemicals from 1975-1985 to control aquatic weeds is expected to be $374,000 to treat 4,461 acres.[83]

Herbicides are also used to "eliminate undesirable vegetation from marshlands and coastal areas, for weed control in agricultural operations" and to "improve the production of waterfowl food plants" by killing off certain fish, such as carp. (Often, when carp and other "nuisance" fish are poisoned off, numerous non-target fish and other creatures are killed as well.) In 1975, 19 states used herbicides such as 2,4-D on 17,227 acres for these purposes. These herbicides are also used to control unwanted vegetation to benefit such fish as trout and bass.

There is almost no discussion in the DEIS of the harmful effects on wildlife and the environment of the use of these toxicants. The document does admit that "Chemical treatment kills fish and some invertebrates." And under the section on "Unavoidable Adverse Environmental Impacts," it states that "FDA approved pesticides for plant, insect, and animal control will be applied to 520,000 acres between 1975 and 1985. Short-term contamination of the soil will result." Most of the adverse effects of pesticides on wildlife are simply ignored in the DEIS. And nowhere is it adequately explained how the government can justify taking millions of dollars of funds set aside for conservation purposes and spending it on toxic chemicals to poison hundreds of thousands of acres of fish and wildlife habitat.[84]

Ignoring The Law

The "poisoning of America" has not taken place accidentally or in secret. On the contrary, it has been the result of a deliberate policy designed to continue and increase the sale and use of toxic chemicals. It has been fostered by a coalition of government and industry officials, many of whom move back and forth, through the classic "revolving door", between the private and public sectors, one year regulating the chemical industry for government, the next year working for that same industry.

For the last few decades, various U.S. government officials have flouted the laws and legal procedures laid down by Congress to ensure that toxic chemicals are properly controlled. And on those rare occasions when the regulatory agencies have attempted to remove hazardous chemicals from the marketplace, powerful members of Congress have intervened to block such action. In addition, the chemical industry has often been able to stymie such efforts through legal delaying actions and lawsuits that have regularly tied up government agencies and environmental groups for years at a time. Yet, there can be no mistaking the clear language and intent of the law.

The U.S. government has clear authority, under several statutes, to keep harmful chemicals off the market. The most important of these is the Federal Insecticide, Fungicide and Rodenticide Act (FIFRA), Section 12 of which bans the marketing of pesticides which are not registered by EPA. Before such a product can be registered, EPA must determine that it will "perform its intended function without unreasonable adverse affects on the environment," which is defined as "any unreasonable risk to man or the environment, taking into account the economic, social, and environmental costs and benefits of the use of any pesticide." EPA may cancel a pesticide's registration if the product "...generally causes unreasonable adverse effects on the environment."

The burden of proving that a pesticide satisfies the registration standard is on the manufacturer and user.[85] Yet over the last decade, EPA has adopted just the opposite interpretation of this statute, requiring citizens and public interest groups to overcome the enormous burden of proof involved in showing that a chemical is hazardous.

The fact that the federal government long ago abdicated its responsibility to protect the public from hazardous chemicals has been thoroughly documented through the years—by books, Congressional hearings, government advisory committees and numerous other investigations. Virtually no effective changes have been made in the way the system has always operated to prevent meaningful action from being taken against pesticides harmful to human health.

Ironically, EPA, which is now under attack from environmentalists for refusing to restrict the use of hundreds of carcinogenic chemicals, was given jurisdiction over pesticides in 1970 largely because the U.S. Department of Agriculture (USDA) had failed to protect the public. But in taking over Agriculture's responsibilities, EPA also adopted many of its personnel and policies as well; therein lies the problem.

USDA's Sellout To Industry

The U.S. government has had authority to regulate toxic chemicals for over three decades, since passage of the original Federal Insecticide, Fungicide, and Rodenticide Act (FIFRA) of 1947. In 1964, USDA agreed to permit the Departments of

Health, Education, and Welfare (HEW) and Interior to help determine if pesticides under review constituted a threat, respectively, to human health, or to fish or wildlife. However, USDA never took these provisions seriously nor attempted to enforce them.

In September 1968, Congress' General Accounting Office (GAO) issued a report taking USDA to task for failure to enforce the law. The study cited numerous instances in which Agriculture allowed toxic chemicals to be used in an unsafe manner over the repeated objections of the U.S. Public Health Service and other agencies. The report disclosed that it was departmental policy, when a hazardous product was identified, to seize only that sample and not to remove from the market other quantities of the same product located elsewhere. Perhaps the most damning indictment was the revelation that USDA had never initiated a single criminal prosecution in 13 years under FIFRA despite overwhelming evidence of numerous and repeated illegal acts by some shippers.[86]

In November 1969, following extensive hearings in May and June by subcommittee Chairman L. H. Fountain (D-N.C.), the House Government Operations Committee issued a scathing report detailing how USDA had ignored its responsibilities to regulate the use of potentially lethal chemicals. The report set forth an almost unbelievable account of collusion between the pesticide industry and the government agency with responsibility for regulating that industry's products, the result being that USDA "failed almost completely to carry out its responsibility" to enforce the law. The report described how the registration of pesticides was granted almost automatically: over 45,000 such products were registered as "safe and effective when used as directed" without any meaningful review or the submission of adequate information. The report listed the following "Findings and Conclusions," among others:

> Until mid-1967, the USDA pesticides regulation division (PRD) failed almost completely to carry out its responsibility to enforce provisions of FIFRA intended to protect the public from hazardous and ineffective pesticide products being marketed in violation of the act...

Numerous pesticide products have been approved for registration over objections of the Department of Health, Education, and Welfare as to their safety...

HEW objected to proposed registrations for more than 1,600 pesticide products during the 5-year period ending June 30, 1969.

...PRD does not keep records of products registered over HEW objections and has failed or even refused to inform HEW of action taken with respect to its objections.

PRD has approved pesticide products for uses which it knew or should have known were practically certain to result in illegal adulteration of food...

PRD has no procedures for warning purchasers of potentially hazardous pesticide products.

The Agricultural Research Service failed to take appropriate precautions against appointment of consultants to positions in which their duties might conflict with the financial interests of their private employer. Facts disclosed by the subcommittee investigation raised a number of serious conflict of interest questions.[87]

The report discussed in detail these questions of "serious conflicts of interest" on the part of government and industry officials, with the latter often being directly involved in drafting the administrative regulations and policies on pesticide use for products made by their employers. For example, in 1965, Dr. T. Roy Hansberry was appointed to serve on a seven-member task force to advise PRD on pesticides. At this time, Dr. Hansberry was an agricultural research official for a company owned by Shell Chemical (a subsidiary of Shell Oil). In addition, Dr. Mitchell R. Zavon served from 1963-1969 as a consultant to PRD while at the same time serving as a paid consultant to Shell, and holding stock in a Shell affiliate.[88] But the most flagrant situation involved John S. Leary, Jr., who served as PRD's head of pharmacology. After overruling objections from one of his subordinates and the Public Health Service (PHS), he permitted Shell to market its "No-Pest Strip" in a way that had been shown to be hazardous. Shortly thereafter, and 25 days after rejecting a report critical of the product from a PHS Committee, he left PRD to go to work for Shell.[89]

Indeed, Shell's ability to market this hazardous product despite objections from scientists and government officials characterized the firm's cozy relationship with Agriculture. In 1963, the "No-Pest Strip" was first registered by PRD for use in restaurants and other places where food is left uncovered, despite the near certainty that food would be contaminated by poisonous residues. In fact, tests performed by Shell confirmed this contamination, and USDA meat inspectors refused to allow the strips to be used in areas where meat was processed.[90]

From 1965 until early 1969, the Public Health Service strongly objected to registration of this product since it resulted in humans continually being exposed to the pesticide. In late 1967, PHS agreed to withdraw its objections if an adequate label was placed on the product warning against exposure of children, the aged and the ill. Thereafter, PRD repeatedly told Shell that the warning label had to be added, but Shell ignored the demand until January 1969, when it agreed to comply.[91]

Shell's influence with the government is hardly surprising in light of the fact that three of the company's employees—Hansberry, Zavon and Leary—were intimately involved in USDA's decision-making process, with the latter two specifically working on the "No-Pest Strip." The most unabashed and active advocate of Shell's views was Leary of PRD. On 6 March 1963, representatives from Shell visited PRD to ask the agency to withdraw its order that Shell's proposed pesticide strip label be changed to bear the word "Poison" and a skull and crossbones. On the same day, PRD's order was withdrawn, and the following day registration was granted for the strip.[92]

In 1965 and 1966, PHS continued to express, orally and in writing, its opposition to the registration of Shell's pesticide strip products, but with Leary's help, the registration was maintained. On 14 November 1966, Leary resigned from USDA, effective on 31 December, "to accept a position with Shell Chemical Company." Before leaving, Leary did one last favor for his employer-to-be. On 6 December, three weeks after giving notice of his planned departure, Leary drafted a memorandum attacking PHS' arguments and stating that its report "serves no useful purpose" and "should be set aside."[93]

Shell's close relationship with USDA also had one other remarkable result. When Congressman Fountain held his 1969 hearings, five years had passed during which PRD had authority under the 1964 pesticide act, FIFRA, to suspend immediately any pesticide when such action was "necessary to prevent an imminent hazard to the public." But during this period, PRD had used this authority on only one occasion, and did so to suspend an insecticide strip made by another firm that was competing with Shell! This was done even though the rival strip had the identical active ingredients and directions for use as Shell's. The only significant difference was in the material used to hold the active ingredient. Lest anyone wonder why this one action was taken, the committee report notes that "the case began when representatives of Shell visited the PRD to complain about" the rival product.[94]

EPA: A New Name For Old Policies

The incompetence, mismanagement and conficts-of-interest that characterized USDA's dealings with the chemical industry—and which helped lead to jurisdiction over this area being transferred to EPA—are little different from present-day practices at that agency.

Shortly before it set up shop, EPA was given a mandate by both Congress and the scientific-community to act expeditiously to control toxic chemicals. The Passage in 1969 of the National Environmental Policy Act (NEPA), and the establishment shortly thereafter of EPA was hailed as ushering in a new era, one in which protecting the environment would be a concern of all federal agencies. EPA would supposedly be the lead agency in acting to clean up pollution and prevent further degradation of the environment.

It soon became apparent that the control of toxic pesticides would have to be a priority goal of the new agency. In December of 1969, the Department of Health, Education, and Welfare (HEW) issued its landmark *Report of the Secretary's Commission on Pesticides and Their Relationship to Environmental Health,* warning of the hazards of chemical pesticides. The group that drafted the report, called the Mrak Commission after its Chairman, Dr. Emil Mrak, Chancellor Emeritus of the University of California at Davis, made a number of recommendations concerning the future use of several dangerous pesticides. Among other things, the Commission urged that "corrective action" be taken on pesticides containing arsenic, inorganic lead and mercury, as well as chlorinated hydrocarbons such as aldrin and dieldrin, benzene hexachloride (BHC), chlordane and heptachlor, endrin and lindane. Citing research findings that many of the chemicals involved cause tumors and/or birth defects, the Commission recommended that these pesticides, as well as captan, carbaryl, 2,4,5-T, forms of 2,4-D and others, "should be immediately restricted to prevent risk of human exposure." Concerning the prevalence of pesticides in our society, the Mrak report referred to "the absurdity of the situation in which 200 million Americans are undergoing lifelong exposure, yet our knowledge of what is happening to them is at best fragmentary and for the most part indirect and inferential."[95]

Today, over a decade later, with most of these chemicals still in widespread use, the Mrak Commission's recommendations are as relevant as ever, but we seem not much closer to heeding the warnings now than we were then. The main reason for this has been the inaction of EPA.

The Environmental Protection Agency was officially set up on 2 December 1970, having been given the responsibilities for pesticide regulation that the USDA had failed to fulfill. But action to limit the use of the most toxic of these poisons, if taken at all, was incredibly slow in coming. Even the strengthening amendments to FIFRA passed by Congress in 1972, giving EPA additional power to control toxic chemicals, did not succeed in stimulating the agency to take the required action.

From its start to the present, EPA has not used its authority effectively to remove toxic chemicals from the marketplace. EPA did not even *propose* to restrict lindane, for example, until July of 1980. It was not until six years after the Mrak Commission Report, and five years of litigation by the Environmental Defense Fund, that the use of aldrin and dieldrin was partially banned in the U.S. Many other chemicals cited by the Mrak Commission, such as 2,4-D, 2,4,5-T and captan, remain in widespread use, with chlordane and heptachlor also still in use on a more restricted basis. By 1979, a full decade later, EPA had acted on only six of the nineteen pesticides cited by the Mrak Commission as suspected hazards to human health.[96] And as of 1980, only about 60 chemicals were being appraised for re-registration.

In December 1976, after holding extensive hearings, Senator Edward M. Kennedy's Subcommittee on Administrative Practice and Procedure issued a devastating report on EPA's enforcement of the pesticide program.[97] The Subcommittee's conclusions were remarkably similar to those of the 1969 LaFountain report, except that the focus of the inquiry was EPA instead of USDA.

Few improvements had taken place in policy, competence or effectiveness since EPA had taken over the program, according to the Subcommittee, in part because "many of the personnel and much of the philosophy from USDA were simply transferred to the EPA pesticide program in 1970." Senator Kennedy concluded in the report:

> EPA has largely failed in its responsibility to assure the safe use of pesticides as mandated by Congress. As a consequence, EPA has failed the consumer and the farmer, as well as the pesticide industry.[98]

In the report's introduction, Subcommittee Chairman Kennedy made it clear that "EPA must shoulder the major share of the blame" for the intolerable situation that had developed:

> I find it incredible that a regulatory agency charged with safeguarding the public health and the environment would be so sluggish to recognize and react to so many warnings over the past 5 years. The EPA was warned and certainly should have known that testing data, submitted by industry as long as 25 years ago, should not be accepted at face value in the re-registration of thousands of pesticide products presently being used on our farms and in our homes. But EPA by and large ignored these warnings.[99]

Senator Kennedy also expressed amazement and dismay that EPA, upon learning of the invalidity of much of this data, did nothing about it.

The extensively documented report contains an entire section on how "EPA has misled the Congress, the General Accounting Office, and the public regarding its pesticide programs."

The conclusions of the Subcommittee report includes a scathing indictment of EPA, calling the agency's policies "fundamentally deficient:"

...the severe inadequacies of pesticide regulation are not attributable in any significant way to deficient legislation. Rather, the principal cause lies with EPA's poor administration of the program...[100]

A month after Senator Kennedy's report was issued, the Carter administration came into office and a new team took over at EPA, promising to correct the mistakes of the past. But four years later, at the end of Carter's term, the situation at EPA had changed little if at all. In some ways, the agency's policies had become worse than ever before, as is discussed throughout this book. And under the Reagan administration, the entire regulatory structure is being dismantled (see "Epilogue").

Cheating On Safety Tests

New pesticide products are still evaluated by the government for registration and use on the basis of information supplied by the products' manufacturers—hardly disinterested parties.

As the Senate Subcommittee on Administrative Practice and Procedure reported in December 1976;

> Regulation of pesticide use by the Federal Government is critically dependent on the safety testing data submitted by the firms that manufacture and market pesticides. EPA's pesticide program cannot even begin to fulfill its purpose if the data upon which it relies for regulatory decisions are deficient, incomplete, or otherwise faulty.[101]

That statement remains as true today as it was then. Time and again, it has been shown that such industry-financed tests are susceptible to manipulation and fabrication by those wishing to "prove" the safety of a chemical. This has happened with aldrin, dieldrin, chlordane, heptachlor, captan, leptophos and other chemicals that were approved on the basis of bogus data and later found to cause cancer, birth defects, neurological disorders and other serious health effects.

In June 1981, Dr. Joseph Calandra, former president of a major testing laboratory, Industrial Bio-Test (IBT) of Northbrook Illinois, and three of his employees were indicted by a federal grand jury in Chicago on charges of falsifying research results to make it appear that certain chemicals did not cause cancer or birth defects. The investigation called into question the safety of hundreds of pesticides, drugs, and food additives now in use. Over 20,000 tests for over 200 corporations and several federal agencies had been carried out by the lab over the last decade, and more than 100 chemicals were approved for use in the U.S. as a result of its tests.[102]

When EPA analyzed IBT's tests on some 200 pesticides, 75 percent of the tests were found to be unreliable and were declared worthless. However, it will take several years to re-test the chemicals, during which time they will remain on the market, even though in such cases EPA has the power to revoke their registration.

In one case mentioned in the indictment, during IBT's tests of TCC, a compound used in deodorant soaps, the lab replaced rats that had died with live ones, and did not report that the chemical affected the animals' testes. In tests on two pesticides used on a wide variety of fruits and vegetables, experiments were cut short and results were fabricated. When EPA tried to investigate IBT's tests on two other widely used pesticides—2, 4-D and captan—the lab admitted that the data had been shredded.[103]

Fortunately, there are often effective alternatives to using animals in such tests, perhaps the best being the Ames test, which takes advantage of the fact that bacteria contain genetic material similar to that of humans. Such bacteria-based tests are not only reliable but are also much quicker and cheaper to perform than animal tests. Developed by Dr. Bruce Ames and Joyce McCann of the University of California at Berkeley, the Ames test (also called the *Salmonella* test) can be completed in two days or so at a cost of $250 to $1,000 per substance tested, as opposed to animal tests which average at least two to four years, and cost $200,000 to $400,000. (A full set of tests on a pesticide can cost an average of $6 million.)

In test sampling of several hundred chemicals, the Ames test proved to be 90 percent accurate in determining whether or not a chemical was able to cause cancer. For example, the Ames test was able to immediately prove that the chemical Tris (used as a fire retardent in childrens' nightwear) was mutagenic (gene altering) and probably carcinogenic almost a year-and-a-half before animal tests confirmed Tris' carcinogenicity. By the time the garments were finally removed from the market, some 60 million children had been exposed to this potent cancer-causing chemical, and hundreds of thousands of children may contract cancer as a result. (See the chapter on Tris.)[104]

The Hopeless Ineffectiveness Of The Present System

How did we get ourselves into this situation, one might ask, in which the human environment is so pervasively saturated with dangerous pesticides that threaten not only our lives and health but those of unborn generations? The answer is that a combination of industry pressure and governmental inaction has resulted in denying the public its legal and moral rights.

With the government unable or unwilling to fulfill its responsibilities, the job of protecting the public from lethal chemicals has largely fallen to unions and public interest groups. Such organizations were responsible for initiating action on 22 out of 26 carcinogenic chemicals that had been regulated by the government as of 1979.[105] But citizen groups such as the Natural Resources Defense Council and the Environmental Defense Fund (EDF), consisting of overworked, underpaid scientists, lawyers and staffers, and operating on shoe-string budgets, can take on only a very limited number of projects. This is especially true in light of the fact that environmentalists are usually fighting an uphill battle against not only the U.S. government but also the giants of American industry, the chemical corporations. These firms have virtually unlimited resources, countless high-powered lawyers and lobbyists and huge advertising budgets, to say nothing of the many powerful politicians to whom they make generous campaign contributions, and who lend them their support.

While environmentalists have enjoyed some success in having a few of the most dangerous chemicals banned or restricted, the costs have been enormous to the private organizations involved, which are spread thin fighting for the public's welfare on a variety of fronts. And while the litigation to accomplish these victories was taking place—seven years for DDT, five-and-a-half years for aldrin and dieldrin and over three years for heptachlor and chlordane—thousands of new, untested and potentially carcinogenic chemicals were being produced and marketed. While a few Americans are trying to plug up the holes in the dike, an oncoming tidal wave of chemicals is threatening to engulf us.

Unfortunately, the outlook for citizens' groups helping to protect the public is not bright. On 17 July 1980, a federal appeals court ruled against EDF in a crucial decision and greatly restricted the right of public interest groups to be involved in pesticide safety decisions made by EPA.

As EDF pointed out in its 31 July 1980 brief:

> the Panel held in effect that economic interests wishing to challenge EPA pesticide regulation decisions as too restrictive are entitled to an administrative hearing under...FIFRA..., but those supporting more protective safety and health regulation of pesticides do not have a parallel right...
>
> It eliminates meaningful participation by *all* environmental, consumer, labor, and migrant worker organizations in *any* EPA pesticide regulatory decision involving proposed suspension, cancellation, or a change in a pesticide's use classification...EDF and other organizations concerned about the safe use of pesticides will be almost entirely excluded from effectively challenging EPA's regulatory decisions...

This decision, EDF observes, "will effectively eliminate all challenges to the adequacy of the safety and health aspects of a cancellation decision."

Interestingly, EPA, in seeking to prevent the participation of public interest groups in this case, reversed its policy of eight years. The Department of Justice was so appalled by EPA's legal position that it took the extraordinary step of refusing to represent the Agency in the case, and even withdrew from appearance on its behalf. EPA's decision to reverse its position was made by Attorney David Menotti and was personally concurred in by its Administrator Douglas Costle over the objections of some officials in the agency.

Meanwhile, powerful vested interests profiting from the spread of poisonous chemicals continue to dominate the legislative and administrative processes. The chemical lobby consistently succeeds in overwhelming the few political and governmental leaders who seem to be aware of and concerned about the current toxic chemical crisis.

Speaking of his 175 appearances before 66 different Congressional committees during his 4 years as head of EPA, Douglas Costle said in December 1980, shortly before stepping down as the agency's administrator, "You win very few victories; it's just varying degrees of defeat."[106]

The way special interests are able to manipulate our economic and political system virtually guarantees that we will persist in our present ruinous course of action. Yet the consequences of failing to confront immediately and bring the toxic chemical problem under control are unthinkable.

Thus, our present system gives us little opportunity to take meaningful action until the inevitable public health catastrophe looming ahead forces us to face the problem. But by then, it may be too late to remedy the situation.

Integrated Pest Management:
Controlling Insects Without Poisons

Perhaps the greatest irony in "the poisoning of America" is that the massive use of chemical pesticides has not only been unnecessary but self-defeating as well: it has *increased* crop losses to insects rather than reduced them.

A U.S. government study released in April 1980 documents this incredible fact. *Integrated Pest Management,* written by Dale R. Bottrell and published by the President's Council on Environmental Quality, details our folly in relying on synthetic pesticides and offers a sensible way to end our addiction to these toxic chemicals.

Fortunately, there is a reliable and cheaper alternative to the massive use of these chemicals. In recent years, scientists have focused on a method of insect control called Integrated Pest Management (IPM), which makes maximum use of natural insect controls, such as predatory insects, parasites, weather, crop rotation, pest resistant crop varieties, soil tillage, insect traps, and other techniques designed to reduce insect damage to minimal levels. IPM also recognizes that tolerable levels of a "pest" may even be desirable since this provides food for beneficial insects. Under this IPM system, chemicals can still be used when necessary and appropriate, but do not form the basis of the control method.[107]

There is nothing new about the use of natural, biological controls to protect crops; indeed the large scale use of chemicals is a relatively new phenomenon of the last 3 decades. In 2,500 B.C., the Sumerians were employing sulfur compounds to control insects and mites, and in 300 A.D. the Chinese were using colonies of insect-feeding ants to control caterpillars and beetles in citrus orchards.

Today, many farmers are returning to the effective, non-chemical IPM techniques widely used before World War II, when American farms were able to bring in huge harvests without reliance on pesticides. One of the most promising and successful new control devices consists of rearing and releasing into an area large numbers of male insects that have been sterilized through irradiation. They then mate with females in the wild, resulting in a drastic decrease or eradication of the insect population over a period of time. Other methods of biological control include the release of wasps in crop fields or gardens to devour insects, worms, and their eggs. CEQ estimates that IPM systems for cotton, citrus, soybean, alfalfa and deciduous fruit crops (which account for 70 percent of insecticides used) could reduce pesticide use by as much as 50 percent in the next five years, and 70 to 80 percent in the next ten years, *with no reduction in present crop yield levels.*" (Emphasis added.).[108]

The U.S. Agency for International Development (AID), until recently a vigorous promoter of the use of pesticides in undeveloped nations, now stresses IPM techniques as well. A 12 June 1980 AID publication reports that this system holds great promise for the future:

> The IPM approach could cut U.S. pesticide use by up to
> 75% on some crops, reduce pre-harvest losses by 50% and
> result in significant savings for overall pest control,
> according to the Congressional Office of Technology
> Assessment.
>
> Since 1971, the Agriculture Department's Cooperative
> Extension Service has been demonstrating this approach
> on a variety of crops and cattle feedlots. For almost every
> crop trial in more than 30 states, pesticide use dropped

sharply, yields and quality were maintained and profits went up. Mosquito control districts in California have used IPM for several years, cutting pesticide use by 90%, reducing costs for labor and material, and almost eliminating environmental pollution. Equally encouraging results have also been reported for IPM in forestry and urban uses.[109]

In contrast to IPM, chemical pesticides kill and drive away birds, and thus remove some of nature's most effective insect control agents. The Garden Club of America has distributed a brochure describing "What One Bird Can Do" in terms of insect control:

- A brown thrasher can eat 6,180 insects in one day.

- A swallow devours 1,000 leafhoppers in 12 hours.

- A house wren feeds 500 spiders and caterpillars to its young during one summer afternoon.

- A pair of flickers considers 5,000 ants a mere snack.

- A Baltimore oriole consumes 17 hairy caterpillars a minute.

Most insects are harmless or beneficial to man; if not poisoned off, they help control the few that are "pests." There are more than a million different species of insects, and they comprise somewhere around three-fourths of all the world's indentified species. Approximately one percent of them are considered pests, with many of the rest performing such beneficial tasks as pollinating flowers, plants, and crops; scavenging and cleaning up carrion, rotting plants, and other "garbage," and, of course, preying on other insects. Even if one assumes that 10 percent of the insects compete with us in one way or another, that leaves 90 percent that are harmless or useful to us. Yet, insecticides make no distinctions, killing off the helpful ones along with the pests, and poisoning us in the process. IPM methods take advantage of the insects' natural tendency to keep pest populations in check.[110]

The use of naturally-occurring insecticides and repellants is also showing promise. Many plants of the chrysanthemum family contain pyrethrum, which is considered a safe and broadly effective natural insecticide. The pungent-smelling marigold is a traditional insect repeller, and gardeners in the 1800's planted these flowers around their beds to keep out insects and animals. Modern-day urban dwellers have reported that marigolds planted in window boxes keep bugs away during the summer. Two Puerto Rican plants with a strong odor, cockroach grass and poleo, also are reported to repel bugs, and can be planted in window boxes to keep insects away. Dried rosemary flowers are reputed to repel moths, and a combination of cinnamon and clove oil has been used for this purpose for centuries.[111]

Other scents and fragrances with reputed insect-repelling qualities include citronella, vetiver, osage orange, bitter orange and bigarade, all of which can be purchased in the form of incense sticks and candles. Garlic, onions, hot peppers and nasturtium have long been used to keep bugs away, and can be blended together with water to form an organic repellent. And that old standby remedy for pests—soapy water—often works as a way of ridding plants of aphids and other bugs. Additional natural remedies include mixtures of water with such substances as cigarette, pipe, or chewing tobacco; fine clay, such as diatomaceous earth; liquid paraffin; glue; buttermilk and wheat flour; equisetum (horsetail grass or meadowpine); and chamomile.

In addition to those using IPM techniques, an increasing number of farmers (30,000 by one estimate)[112] and gardeners are raising crops and plants without the use of even small amounts of synthetic chemicals. "Organic" farming—growing crops without chemical fertilizers and pesticides—has long been the subject of ridicule by agribusiness interests and government agencies. Earl Butz, Secretary of Agriculture under President Nixon, used to say that before the U.S. could consider organic farming, it would have to decide which 50 or 60 million Americans were going to be allowed to starve.[113] This view typified the closed-minded attitude of many agricultural officials toward ending the farmer's reliance on synthetic chemicals.

But various studies of the feasibility of organic farming have come up with encouraging conclusions. An Agriculture Department task force of scientists and economists conducted such a study in 1979 and drew "positive conclusions on the importance of organic farming and its potential contributions to agriculture and society," according to one of the task force's members. It found that some farmers experienced no reduction in yields when they gave up the use of chemicals, and those who did lose some production made up for it through savings in chemical costs. (Most pesticides and other farm chemicals are made from petroleum and natural gas, the prices of which have soared in recent years. Pesticides which cost 25¢ an acre to apply in 1950 reached $4.50 an acre in 1980—an increase of 1900 percent! The overall cost of pesticides has increased 35 fold during this period—from $87 million to $3.5 billion.)[114]

Another research project, a long-term, five year study by the Center for the Study of Biological Systems at Washington University in St. Louis, matched a group of farms with similar crops, soil conditions, and acreage, half of which used chemicals, the other half of which did not. According to the Center's director, the highly respected environmentalist Barry Commoner, "A five-year average shows that the organic farms yielded, in dollars per acre, exactly the same returns. In terms of yield, the organic farms were down about 10 percent. The reason why the economics came out is the savings in chemicals made up for the difference."[115]

There is, thus, a feasible, realistic alternative to the massive use of toxic pesticides; it is not necessary to poison our food in order to grow it. But whether we begin to move down this road may depend on factors not related to the merits of the method. The political and economic power of the chemical industry can be expected to be used to the fullest to oppose any government proposals that encourage farmers to decrease their reliance on chemical pesticides. Whether or not this political obstacle can be overcome remains to be seen, and will depend to a large extent on how loudly the public demands that changes be made. Our lives, and the future of American agriculture, may well depend on the outcome of this struggle.

Voices In The Wilderness

For decades, a few voices here and there have tried to warn us of the dangers of relying on the indiscriminate use of chemical poisons to grow our food. Many of the things they warned would happen have now come to pass, and the highest officials of our government have confirmed the validity of the arguments of these early "prophets of doom."

Rachel Carson's *Silent Spring*, which appeared in 1962, is generally regarded as the first widely-publicized attack on pesticides. But it was preceded by a few years by a June 1959 article in the mass-circulation *Reader's Digest*, which related to its tens of millions of readers a devastating case against the massive and increasing use of these chemicals. Entitled "Backfire in the War Against Insects," and written by editor Robert S. Strother, the article cited several incidents of pesticide misuse, and warned:

> The new insecticides, often used as massive sprays from planes, kill birds, fish, and animals along with insects of all kinds, good as well as bad. There is mounting evidence that massive aerial spraying of pesticides may do more harm than good. Until the full results are known, all concerned should heed the warning: "Proceed with Caution."[116]

This article, written over two decades ago, described the ruinous impact pesticides were having on humans, wildlife and the environment, and forecast many of the problems these chemicals would cause in the subsequent 22 years of their massive use.

Even in the 1950's, before numerous laboratory tests had overwhelmingly proved a link between many pesticides and cancer, the elevated incidence of various diseases paralleling the increased use of insecticides was being noticed. Anticipating what we know today—that many of these chemicals cause cancer, neurological disorders, and break down the body's immunological system—Strother wrote:

> And finally, there is the greatest question of all: how serious are the hazards to human health? Doctors are increasingly troubled by the possibility that DDT and its much more poisonous descendants may be responsible for the rise in Leukemia, hepatitis, Hodgkin's disease, and other degenerative diseases.[117]

Today, of course, that question has been answered, in the affirmative.

In 1962, Rachel Carson's epic work on the dangers of pesticides, *Silent Spring*, sounded another warning of how our society was poisoning itself. Citing voluminous scientific studies, Carson described how the massive and indiscriminate use of insecticides was wiping out wildlife, destroying nature's ecological balance, and contaminating our food, air and water, as well as our bodies, with chemicals that cause cancer, birth defects and death. While a few of the most obviously toxic pesticides have since been restricted, almost all of the chemicals whose horrors she documented remain in widespread use, and a multitude of new and equally toxic ones have been introduced.

Although she is best remembered for linking the use of DDT and other insecticides to the decimation of birds and other wildlife, Rachel Carson was also one of the first to draw public attention to the longer-term, subtler effects of pesticides. Even in the late 1950's, startling evidence was accumulating linking the rise in the use of pesticides to the simultaneous rise in the American cancer rate:

> ...malignant growths...accounted for 15 percent of the deaths in 1953 compared with only 4 percent in 1900...The situation with respect to children is even more deeply disturbing. A quarter century ago, cancer in children was considered a medical rarity. *Today, more American children of cancer than from any other disease ...* Within the period covered by the rise of modern pesticides, the incidence of leukemia has been steadily rising. In the year 1960, deaths from all types of malignancies of blood and lymph totalled 24,400, increasing sharply from the 16,690 figure of 1950. In terms of deaths per 100,000 of population, the increase is from 11.1 in 1950 to 14.1 in 1960.[118]

Carson also reported that doctors at the world-renowned Mayo Clinic in Rochester, Minnesota found that in treating these cancerous diseases of the blood-forming organs, "almost without exception these patients have had a history of exposure to various toxic chemicals, including sprays which contain DDT, chlordane, benzene, lindane and petroleum distillates."[119]

Although there have been numerous attempts by industry to discredit Rachel Carson and her book, its facts, thesis and conclusions have never been successfully refuted. Indeed, in 1963, the year after it was published, the President's Science Advisory Committee supported Carson's findings and stated that "elimination of the use of persistent toxic pesticides should be the goal."[120] Since then, the use of such toxins has dramatically increased, as has the cancer rate (see pp. 247-49).

The accuracy and validity of *Silent Spring* was no inhibition to the chemical industry's attacking and attempting to discredit it, a vicious campaign which started even before the book had been published, and continues today. In 1962, Velsicol Chemical Corporation even tried to prevent the publication of *Silent Spring* because of the book's criticism of two of its best-selling pesticides, chlordane and heptachlor, which have since had their major uses banned. Velsicol wrote to the publisher and charged that the book's arguments on pesticides might be part of a "sinister" Communist plot to destroy American farms (see p. 365). But one can just as easily invoke the same argument on the other side. What could be more pleasing to this country's enemies than to see America's people poisoned into sickness and death, their food contaminated with cancer-causing chemicals, and large segments of the population alienated from and embittered against the government and much of the corporate system.

Nor was Velsicol's letter and its charges an aberration; it was, in fact, consistent with the actions, policies and propaganda the company has undertaken before and since. For example, in September 1967, Velsicol's chief spokesman, Louis A. McLean, authored an article in *BioScience* magazine, the journal of the American Institute of Biological Sciences, the opening paragraph of which compared opponents of pesticides to Marx, Lenin and the Nazis. McLean explained that pesticide critics were "neurotics, driven by primitive, subconscious fears ... they are actually preoccupied with the subject of sexual potency to such an extent that sex is never a subject of jest."[121]

Following publication of *Silent Spring*, the pesticide lobby countered with a book of its own extolling the value of these chemicals: *That We May Live*, ostensibly authored by Congressman Jamie Whitten, the powerful Chairman of the House Appropriations Subcommittee on Agriculture, which exercised jurisdiction over the budget of EPA and the Agriculture Department. But such relatively amateur efforts as these pale in comparison to the slick, sophisticated multi-million dollar public relations campaigns now waged by the chemical industry. Despite all this, Rachel Carson's arguments remain as valid today as when they were first presented two decades ago. Only now, it may be too late for us to undo the harm we have done to the environment and to ourselves.

Sealing Our Doom

It is ironic that in so casually trying to wipe out other forms of life that were not perceived as immediately useful or convenient to us, we may well have sown the seeds of our own destruction. It was naive of us to think we could eradicate the insects; they are older, more adaptable and much more abundant than we are, and comprise 75 percent of all known species on earth.[122] It is estimated that for each human on earth, there are a million insects.[122] While man first evolved only about a million or so years ago (give or take a few hundred thousand), insects have been around for well over 300 million years. Insects will probably still be around long after we are gone.

Our greatest mistake was to assume that our chemical poisons could be selective, that we could kill off numerous other forms of life and not be affected ourselves. We have been duped by our smug but ill-advised confidence, by our belief that we, as a "superior" life form, could be immune to the natural laws that govern the way poisons interact with, and terminate the life-sustaining processes of, creatures similar in chemical and biological structure to ourselves. As writer Cleveland Amory, President of the Fund for Animals, puts it, "It is this supreme arrogance, combined with our utter indifference towards the needless destruction of other living things we find inconvenient or dispensable, that may have already sealed, for many of us, our own doom."[123]

Our use of pesticides has resulted in the poisoning of much of humanity. All Americans carry in their bodies a variety of cancer-causing insecticides. Perhaps sooner than we imagine, the time may come for the eradication of the greatest and most destructive pest species ever—man himself. And when we go, we surely will not be mourned by whatever creatures, if any, remain on this once pure but now poisoned planet.

What You Can Do

Employment of the IPM techniques and other biological controls discussed at length in this chapter (see pp. 125-29) can help the home gardener achieve greater success at lower costs while eliminating the large-scale use of chemicals. A fundamental rule to remember, as the Rachel Carson Council points out in its publication *How to Control Garden Pests Without Killing Almost Everything Else*, "Without deliberate manipulation by man, insects...are naturally controlled by a variety of factors." [124] Some of these natural controls include weather, food supply (small stands of diversified plants can prevent the buildup of a pest, as opposed to "monoculture," large stands of the same species), and biological controls (fungus diseases and natural enemies, such as spiders, toads, birds, ladybugs, praying mantises, lacewings, and wasps).

Cultural controls include selecting species of plants that are resistant to certain pests. Controlling the watering and fertilizing of plants can also affect the population levels of certain insects. The use of sawdust between flower beds can keep insects away. So can tin cans opened at both ends and stuck into the ground as plant barriers. Sprinkling and hosing can wash insects off plants. And planting flowers in your vegetable garden can also help keep pest populations in check, since the pollen and nectar attract beneficial insects.

If intolerable damage is being done to your plants and some lethal controls must be used, the use of biological insecticides can selectively control "pests" without harming their natural enemies or other creatures. Some of the most effective are naturally-occurring disease organisms that are available commercially, such as *Bacillus thuringiensis* (BT), which is available as a powder that can be wettened and sprayed on a plant under attack by caterpillars. *Bacillus popilliae* can control Japanese beetles by causing milky spore disease. Chemicals that create a sexual scent (pheromones) can be used to lure insects into traps, as well as to confuse the males and prevent reproduction. [125]

Often, the best insect control strategy is to leave your plants alone. A garden full of bees and other beneficial insects is a healthy one, and it is almost impossible to use chemicals without killing off the helpful insects along with the undesirable ones. Ants in the yard rarely need to be controlled, and are extremely valuable since they perform beneficial tasks such as aerating and turning the soil, cleaning up carrion, excrement, and other such "garbage," and even killing fly larvae and termites.

Without insects to pollinate our crops, we would quickly starve. Once we accept the truth that insects are here to stay, we will have come a long way towards halting the poisoning of our society.

Chapter Three

Toxic Wastes: 80 Million Tons A Year

"I believe that it is probably the first or second most serious environmental problem in the country. One of the difficulties is that we really do not know what the dimensions of the problem are. Essentially, there is very little downside risk to anybody who illegally disposes of chemicals in such a way as to be harmful to the public health.

We do not know where the millions of tons of stuff is going. We feel that the things that have turned up like the Love Canal are simply the tip of the iceberg. We do not have the capacity at this time really to find out what is actually happening. In my view, it is simply a wide open situation, like the Wild West was in the 1870's, for toxic disposal.

The public is basically unprotected. There just are not any lawmen out there, State or Federal, policing this subject."

**James Moorman, Assistant Attorney General
for Land and Natural Resources, U.S.
Department of Justice, in testimony before
the House Subcommittee on Oversight and
Investigations, May 16, 1979[1]**

Each year, American industry generates some 57 to 80 million tons of toxic wastes, which is more than the combined weights of all the cars now on U.S. roads. (Some estimates are double these figures.) The lower estimate of this staggering amount of hazardous material amounts to 500 to 600 pounds *each year* for *each* American.[2]

Moreover, the Environmental Protection Agency (EPA) estimates that only about 10 percent of the hazardous wastes of these thousands of chemicals are disposed of in a safe, legal and acceptable manner[3]; 90 percent are dumped illegally or disposed of in a manner that represents a potential threat to humans or the natural environment. EPA has called the situation "the most serious environmental problem in the U.S. today."

According to EPA, there are from 32,000 to 50,000 disposal sites in the U.S. containing hazardous wastes, of which 1,200 to 2,000 may pose "significant risks to human health or the environment."[4] Only about 200 of these dump sites are licensed for the disposal of toxic wastes.[5] In 1980, new chemical waste sites were being discovered at the rate of 200 a month![6] Like malignant tumors on the earth's flesh, these sites continue silently to spread their poisons throughout the environment, infecting an ever-growing area with their lethal toxins.[7]

It may well prove impossible to clean up the nation's dump sites and polluted areas at any price, but the anticipated costs for steps that must be taken in the next few years are staggering. One EPA study estimated that it would cost over $44 billion just to clean up the most dangerous of the waste sites, with the public being forced to pick up half the tab.[8]

The magnitude of this hazardous waste problem can hardly be exaggerated. EPA Administrator Douglas Costle said in late 1980 that "every barrel stuck into the ground (is) a ticking time bomb, primed to go off." His Deputy Administrator Barbara Blum called the situation "one of the most serious problems the nation has ever faced." And Dr. Irving Selikoff, director of the Environmental Sciences Laboratory at Mount Sinai Medical School, says of toxic waste dumping, "we can't survive if we continue." He predicts that toxic waste will be "the major environmental and public health problem facing the U.S. in the '80's.[9]

The Health Disaster At Love Canal

The best known, certainly the most infamous, of the nation's toxic dump sites is the Love Canal area of Niagara Falls, New York. There, the Olin Corporation and the Hooker Chemical Corporation buried over 20,000 tons of toxic chemicals, many of which are known to cause cancer, birth defects, and other disorders.

The chemicals thought to be responsible for these problems were dumped into the canal between 1942 and 1953, and in the latter year, Hooker deeded the land to the local board of education for a token payment of $1.00.[10] But the seriousness of the situation did not become apparent until 1976, after years of unusually heavy rains raised the water table and inundated basements. Houses began to reek of chemicals; trees and garden vegetables died; children and pet dogs began to come home with chemical burns on their feet and limbs; some pets even died.

After residents of the area began to experience an unusually high number of miscarriages, deformities in their newborns and incidents of blood disease, cancer, epilepsy, hyperactivity and other ailments, it became apparent that the community was unfit for human habitation. In August 1978, New York State Health Commissioner Dr. Robert P. Whalen recommended that pregnant women and children under two be evacuated, saying that there was "growing evidence . . . of subacute and chronic health hazards, as well as spontaneous abortions and congenital malformations" at the site.[11]

When state authorities tested the air, water, soil and homes in the area for toxic chemicals that month, at least 82 separate compounds were identified, a dozen of which were known or suspected carcinogens. Pollution in the air was measured at 250 to 5,000 times the levels considered safe. The 1978 State Health Department survey found that there had been an abnormally high miscarriage rate of 29.4 percent, and that five children of the 24 born in the area had birth defects.[12] State health officials reported a 50 percent higher risk of miscarriage among women in the area. And in 1979, of 17 pregnant women in Love Canal, only two gave birth to normal children. Four had miscarriages, two had stillbirths, and nine had babies with defects.[13]

On just one road in the area, 96th Street, where 15 homes are located, eight people have developed cancer in the last 12 years: six women had cancerous breasts removed; one man developed throat cancer and another man got bladder cancer. A 7-year old boy had convulsions and died of kidney failure; a pet dog developed cancerous tumors and had to be destroyed.[14]

Some of the most alarming health data about the effects of these chemicals on Love Canal residents have been gathered by Dr. Beverly Paigen of the Roswell Cancer Institute. She found, for example, a much higher incidence of health disorders among those living in homes above moist ground or wet areas. In testimony before the House Subcommittee on Oversight and

Investigations on 21 March 1979, chaired by Congressman Bob Eckhardt (D-Tex.), she recounted the astonishing history of several families who have inhabited one such house located directly above where liquid wastes were seeping out of the dump:

> This house is rented, and four families have lived there during a 15-year period. In family number one, the wife had a nervous breakdown, and a hysterectomy due to uterine bleeding.
>
> In family number two, the husband had a nervous breakdown. The wife had a hysterectomy due to uterine cancer. The daughter developed epilepsy and the son asthma.
>
> In family number three, the wife had a nervous breakdown. Both children suffered from bronchitis.
>
> In family number four, who lived there less than two years, the wife developed severe headaches after moving in. She also had a hysterectomy due to uterine bleeding and a pre-malignant growth.[15]

Dr. Paigen also found that of the 16 children born to residents of the wet areas between 1974 and 1978, nine had birth defects. The overall incidence of birth defects was set at 20 percent, and the miscarriage rate at 25 percent, compared to 8.5 percent for women moving into the area. Eleven out of thirteen hyperactive children lived in wet areas, and there was 380 percent more asthma than in dry areas of the canal. The incidence of urinary disease and convulsive disorders was almost triple that of dry areas, and the rates of suicides and nervous breakdowns almost quadruple.[16]

Dr. Paigen also found that the health of the Love Canal victims dramatically improved when they moved out of the area. As she testified before Congressman Eckhardt's subcommittee,

> Of the nine families who reported that frequent ear infections were a major problem living on the canal, all nine reported a major improvement in this problem, including one child whose loss of hearing returned to normal. Of the 50 families who reported that colds, pneumonia, bronchitis, and sinus infections were a major

problem while living on the canal, 49 reported an improvement.

Of the 12 asthmatics, 11 reported an improvement. Some of these have not had a single attack since moving. Of the 17 families who reported skin rashes as a problem, 14 have experienced improvement since moving. Of the 12 families who reported that severe depression or a nervous condition were health problems, 11 have reported major improvements.

Of the 39 families that reported migraine or frequent headaches as a problem, 38 reported a major improvement.

One individual case is illuminating. One child had been extensively studied in Buffalo Children's Hospital for severe growth retardation. At age 3, she had a bone age of 1 year. Her doctors told the parents that the child would probably be a midget.

Since leaving the canal, this child has begun to gain weight and to grow rapidly.[17]

Dr. Paigen's study has been criticized by a panel of scientists appointed by New York Governor Hugh Carey, which in October 1980 charged that she, EPA and the State Health Department had "fueled rather than resolved public anxiety." Dr. Paigen freely admits that aspects of her study, conducted on a shoestring budget and using volunteer help, had shortcomings. Her defenders point out that epidemiological studies are enormously expensive, and that, in a classic "Catch 22" situation, such large sums are difficult to come up with unless there can be prior demonstration of possible health problems.

Eventually, New York state authorities termed the area "a grave and imminent peril" to the health of those living nearby. Several hundred families were evacuated; the school was closed, and the site enclosed by a barbed wire fence; thousands of additional families were advised to leave and President Carter declared the Canal a disaster area.[18]

In May 1980, further testing of the residents revealed alarmingly high levels of genetic damage among people living near the area. Following this revelation, an additional 710 families were evacuated at a cost estimated to run between $3 million and $60 million.[19]

The residents of Love Canal may be serving as an "early warning system" for our society. Countless other Americans throughout the country have been exposed to similar concentrations and combinations of these chemicals. Thus, similar heath affects can be expected to show up elsewhere, especially with toxic wastes being produced and disposed of at a rate of up to 80 million tons a year.

Hooker's Cover-up Of The Hazards

As with so many other health disasters involving toxic chemicals, the one at Love Canal could have been avoided if the company responsible had not concealed the facts from the government and the public. When the Hooker Chemical Company deeded the land to the Niagara Falls Board of Education in 1953 for a token $1.00, the company made no real effort to warn the Board about the toxic nature of the chemicals buried there. But Hooker did protect itself: the legal document covering that sale disclaims liability for any death or injuries that occur on the property, and has the school board assume liability for any claims that result from the buried chemicals. Indeed, in ridding itself of this poisonous cesspool, Hooker adopted the stance of a good and generous corporate citizen acting in the highest traditions of civic responsibility.

In its letter agreeing to donate the land for a new school, Hooker's executive vice president wrote to the school board,

> We are very conscious of the need for new elementary
> schools and realize that the sites must be carefully selected
> so that they will best serve the area involved ... we are
> anxious to cooperate in any proper way. We have,
> therefore, come to the conclusion that since this location is
> the most desirable one for this purpose, we will be willing
> to donate the entire strip ... to be used for the erection of
> a school at a location to be determined ..."[20]

The company never warned the public of the dangers to school children or to people who decided to build their homes in the area—information that was known only to Hooker.

On 18 June 1958, a company memo observed that "the entire area is being used as a playground," and "that 3 or 4 children had been burned by material at the old Love Canal property." But Hooker remained silent to avoid legal repercussions.

Again, in 1968, Hooker was made aware of rotten drums, leaking chemicals and the presence of lethal materials at the site when such wastes were discovered by a Department of Transportation team working on a highway adjacent to the area. In order to identify the chemicals found in the drums, Hooker had them analyzed, and discussed the presence of these dangerous chemicals in a 21 March 1968 internal company memorandum entitled "Residue Sample from the Old Love Canal."[21] Again, no action was taken to inform or alert the public about these hazards.

Recently, a Hooker official acknowledged that the company was aware that a critical health hazard existed from leaking chemicals but did not warn the residents. This remarkable admission was brought out in testimony given on 10 April 1979, before the House Subcommittee on Oversight and Investigations, by Jerome Wilkenfeld, a former official of Hooker who by then had been promoted to director of Health and Environment for its parent company, Occidental Petroleum. Questioning from Congressman Albert Gore, Jr. (D-Tenn.) produced the following exchange:

> *Congressman Gore:* Did you take any steps to inform the people who lived adjacent to the Love Canal dump site ...of what kinds of chemicals were in the dump site and what the hazards to their health were?
>
> *Mr. Wilkenfeld:* No, we did not.
>
> *Congressman Gore:* Why not?
>
> *Mr. Wilkenfeld:* We did not feel that we could do this without incurring substantial liabilities for implying that the current owners of the property were doing an inadequate care of the property.[22]

Yet, even with a situation such as the Love Canal, where damage to human health has been so thoroughly studied and well-documented, the attitude of the chemical industry remains one of soft-pedalling the harm, denying guilt and even shifting the blame to the victims. For example, during a 2 July 1980 television interview on the MacNeil/Lehrer Report, Geraldine Cox, vice president of the Chemical Manufacturers Association, compared Love Canal residents to "hypochondriacs" and suggested that their illnesses were imaginary or exaggerated.[23]

On 14 October 1979, Armand Hammer, the multi-millionaire businessman and philanthropist who is Chairman of the Board of Hooker's parent company, Occidental Petroleum, said on "Meet the Press" that the Love Canal problem "has been blown up out of context." And when Hammer chaired Occidental's annual meeting in May 1980, the company rejected a stockholder's resolution calling on the firm to establish policies to prevent future environmental tragedies. At one point, when supporters of the resolution were trying to speak in its favor, their microphones were cut off, and a nun asked Hammer, "Are you refusing to hear?" "Yes, I am refusing to hear," he replied according to the Buffalo *Courier-Express:* "Go back to Buffalo."[24] (In October 1981, Hammer was appointed by President Reagan to be Chairman of the President's Cancer Panel.)

The State of New York had spent over $35 million and the federal government over $6 million on the Love Canal problem as of early 1981; in addition, they committed $20 million more in loans and grants to purchase the homes of several area residents. Temporary housing for relocated residents in hotels and apartments cost $1.3 million a month.[25]

Hugh Kaufman, head of hazardous waste assessment for EPA, described the situation thus: "What Hooker did at Love Canal, putting these wastes there, knowing about the potential for poisoning people, and then finding people were being poisoned and not doing anything about it, was just like taking a gun and pointing it at the heads of those people."[26]

1,000 Other Love Canals

In mid-1980, Hugh Kaufman, of EPA, informed the Senate Judiciary Committee,

> The amount of hazardous wastes generated in the U.S.
> every day would fill three professional baseball
> stadiums... Perhaps tens of thousands of Americans living
> in the northeastern part of the United States are either
> getting sick or dying early as a result of hazardous waste
> management practices.[27]

According to Kaufman, "there are at least 1,000 Love Canals around the country. We know of 4,000 to 5,000 potential ones right now."[28] "The only thing unusual about Love Canal," he once said, "is that it was discovered."[29]

Indeed, Love Canal is only one of at least 15 dump sites discovered in the Niagara Falls area that are considered "imminent hazards" to public health. Three of Hooker's sites, which are estimated to contain some 352 million pounds of industrial chemical wastes, are considered worse than Love Canal.

One of these sites, at Hyde Park, is reported to contain 2,000 pounds of dioxin, 6,600,000 pounds of TCP (which often contains dioxin), 400,000 pounds of the highly-toxic carcinogen mirex and 9 million pounds of extremely lethal C-56 derivatives used to make kepone and other pesticides, as well as large amounts of chlorine, benzene and other deadly chemicals. In all, it contains over 80,000 tons of the same wastes dumped into Love Canal.

Hooker's S-Area site contains some 149 million pounds of industrial chemicals, including over 36 million pounds each of C-56 and chlorobenzene. This site is located only a few hundred feet from, and appears to have contaminated, Niagara Falls' water treatment plant, endangering the main water supply for the city's more than 100,000 people.[30]

In fact, there are a total of 161 old chemical dump sites that have been identified in the Niagara County area; the composition and dangers of almost all of them have not yet been adequately assessed.[31]

Indeed, Love Canal has generated tragedy and publicity out of proportion to its size. The canal itself is only 60 feet wide and 3,000 feet long. Thousands of other dump sites across the country are much larger and potentially much more dangerous.

In May 1980, the Department of Health and Human Services (formerly HEW) released a report estimating that over 50,000 dump sites for hazardous wastes were being improperly operated, and that "more than 30,000 dump sites pose a significant health risk ... (They) contain a large number of variety of chemicals which may produce cancer, disorders of the central nervous system, reproduction disorders, and many other illnesses." The report also concluded that "the greatest risk to large populations may be through contamination of drinking water."[32] The report was made public by Dr. David P. Rall, Director of the National Institute of Environmental

Health Sciences (NIEHS), at a Congressional hearing on the subject.

On 5 June 1980, EPA released a report estimating that over 1.2 million Americans may be exposed to serious health hazards from just the 645 waste disposal sites it had studied in detail. The 108 highest risk sites potentially affect 629,603 people. As other sites are studied, the figures will rise accordingly and amount to millions upon millions of Americans in potential danger.[33] (In 1977, when EPA surveyed 50 landfill sites that had no previous history of problems, it found that 47 were leaking into the ground cyanide, mercury and various other toxic chemicals.)[34]

At a 6 June 1980 Senate hearing chaired by Senator Edward M. Kennedy (D-Mass.), Dr. Rall warned that the results of 50 years of careless dumping of chemicals will threaten the health of Americans for the indefinite future. The most serious threat, he said, may be to the nation's ground water, from which "40 percent of the American people get their drinking water," and into which is seeping much of "the 300 billion pounds of chemicals we dispose of each year." Since "75 percent of the waste dumps are either over aquifers (water-bearing layers of rock) or wet areas," he pointed out, there is a strong chance of "contaminating the water we drink," even in major cities and towns "10 or 20 miles or further" from these sites."[35]

Long Island: Hooker Strikes Again

As the Love Canal situation became more widely publicized, more and more dump sites of a similar nature began to be discovered. For example, on 9 July 1980, it was revealed that toxic chemicals had polluted the sole water source (underground aquifers) for, and jeopardized the health of, three million people on Long Island. After two years of research, the New York Public Interest Research Group (PIRG) released a study entitled "Toxics on Tap," reporting that a wide variety of deadly chemicals had turned Long Island's water source into "an industrial sewer." The pollution, from industrial discharges, was found to be most concentrated in the heavily-populated industrial areas on the western part of the island, which also has an abnormally high cancer rate.[36]

The island's drinking water is further endangered, PIRG also charged, because some 66 companies were dumping almost 10 million gallons of polluted waste water *each day* into Long Island's eleven sewage systems, none of which is equipped to treat toxic wastes.[37]

The PIRG report blamed the Hooker Chemicals and Plastics Corporation for the pollution at many of the sites, since Hooker dumped hundreds of thousands of pounds there of carcinogenic vinyl chloride, as well as other cancer-causing chemicals, between 1946 and 1975. The dumping apparently violated no laws.

Officials warned that the damage that had been done was irreversible for the foreseeable future, and that since New York has some of the strongest environmental controls in the country, the implications for the rest of the nation were ominous. One scientist said of the situation, "If this is happening to Long Island's water, it's happening everywhere." Walter Hang, a molecular biologist and co-author of the report, pointed out:

> Even if all pollution discharges were halted today—which our study shows is impossible—the toxic chemicals that have been discharged in previous years all over Long Island will not bleed out into the surrounding seas for approximately the next 3,000 years... What we have shown today, with this study, is that the nationwide program to protect drinking water, to safeguard public health, and to control industrial discharges is totally ineffective.[38] [39]

Cancers On The Landscape

As more becomes known about our nation's hazardous wastes situation, the larger the dimensions of the problem grow. A few of the recently-discovered dump sites that have been studied give an alarming indication of what we can expect to turn up as investigations proceed of the up to 50,000 hazardous waste sites scattered throughout the country.

One of the nation's most hazardous dump sites is in Woburn, Massachusetts, which has 60 acres of contaminated land and lakes, as well as the highest cancer rate of any large community in the state. The childhood leukemia rate is over double the national average; the incidence of kidney cancer is 50 percent higher than normal.[39B]

In Toone, Tennessee (Hardeman County), well water has been contaminated by leaks from some 350,000 fifty-five gallon drums containing some 16.5 million gallons of such toxic, carcinogenic chemicals as aldrin, dieldrin, endrin, benzene and heptachlor, dumped there by Velsicol Chemical Company. Local drinking water was found to contain at least a dozen dangerous pesticides; the carbon tetrachloride level in one well was 48 times that found in Cincinnati's Ohio River when authorities warned against drinking water from it. Other well water was found to be contaminated with at least half a dozen cancer-causing chemicals at levels 2,000 times the allowed limit.

Residents of the area complained of a wide variety of illnesses, including dizziness, nausea, liver and urinary tract problems and rashes.[40] Dozens of people developed various diseases, and a wave of unusual birth defects swept the area, including one baby born with its stomach outside of its body.[41]

In Shepardsville, Kentucky's 23-acre "Valley of the Drums," authorities discovered, but were baffled as to what to do about, 17,000 deteriorating, leaking barrels of chemical wastes illegally dumped there. Later, several hundred of the drums stored there were seen floating down the Ohio River after it flooded its banks.[42]

In 1979, in Luzerne County, Pennsylvania, an abandoned coal mine into which up to a million gallons of toxic chemicals and oil had been dumped was leaking 1,000 gallons *a day* into the Susquehanna River.[43]

An Elizabeth, New Jersey company, Iron Oxide, specialized in hard-to-neutralize acid wastes. When the state raided the facility—using helicopters, frogmen and police—it found that instead of being treated, the acids were being dumped into an adjoining river.[44]

In 1979, also in Elizabeth, investigators found that a site occupied by the Chemical Control Company contained over 50,000 barrels of extremely toxic, explosive and flammable waste, stored within a few feet of the firm's waste incenerator, and a quarter mile from a huge complex containing millions of gallons of liquified natural gas and propane storage tanks. The state warned that a fire or explosion could have produced a cloud of chemicals that could travel for miles and endanger hundreds of thousands of people in northern New Jersey and New York City. A 30 May 1979 report by the federal Bureau of Alcohol, Tobacco, and Firearms stated, "Health officials reported that enough poisons and pesticides had been noted on the premises to provide a minimum lethal dose to all of Staten Island and lower Manhatten in the event of a fire at Chemical Control Corporation."[45]

Although the firm was paid millions of dollars to treat and dispose of chemical wastes, all it really did was store the steel drums in large piles. Some of the barrels of waste were so dangerous that they exploded while being hauled there, according to the company's former president William J. Carracino. He also revealed that the site contained one thousand 55-gallon drums laden with dioxin, three ounces of which in its pure form in New York's water supply could kill everyone in the city.[46]

When the state did begin to clean up the dump, it found over 500 pounds of high explosives there, including TNT and nitro-glycerine, which fortunately were hauled off to a safer location — and not a moment too soon. Many of the barrels of wastes were clearly marked with the names of some of the country's largest firms, such as the 3M Company. The clean-up costs were estimated at between $11 and $13 million, including $50 to $150 a barrel for disposal, and an additional $600 a barrel to analyze the contents.[47]

In April 1980, what many people feared might happen, did. The site exploded and burned, releasing clouds of toxic fumes throughout the area, causing schools to be closed, and forcing the sick and elderly to remain in their houses.[48] The shocks from the explosion rattled windows of buildings in New York ten miles away. The fire burned out of control for ten hours, and the huge clouds of toxic smoke were prevented from drifting into New York City only by favorable wind currents.[49]

As *Time* magazine New York Bureau of Chief Peter Stoler watched the fire buring at 3:00 in the morning, he looked down to see his nylon sneakers "being eaten by chemicals in the soil. They were literally dissolving off my feet."[50] Fortunately, the most dangerous compounds had already been removed. The wastes were finally taken away after the fire and burned at New Jersey's sole toxic waste incinerator.

Six months after the fire, firemen and waste disposal workers were reported to be still suffering from dizziness and diarrhea. It is not known if or how local residents were affected by exposure to these chemicals, since no epidemiological studies have been conducted. But as Michael Brown recently wrote, the earlier removal of the explosive may have averted a major disaster:

> Had these chemicals and explosives still been present last spring, they would have touched off a much bigger conflagration, blown up natural gas tank farms nearby, and caused carcinogens to rain down upon an even wider territory, probably reaching the crowded streets of Manhattan.[51]

One of the people that New Jersey officials say was an owner of the dump — John Albert — had been indicted in Massachusetts for illegal dumping. He was arrested and indicted in New Jersey for allegedly running an illegal million dollar gambling ring,[52] and for mail fraud charges in connection with an illegal dumping scheme. Albert, of North Brunswick, New Jersey, has been identified by organized crime investigators as a "soldier" in the "crime family" of Frank Tieri. The charges against Albert reportedly resulted from information obtained from wiretaps used to investigate gambling and narcotics rings operating in the state.[53] (Such investigations by federal authorities have revealed that organized crime is moving into toxic waste disposal in other states as well.)

In 1979, the former president of the now-bankrupt firm, William J. Carracino, expressed to *The Philadelphia Inquirer* his philosophy on how to run a successful waste disposal enterprise: "There are millions to be made in this business. The overhead is nothing. You find a place — don't buy it, don't buy anything, just rent it — and dump. When they catch up to you, just declare bankruptcy and get out."[54]

Carracino was later indicted on illegal dumping and federal mail fraud charges, and was sentenced to serve two to three years in prison for the former charge. On 25 November 1980, he testified before the New York State Senate Select Committee on Crime. He told the Committee that organized crime elements had taken over the Chemical Control Corporation, and that he believed the Elizabeth fire might have been set deliberately after the state took over the warehouse in May

1979. Carracino testified that a group of six men, armed with guns and led by Albert, forced him out of the company and took it over in September 1977. He also estimated that the owners made $7 million in profits during this 20 month period by storing the wastes and not going to the expense of disposing of them.[55] At a December 1980 Congressional hearing, Albert denied the takeover account, but repeatedly invoked the Fifth Amendment when asked about his ties to the corporation.[56]

Shortly after the Elizabeth explosion, New Jersey experienced two more blowups at sites where chemicals were stored. On the Fourth of July, 1980, fireworks were provided when chemical wastes stored behind a plant producing industrial paint went up in flames, spreading poisonous fumes over the town of Carlstadt.[57] Three days later, in Perth Amboy, steel drums at a chemical waste facility exploded, "shooting up in the air like jets." The blast caused a huge fire that burned out 16 business firms and 15 buildings in the industrial complex, and forced the evacuation of several hundred people from nearby homes.[58]

Deep Well Injection

Sometimes the earth violently objects to the poisons we casually inject into it. In the 1960's, army personnel at the Rocky Mountain arsenal in Denver, Colorado, having already polluted the groundwater with wastes from the production of nerve gas and closely related pesticides, dumped some 150 million gallons of wastes into a deep well. This had the effect of disturbing and intensifying underground pressures, and caused Denver's first earthquake in 80 years, followed by some 1,500 tremors and quakes of varying intensities over the next five years. The pressure sometimes forced poisonous gases and wastes to shoot up like geysers out of abandoned wells, and even oil well rigs were toppled. When the so-called deep-well injection of wastes stopped, so did the earthquakes.[59]

The deep-well injection techniques practiced by the estimated 650,000 such wells operating across the country have sometimes caused similar problems. In 1968, for example, a well in Erie, Pennsylvania, into which 150,000 gallons of wastes a day were being pumped, suddenly shot a geyser 20 feet into the air. Some four million gallons of wastes gushed out in the three weeks it took to cap the well.[60]

In 1976, EPA proposed regulations to put some controls on these operations, but pressure from industry killed the proposal (see pages 155-58). Four years later, on 8 July, 1980, EPA issued a watered-down version of the rules, giving the states nine months to draw up procedures to regulate these wells somewhat.

These and other half-way measures are akin to putting a band-aid on a hemorrhaging jugular vein. While government bureaucrats mull over their sets of regulations, industry carries on business as usual, generating some 125 to 300 billion pounds a year of hazardous wastes whose disposal remains a critical problem.

The Midnight Dumpers

According to EPA, 90 percent of all toxic waste is illegally and/or improperly disposed of — "dumped" wherever it is convenient or profitable. This is an inevitable consequence of the lack of proper waste sites to handle the enormous volume of chemical wastes, a situation which guarantees that such poisons will continue to be dumped in ways that will contaminate the environment and threaten the health of millions of Americans.

An investigation of illegal dumping by Georgette Jasen and Jonathan Kwitny of *The Wall Street Journal* found that the practice was becoming a burgeoning and lucrative industry:

> Recognizing that it's easier and cheaper to dump illegally than by the book, unscrupulous operators — some reportedly linked to organized crime — have taken over landfill and waste-hauling concerns. Manufacturers, desperate to unload their toxic byproducts, often don't press for details about how or where they are dumped. And detection is unlikely, at least until environmental damage becomes obvious ... For unscrupulous operators, hazardous waste disposal is a low-overhead, high margin business. They can underbid legitimate competitors because they don't bother with costly de-toxification procedures or licensed storage facilities. Fly-by-night haulers can maximize profits by purporting to be legitimate, charging top rates, and then dumping poisons down the nearest sewer.[61]

The proper disposal of common industrial solvents and de-greasing agents, which often includes complete incineration, leak-proof storage, and even re-cycling, can cost $10,000 per tank car load. Such costs can be much greater for the more toxic and hazardous chemicals, especially if they are extremely stable and do not easily break down, such as PCB's and dioxin.

Nor are shady, fly-by-night operators the main culprits. Many of the nation's biggest corporations get rid of their toxic wastes through the lowest bidder and ask no questions about the destination of these poisons. For all they know or care, the chemicals may be dumped in the nearest river, lake or sewage system. Often, this is precisely what happens. As *The Philadelphia Inquirer* noted in its detailed six-part series on chemical dumping, "For the industry that produces toxic wastes, the choice is this: Pay large sums to transport millions of gallons of toxics halfway across the country for safe disposal; pay for costly treatment to detoxify the wastes; or take the cheap way out and pay a local midnight dumper."[62]

Only a fraction of the illicit dumping incidents ever comes to light, but those that do indicate that the problem is widespread. In early 1979, the nation's biggest waste disposal firm, Browning-Ferris Chemical Services, was accused in Houston, Texas of illegally mixing the toxic wastes cyanide and nitro-benzene with motor oil, which was then used on the surface of six dirt roads as a dust suppressant. A company chemist was forced out of his job after he resisted orders to make the poison-ous mixture, which was offered free of charge — or warning — to construction companies and municipalities. People living in the vicinity of the roads on which it was used experienced nausea and other symptoms, and the death of livestock also occurred.[63]

Two years after the dirt roads were treated with the oil, nitrobenzene could still be found in significant quantities. This toxic chemical gives off an extremely noxious odor, even in small quantities. A Congressional report on this incident includes the following account:

> The roads near Corrigan contained a stench that could
> only be described as overpowering six months after they
> were oiled. Both investigative reporters from the Port
> Arthur News and the staff person from the House Sub-

committee on Oversight and Investigations, who visited the subdivision prior to the removal of the roads, reported headaches and nausea from just a few hours exposure.[64]

Later, eight roads in east Texas were ordered dug up and removed by the Texas Department of Water Resources. Browning-Ferris officials disclaimed any knowledge of the ultimate disposition of the contaminated waste oil and toxic chemicals that were supplied to oil recyclers in Louisiana. However, investigators from the Subcommittee learned that the Louisiana firms do extensive business with asphalt contractors in that state and in Mississippi, making it probable that roads in these states are similarly contaminated.[65]

This was by no means an isolated incident; oil waste or sludge is often mixed with toxic chemicals and disposed of on roads to keep down dust. Such improper dumpings of wastes are known to have taken place in Kansas, Alabama, Illinois, Missouri, Minnesota, and South Dakota, but most doubtlessly go undetected and unreported.

One aspect of this waste oil ploy has become a multi-million dollar business: toxic wastes are mixed with heating oil and sold for fuel. When the oil is burned, the toxins are released into the house or building and the outside air, causing potentially deadly pollution.

In 1981, ABC Television's 20/20 show uncovered and filmed a New Jersey plant that was selling such oil to firms in at least four states, as well as to apartment houses and office buildings in New York City. After New Jersey officials made an unannounced inspection visit to the facility and found that the oil being sold was mixed with such hazardous compounds as PCB's, PCE and TCE, the plant soon closed down and later filed for bankruptcy. This left the state with the task of cleaning up the site, at a projected cost to the taxpayers of some $1 million.

Nor was this an isolated case. Waste oil mixed with chemicals has shown up in a dozen different states. And when EPA sampled products from nine waste oil companies, six turned out to be selling oil contaminated with PCB's![65b]

Another common way of disposing of toxic wastes that is even less expensive than using midnight dumpers is the do-it-yourself method. One reason cities and industries have grown up along rivers is that these waterways provide a convenient waste disposal system. They have been extensively used, and abused, by American industry, and today most of the country's major river systems are seriously contaminated (see pp. 169-70).

There are countless examples of this practice that could be cited, but one of the most recent typifies the approach taken by many segments of industry. In November 1980, the FMC Corporation, one of the country's biggest chemical and machinery manufacturers, pleaded guilty to concealing the massive secret dumping of carbon tetrachloride into rivers in Ohio and West Virginia. As a result of the dumping, the deadly chemical contaminated drinking water supplies in such cities as Huntington, West Virginia; Cincinnati and Louisville. When EPA investigated the situation and traced the pollution to an FMC plant, the company assured the agency that only 100 pounds a day of the chemical were being dumped; in actuality, 3,000 pounds were being discharged daily. The firm was fined $35,000 and agreed to pay $1 million into an environmental trust fund.[66]

The massive illegal dumping of poisonous wastes by industrial and "midnight" haulers has been caused by two major factors: the absence of laws and regulations requiring that records be kept of the location of chemicals from production to disposal, and the acute lack of properly maintained waste disposal and treatment sites. The shortage, in fact, is so critical as to guarantee that toxic wastes will continue to be improperly disposed of in streams, swamps, rivers, lakes, the ocean, forests, fields, farms, sewers, along roadways and in other isolated areas. As Michael Brown notes in his excellent book on the hazardous waste problem, *Laying Waste: The Poisoning of America by Toxic Chemicals*, in the late 1970's, the approved chemical dump sites in Illinois received some 2 million tons a year, but 14 million tons of waste were produced in the state. It is not known where the other 12 million tons ends up. California produces over 100,000 tons of hazardous wastes *each month*, 92 percent of which goes "straight into the ground." At one time, the island chain of Hawaii had only one firm licensed to dispose of chemical wastes, which processed some 20,000 gallons a month of waste oil. But five times that amount was being generated just on the island of Oahu. Again, no one knows where the unaccounted-for portion was, and is, ending up.[67]

One of the most innovative approaches to the toxic waste problem is being taken by the State of California, where

Governor Jerry Brown has ordered a halt by January 1983 to the dumping on land of the most hazardous wastes. In taking this action, Brown relied on an extensive, two-year study by his Office of Appropriate Technology, which concluded that almost 75 percent of the toxic wastes dumped in landfills could be recycled, safely incinerated, baked into hard blocks, or chemically altered into harmless substances, in each case using techniques already proven to be effective. Under some circumstances, such waste materials that are combustible can even be used as supplemental fuels, such as in cement kilns, thus disposing of dangerous chemicals while cutting fuel bills.[67b]

EPA's Non-Enforcement Of The Law: The Resource Conservation And Recovery Act

While new laws have long been needed to deal effectively with the hazardous waste situation, EPA has not even used the laws that are on the books to address the problem. A glaring example of this failure is EPA's refusal to implement and enforce the 1976 Resource Conservation and Recovery Act (RCRA). This law gave EPA authority to regulate to some extent the disposal of toxic wastes, but the agency was two years late in promulgating the regulations to implement the act! As a result of EPA's unwillingness or inability to take firm and timely action to protect the public, millions of Americans will continue to be threatened by the improper disposal of poisonous chemicals.

In 1978 and 1979, the House Subcommittee on Oversight and Investigations, chaired by Congressman Bob Eckhardt (D-Tex.), conducted an extensive investigation of the problem of hazardous waste. In its September 1979, report, the Subcommittee concluded that "the hazardous waste disposal problem cannot be overstated," and issued a devastating indictment of EPA's failure to properly address the problem:

> This report is being published on the third anniversary of the enactment of RCRA. The Act recognized that speedy promulgation of regulations was the cornerstone of effective regulation of the disposal of hazardous waste; therefore, it required EPA to promulgate regulations within 18 months of the date of enactment of RCRA. April 21, 1978 passed without any final regulations and with precious few

proposed regulations. Now, 3 years later, we are still awaiting the promulgation of the first hazardous waste disposal regulations.[68]

EPA's inaction has also discouraged states from improving their waste disposal programs, since "those states that do act vigorously may have relatively little impact on the problem because many companies will simply ship their wastes to states with fewer restrictions."[69] "As a result," the Subcommittee reported, "even an extraordinary effort, commenced immediately, cannot achieve adequate protection for the American public for years to come."[70]

The effect of EPA's policies, as the Environmental Defense Fund testified on 30 October, 1978, has been to render RCRA ineffective:

> Approximately 260 million pounds a day of chemicals that cause cancer, birth defects, nerve damage, and that destroy rivers and wildlife, are being disposed of without Federal regulation, just as they were in 1976 when this legislation (RCRA) was passed.[71]

When EPA finally did act to issue *some* of the required rules in May 1980, the agency declared at the time that "the improper management of hazardous waste is probably the most serious environmental problem in the United States today."[72]

When EPA Administrator Douglas M. Costle announced these new rules, he warned ominously: "Let me predict now that the process we are starting will turn up information and situations which will shock our nation."[73]

Even these partially-complete regulations, coming two years late, were attacked by environmentalists and even other EPA officials as so riddled with loopholes and exemptions as to be largely ineffectual. Critics pointed out that under these regulations, future "Love Canals" would not only be likely to occur, but were in fact probably already happening across the country. Most of these deficiencies were contained in EPA's proposed version of the regulations (issued only after several lawsuits were filed), and were severely criticized in the 1979 hearings and report by Congressman Eckhardt's Subcommittee.

As the Assistant Attorney General of the State of New York, John Shea, testified before the House Subcommittee on 16 May 1979 on EPA's proposed rules:

> The two major flaws in those regulations are that the standards for landfills would allow migration of chemicals to occur after only 50 years, which is merely a fraction of the period of toxicity for these materials. Second, the operators of dump sites would be required to care for the sites for only 20 years after the dumps are closed. Love Canal has proven that to be far too short a time. What this means is that EPA's regulations will permit future Love Canals.[74]

Even the head of EPA's Hazardous Waste Implementation Branch, William Sanjour, admitted that the regulations would not be effective in protecting the public and would not even have prevented the tragedy at Love Canal:

> Hooker would have had no trouble complying with these regulations. They may have had a little extra paperwork, but they wouldn't have had to change the way they disposed of the wastes.[75]

Other glaring weaknesses in the regulations include the licensing procedures for firms or individuals handling hazardous wastes, which allow those applying to receive automatic temporary permits for up to five years, if EPA does not act on the application. Moreover, many extremely toxic pesticides are not included in the list to be regulated. And in ascertaining the toxicity of a chemical, EPA refused to include such obviously relevant properties as infectiousness, radioactivity, or teratogenicity and mutagenicity — the ability to harm unborn children or cause genetic alterations.

Moreover, one of the toughest rules — requiring that toxic wastes be kept track of "from cradle to grave" and holding recipients of such wastes accountable for their disposal — did not go into effect until November 1980, thus creating an upsurge of careless and improper dumping before it became illegal. As Ed Magnuson reported in *Time* magazine's 22 September 1980 cover story on "The Poisoning of America:"

> To beat the deadline, some companies have been taking chemical refuse they have stored on their property for

months or even years and simply getting rid of the stuff as swiftly and as surreptitiously as they can, often dumping by night and running.[76]

The rush to dump toxic wastes, made inevitable by EPA's actions, was widely publicized while it was taking place, but there was little that could be done to prevent it.

On 16 November, for example, *The New York Times* reported in a front page story by Michael Knight:

> Thousands of tons of hazardous and toxic wastes are being hurriedly dumped into city sewer systems, spilled from moving trucks onto busy interstate highways, and abandoned in shopping center parking lots around the country in a last minute rush to dispose of the chemicals before a Federal "cradle-to-grave" waste monitoring system begins next week.

> Environmental protection officials say...the short term effect has been to increase to a frantic pace the improper and illegal disposal that the law was designed to halt.[77]

Hugh Kaufman, EPA's chief toxic investigator, noted at the time:

> The pressure is on the companies that make and use these chemicals to get the stuff off their site and onto somebody else's court before November 19, when they have to accept responsibility for them. Nobody wants to have a paper trail back to them so that they can be sued some day, not if they can help it.[78]

Kaufman observed that a major problem was that EPA had still not fulfilled the 1976 law's requirement to define proper and legal disposal techniques.

One of the human victims of RCRA's crippling was William Sanjour, who, as head of EPA's hazardous waste assessment branch, was instrumental in drafting and working for the passage of the law. Three years after its enactment, Sanjour expressed dismay at how it had been implemented: "I've probably killed more people by passing that act. I worked very hard for it. As a result they're breeding and dumping waste as fast as they can before it takes effect." Sanjour did not willingly acquiesce to the dismantling of RCRA, but he paid a heavy price for his opposition. According to him, "In June of 1978, I was ordered to cut back on the regulations, to soften them to make them easier on industry because of President Carter's fight on inflation." After Sanjour protested the order, he was transferred to an office that was being abolished and given no staff and no work.[79]

Hugh Kaufman is equally pessimistic about the outlook for RCRA:

> "What do you expect? EPA lobbied against RCRA in the first place, and it has no intention of ever aggressively enforcing it.[80]

EPA's lack of enforcement of RCRA has changed little, except to deteriorate further under the Reagan administration. On 30 September 1981, Congressman James J. Florio (D-N.J.) released a study by Congress' General Accounting Office (GAO) entitled "Hazardous Waste Facilities with Interim Status May Be Endangering Public Health and the Environment." It described how EPA was allowing thousands of such sites to operate with temporary permits without determining if they were safe. Over 8,300 known sites had not been registered under the law and were not being pursued. In analyzing samples of EPA's inspection reports, GAO found that 122 out of 127 facilities were in violation of current regulations.[81]

The report detailed how it had taken EPA four years to issue regulations for such dumps since passage of RCRA, and projected that it would take another eight years for the agency to develop final regulations and issue permanent operating permits, or even longer if staffing or funding were reduced. "We are facing a national chemical nightmare," said Florio, "and this administration is treating it like a joke."[82]

As of early 1982, the Reagan administration had cancelled the modest, loophole-ridden landfill standards proposed by the Carter administration, and were working on new rules to govern the over 80,000 waste sites throughout the country that EPA has identified (over one third of which may contain hazardous wastes). Environmentalists feared that the new regulations would be much weaker than the former ones. But whatever happens, as Hugh Kaufman pointed out at the time, "you still have no enforcement effort, no financial responsibility to the waste disposers, no technical standards. What do you have? Chaos. Nothing." And William Sanjour observed that since all landfills, no matter how well lined, eventually leak, it was impossible to have a toxic waste landfill that was safe: "... the federal government cannot come out with any set of regulations that both protects human health and the environment, and allows existing landfills to continue to function." Thus, new potential Love Canals are still being created. After studying EPA's list of the 115 most hazardous waste sites, Mr. Kaufman and Mr. Sanjour asserted that over 80 percent of the situations would not have been prevented under current regulations.[81b]

Other Environmental Laws Ignored

RCRA is not the only law available to EPA to deal with hazardous waste sites, but the other environmental statutes are rarely invoked in such cases. Although thousands of these waste dumps violate such federal statutes as RCRA, the Clean Water Act, the Refuse Act and the Toxic Substances Control Act (TSCA), the government, for years, largely ignored most of these clearly illegal situations and refused to enforce the law.

Ironically, the problem of toxic waste disposal has in some ways been increased by earlier legislation to protect the environment, such as the Clean Air Act, the Clean Water Act, the Federal Water Pollution Control Act and the Safe Drinking Water Act. In fact, when RCRA was passed by Congress it was specifically designed to "eliminate the last remaining loophole in environmental law, that of unregulated land disposal of discarded materials and hazardous wastes," according to the 1976 RCRA legislative report of the House Interstate and Foreign Commerce Committee:

> At present, the Federal Government is spending billions of
> dollars to remove pollutants from the air and water only
> to dispose of such pollutants on the land, and in an
> environmentally unsound manner.[83]

Thus, the long history of inadequate enforcement of RCRA helps defeat the purpose not only of this statute, but of other environmental laws as well.

Cleanup Legislation — Is Congress For Sale?

The chemical industry would dearly love to have its past actions — and responsibility for them — treated in the same "out of sight, out of mind" way it has treated toxic wastes: bury and forget about them.

But there is a crucial issue that must first be resolved if an attempt is to be made to clean up the nation's hazardous waste sites — who will pay the bill: the chemical industry, which is responsible for the existence of the dumps, or the taxpayers? One EPA study found that it would cost $44 billion to try to clean up just the most hazardous dumps posing the greatest dangers to human health, with the public having to pay for half of these expenses.

The dispute over this issue came to a head in 1980, when legislation that would address the problem was working its way through Congress. The industry initially tried to kill off the legislation, and did succeed in delaying its passage for a few crucial months. Later, realizing that the passage of some law dealing with waste sites might be inevitable, the industry worked to make the final bill — if it could not be delayed to death — as weak and inexpensive as possible. The usual Washington lobbying techniques were employed.

The legislation was prompted by the discovery of numerous hazardous waste sites, such as the Love Canal, that were polluting the land and water, and were serious threats to public health, but for which no one would accept the responsibility for cleaning up. As the discovery of more and more such dumps attracted national publicity, the pressure grew for Congress to act. The industry responded by making massive financial contributions to key members of Congress.

The chemical and oil industries were more than generous in spreading money around in an attempt to defeat or cripple this so-called "Superfund" legislation, especially the bills that would create a pool of money, financed mainly by the chemical industry, to clean up chemical waste sites. The strong Senate version, which would have made it easier for victims to obtain compensation from polluters, was the main target of the industry campaign. This bill would have allowed citizens to bring suit against dumpers to recover personal damages, medical expenses, economic losses, and damage to natural resources.

Primary sponsors of strong Superfund legislation were Senators John Culver (D-Iowa), John Heinz (R-Pa.), Robert Stafford (R-Vt.), Edmund Muskie (D-Maine) before he left the Senate to become Secretary of State, and Congressmen Bob Eckhardt (D-Tex.), James Florio (D-N.J.), Albert Gore, Jr. (D-Tenn.) and Edward Madigan (R-Ill.).

Congressman John F. LaFalce (D-N.Y.), who represents the district where Love Canal is located, called the fund "the most important environmental bill of this decade and of the next decade." But then-Congressman David A. Stockman (R-Mich.) opposed the legislation, warning that it would make EPA "the czar over every waste site in the country."[84] (As President Reagan's Director of the Office of Management & Budget, and "czar" of the federal budget, Stockman has slashed the budgets of EPA, CEQ, and other environmental and consumer protection agencies, thereby crippling many of their programs, including the "Superfund.")

The fight against a strong Superfund bill was led by the Chemical Manufacturers Association (CMA), which has over 200 member companies, including some of the nation's largest and most powerful corporations. Other industry groups opposing the legislation included the American Petroleum Institute, the Business Roundtable, the U.S. Chamber of Commerce, the National Association of Manufacturers, the Fertilizer Institute and such firms as Dow Chemical and Olin.

In public, the industry's main tactic was to deny that the hazardous waste problem was sufficiently serious to require such sweeping legislation. Robert A. Roland, president of the CMA, took the position that "the Association opposes Superfund legislation because it would create an unnecessary new level of federal bureaucracy and squander the nation's resources at a time when sound fiscal management is imperative."[85]

Bill Lowrey, a Washington lobbyist for Shell Oil, admitted that his firm was vigorously fighting the Senate bill, saying "we are opposing it with everything we've got." Monsanto even sent out a computerized mailing to 40,000 to 50,000 shareholders, urging them to send in letters opposing the legislation.[86] Donald L. Baeder, President and Chief Operating Officer of the Hooker Chemical Company in Houston, Texas, sent letters to Senators and Congressmen claiming that despite "sensationalized media treatment...there is no evidence of chemicals from Love Canal having caused health problems."

However, the industry's most effective tactic in opposing the legislation was a subtler one: massive (and legal) payoffs to key politicans. A study released in August 1980 by Ralph Nader's Congress Watch organization detailed how 37 of the top 50 U.S. chemical producers contributed over $2 million to Congressmen and Senators for the 1978 and 1980 campaigns. Among the top recipients of this corporate largesse were — you guessed it — those members leading the opposition to this cleanup legislation. Congress Watch noted that the average contribution to House members who supported two key committee amendments to strengthen the bill was only $1,078, while opponents averaged $4,765.

Nader described the links between the industry's contributions—given through Political Action Committees (PAC's)—and opposition to the bills.

> The major recipients of chemical company contributions reads, with some exceptions, like a who's who of opponents to the Superfund bill. For example, Senator Howard Baker (R-Tenn.), who topped the Senate contribution list with over $27,000, is a major opponent of the key provision of the Senate bill which would give victims a fairer chance to recover from the chemical companies. Senator Alan Simpson (R-Wyo.), who received over $21,000, was the lone dissenting vote when the bill was reported out of the Senate Committee on Environment and Public Works earlier this year. The Senate bill is now going to the Senate Finance Committee, so it will be interesting to see how Senator Russell Long (D-La.), who was the second largest Senate beneficiary with over $25,000, handles the legislation.
>
> On the House side, Representative Jim Santini (D-Nev.), the sixth largest House recipient with $13,250, was the only Democrat to vote against the bill when it was reported out of the Subcommittee on Transportation and Commerce. And Representative Jim Martin (D-N.C.), who received over $14,000, was an opponent of two amendments that strengthened the bill in the Ways and Means Committee.[87]

In fact, Senator Long did his best to hold up final passage of the Senate bill and succeeded in delaying it several vital weeks.

The information released by Congress Watch revealed that the top 50 chemical manufacturing firms in the U.S., with combined 1979 sales of sixty-five *billion* dollars, had made over two million dollars in political contributions between 1977 and 1980.

In the case of members of Congress who would not cooperate, the oil and chemical industries poured money into the congressional campaigns of the opponents of those independent-minded legislators. In 1980, such key sponsors of Superfund legislation as Eckhardt (a primary author of the Toxic Substances Control Act) and Culver (who helped write the Clean Water Act) went down to defeat at the hands of well-financed, pro-industry opponents. According to Peter Harnik of *Environmental Action's,* "Filthy Five Campaign Committee," five of the worst polluting corporations in America — Dow Chemical, International Paper, Occidental Petroleum (parent company of Hooker Chemical), Republic Steel, and Standard Oil of Indiana — contributed over $630,000 to House and Senate candidates between 1 January 1979 and 30 September 1980. Dow alone donated $206,750 through its eight political action committees.[88]

The largest recipient of funds from the Filthy Five was Representative Charles Grassley (R-Iowa), who used the $21,800 to help take away John Culver's Senate seat. The next biggest recipient was Senator Russell Long ($15,500), who helped stall the Superfund legislation at a crucial time, followed by anti-conservation Republican Congressman Steve Symms ($12,500), who defeated Senator Frank Church (D-Idaho). The fifth and sixth largest beneficiaries of Filthy Five largesse were two ultra-conservative Republican Congressmen, Dan Quayle ($9,500) who beat Senator Birch Bayh (D-Ind.), and Jim Abdnor ($8,600), who defeated Senator George McGovern (D-S.D.).[89] Industrial PAC's donated $19 to $30 million to the 1980 Congressional and presidential election campaign; (see pp. 206-7).

Thus, having succeeded in helping to replace a number of key pro-environment senators with pro-industry legislators, the chemical lobby may find that its 1979 and 1980 campaign donations were among the most profitable investments it ever made.

Opponents of the bill succeeded in preventing its passage before the November 1980 elections. This delighted industry, which felt it could get a much more favorable bill—or even better, none at all—the following year from the new Republican—controlled Senate and the Reagan administration. During the campaign, Reagan, the Republicans and their platform stressed anti-environmental, anti-regulation, and pro-industry themes, so the outlook for passage of any effective superfund legislation appeared bleak.

In retrospect it is remarkable that any bill whatsoever was able to pass Congress, considering the emboldened efforts that were made to kill it. After earlier seeming to support a weaker version of the legislation, the CMA came out strongly against any bill whatsoever, prompting Senator Daniel Patrick Moynihan (D-N.Y.) to say, "In the annals of corporate cynicism, I have not encountered anything so brazen."[90]

Senate Republican leader Howard Baker (Tenn.) also tried to kill the bill by suggesting it be put off until the new, Republican-controlled Senate convened in 1981. And ultra-conservative Senator Jesse Helms (R-N.C.), a leading opponent of the bill, threatened to filibuster it to death unless it was greatly weakened.

But, miraculously, in the final days of the lame-duck 96th Congress, a version of the superfund legislation did pass and was signed into law. However, so many compromises had to be made to get the bill through that the final version was substantially weaker than earlier ones.

The original Senate bill called for creating a $4.2 billion fund over six years, and placing strict liability provisions on chemical companies, enabling citizens to sue them for personal and property damage when their health or property was harmed by dumps or spills. The final Senate bill eliminated all such liability clauses, meaning that injured persons will receive no compensation from the fund, and will have to go through the slow, difficult and cumbersome process of seeking relief in state courts. The Senate scaled down the size of the fund to $1.6 billion over five years, with 87.5 percent coming from taxes on chemical products. Coverage of oil spills was also deleted from the bill.

Because only a few days remained before adjournment, the House accepted the watered-down Senate version. And President Carter, on 11 December 1980 signed into law the Environmental Emergency Response Act.

The law does make owners and operators of waste disposal sites, as well as producers and transporters of hazardous wastes, liable for cleanup costs, and for up to $50 million for damage to natural resources owned by the government.

The Act also sets up a new "agency for toxic substances and disease registry" within the Public Health Service to study and register victims of toxic wastes and their health effects.

Both Senator George Mitchell (D-Maine) and Jennings Randolph (D-W.Va.) promised to try to later restore some of the provisions that had been dropped from the bill. Senator Mitchell, speaking before the Senate on 24 November 1980, decried the fact that the bill that had just been passed had been so weakened, pointing out:

> The people of the United States should understand that
> the legislation does not deal with the most serious part of
> the problem ... It provides for the cleanup of places and
> compensation for damage to things, but it provides
> nothing for injury to people ... In effect, under the
> legislation it's all right to hurt people but not trees ...

Asking "By what standard of justice or decency is damage to property more important to persons?," Senator Mitchell strongly condemned Congress for knuckling under to the demands of the chemical industry and letting down the American people:

> Having made the judgment that property is more impor-
> tant than person, we should not delude the people of this
> country about what we have done. Most particularly, we
> must not delude the thousands of people who are victims,
> who were waiting for our help, and who do not get that
> help from this legislation. For them, the Superfund is not
> super.[91]

There is also considerable doubt concerning the Reagan administration's willingness to fund and enforce the Superfund. President Reagan's first budget, submitted to Congress in February 1981, contained cuts of some $50 million in the program, meaning that its impact and effectiveness will be extremely limited.

Additional cuts in personnel and budget at EPA have made it impossible for the agency to properly administer the program. To make matters worse, EPA Administrator Anne Gorsuch announced in mid-1981 that EPA's Office of Enforcement would be disbanded, and regional EPA offices were ordered to cease issuing orders for the cleanup of hazardous waste sites without obtaining the approval of officials in Washington.[92]

And on 25 February, 1982, EPA lifted for at least three weeks a rule banning the placing in landfills of barrels of toxic liquids. Industry immediately began disposing of an estimated 220,000 gallons or more *a day* of such wastes—the same type of dumping that caused Love Canal.

What You Can Do

If the existing laws on the books were effectively implemented and the needed regulations issued and enforced, the hazardous waste problem could be meaningfully addressed.

In addition, there are new techniques that are effective in disposing of toxic wastes, including placement in concrete blocks, recycling, enclosed high intensity incineration, and the chemical alteration of such compounds into harmless substances. New laws and tax incentives to encourage industry to pursue such measures could be important. They would be especially effective if they were combined with taxes on the manufacturers and users of toxic chemicals designed to discourage their production and use. That money could help pay for disposal costs and damages caused. If chemical companies were made legally liable for the pollution and harm to human health and other damages caused by their products and wastes, there would be added economic incentive to adopt policies that might prevent Love Canal-like situations. (Congressman George Miller [D-Cal.] has introduced legislation that calls for fines and even prison terms for corporate executives who are criminally negligent or cover up the dangers of a product or situation that endangers the public, such as what took place at Love Canal.)

But without a system of incentives and controls, in the long run the only real answer to this problem may be to limit our massive production of these deadly poisons.

Chapter Four

Contaminating The Water: Our Nation's "Most Grievous Error"

Our ground waters are threatened by ruinous contamination...this will become the environmental horror story of the 80's...the most grievous error in judgment we as a nation have ever made.

**Eckhardt Beck,
Assistant Administrator, EPA
25 July, 1980[1]**

There is no resource we have that is more precious than water. Yet we are squandering and contaminating our vital supplies of water so rapidly that many people believe that in a few years, the shortage of clean, useable water will dwarf the energy "crisis" of the 1970's.

There is no question that in some areas, progress has been made in cleansing our rivers and lakes of certain pollutants, usually the more visible and obvious types, such as oil and raw sewage. We have come a long way from the days in the 1960's when the Cuyahoga River in Cleveland was so saturated with oil that it would often burst into flames, and fire breaks and fire patrols had to be instituted. But while some of the most noxious pollutants have been removed or reduced in volume, they have often been replaced with colorless and odorless chemicals of a much subtler and more dangerous nature.

The long-term, perhaps permanent poisoning of water tables throughout much of the country appears to have already taken place. From California to New York, wherever wells, water tables, and other water supplies are analyzed, the presence of alarming levels of toxic, cancer-causing chemicals continues to turn up. The main factor limiting the extent of known contamination is the lack of a comprehensive survey of America's water supplies. Such an examination would almost certainly find that clean water has become a very rare commodity.

Repeated warnings by environmentalists for over a decade have recently been confirmed by various government studies and reports, documenting the contamination of America's water supplies and systems with cancer-causing, health destroying chemicals.

A January 1981 report on ground water contamination issued by the President's Council on Environmental Quality (CEQ) states:

> Many compounds once thought safe...can present serious and substantial health risks even in concentrations in the low parts per billion or parts per trillion range. At higher concentrations in drinking water, many toxic organic chemicals are tasteless and odorless, and they cannot be detected without sensitive chemical instrumentation.[2]

The report goes on to point out that such chemicals, when ingested, can cause such health problems as tremors, blindness, nausea, dizziness, skin eruptions and impairment of the central nervous system. And CEQ's 1979 Tenth Annual Report states:

> As many as two-thirds of the nation's lakes may have serious pollution problems... An estimated 80 percent of more than 3,700 urban lakes in the United States are significantly degraded, and, yet, they offer potential aesthetic and recreational value to more than 94 million metropolitan residents.[3]

A March 1980 report prepared by the Library of Congress, and released by the Senate Environment and Public Works

Committee, catalogs the poisoning of America's water resources:

> ...damage to natural resources in the United States by toxic chemicals is substantial and enduring. Historic and invaluable waterways — the Hudson, the Shenandoah, the Delaware, the Susquehanna, and the James River, to name but five — are injured... The most valuable and irreplaceable surface water now known to be contaminated by toxic chemicals (e.g., PCB's, mirex, DDT, mercury, PCDF's, and asbestos) is the Great Lakes system. The Great Lakes contain 20 percent of the world's surface [fresh waters] and 95 percent of surface fresh waters of the United States.... overall commercial fishing has been limited by 50 percent.[4]

The contamination of this body of water is particularly tragic, since some 24 million Americans get their drinking water from the Great Lakes, which comprise the largest single volume of surface fresh water in the world.[5]

Perhaps most depressing of all is the fact that our studies of America's water quality have only scratched the surface. While they have revealed alarming and in many cases permanent damage to our water resources, we may yet learn that there are few safe and unendangered supplies left, and that most Americans have no choice but to consume contaminated water.

Our Nation's "Most Grievous Error"

One of the most severe threats posed by the dumping of toxic chemicals is the increasing contamination of underground water supplies, which provide about half of the American people with their drinking water, and account for a quarter of all fresh water used in the United States. There is thought to be perhaps 50 times more of this subterranean water than what flows through all of the nation's rivers, lakes and streams combined.[6] As a recent EPA publication put it,

> ...groundwater accounts for about two-thirds of the freshwater resources of the world. If we only consider the portion that can be used (minus icecaps and glaciers), then groundwater accounts for almost the total volume.

If only the most "active" groundwater sources are counted, they constitute 95 percent; lakes, swamps, reservoirs, and river channels 3.5 percent; and soil moisture, 1.5 percent.[7]

Once ground water becomes contaminated, there is usually no way to cleanse it again. And unlike river or lake water, underground supplies are not processed at treatment plants before being used by or distributed to the public. We are well on the way to permanently ruining this irreplaceable resource that has been "untainted since the beginning of time."[8]

Some ground water consists of rain that fell to earth a century or more ago, migrating gradually through the ground and permeable rock to accumulate in huge reservoirs of porous rock, gravel and sand known as aquifers. Underground waters flow into rivers and streams through springs, and are in turn replenished by them. Until recently, these water supplies were considered the ultimate in purity; but the pollution of the last few decades may have permanently contaminated much of this precious natural resource. Aquifers do not flow and are not exposed to sunlight and other natural cleansing systems. As former EPA Assistant Administrator for Water and Waste Management Eckhardt Beck observes, groundwater is uniquely susceptible to long-term contamination:

> With the slow patience that only nature knows, ground water inches its way through a maze of infinitesimal cracks and fissures, through compacted beds of glacial sands and gravel, sometimes taking a year to traverse a mere thirty yards, a human lifetime to travel a mile or two. A given drop of today's rain that soaks through permeable soils into one of our aquifers may not see daylight again until the 22nd century or beyond, then to replenish a lake or a stream a few miles away...Ground water, then, is almost everywhere beneath us, a virtual ocean of fresh water, oozing through the earth, an awesome natural resource more susceptible to long term damage than the air we breathe.[9]

Because of the incredibly slow movement of groundwater, contamination in one location may not show up for decades in a well just a few miles away. This means that our present knowledge of water table contamination may just be the tip of the iceberg, and decades ahead may reveal further loss of much of our water supply just from pollution that has already taken place.

In his 25 July 1980 appearance before the House Subcommitee on Environment, Energy, and Natural Resources, Eckhardt Beck began by stating, "I come before you today with the distressing news that one of this nation's most vast and vital natural resources is in serious jeopardy. Our ground waters, long considered virtually pollution-free, are threatened by ruinous contamination." Beck testified that the government's failure to address this problem was "the most grievous error in judgment we as a nation have ever made,"[10] and predicted that "this will become the environmental horror story of the 80's — with after-effects reaching into the next millenium." Stressing that "the problem is serious, for the intruding contaminants are often highly toxic, sometimes cancer-causing," Beck described how utterly dependent we are on ground water:

> More than 100 million of our citizens depend in whole or in part on underground sources of drinking water. Each day, more than ten billion gallons of ground water are withdrawn for use in American homes. Another seventy billion gallons are drawn for agricultural purposes. The prospect that water may contain high concentrations of toxic chemical compounds compels our immediate attention and action.

In his testimony at the hearing, Dr. Robert Harris, then a member of the President's Council on Environmental Quality (CEQ), pointed out that "once contaminated, ground water remains so for hundreds or thousands of years, if not for geologic time... For all practical purposes, ground water contamination is irreversible by natural forces."[11]

Stressing the nation's dependence on ground water, Dr. Harris estimated that there are about 36 quadrillion (36,000,000,000,000,000) gallons of ground water within half a mile of the earth's surface, representing 30 times the amount contained in all the rivers and lakes of America. More than 30 trillion gallons of this are withdrawn and used each year in the United States. Underground sources supply over 40 percent of all irrigation water used to grow vegetables, fruit, grain and other food crops; the amount of ground water used for agricultural and other purposes is growing rapidly.[12] Some 750,000 new wells are drilled each year.[13]

In addition, 20 percent of urban, and 95 percent of rural, drinking water comes from groundwater, primarily wells, and this source is also in serious jeopardy. A 1979 report issued by the House Subcommittee on Oversight and Investigations alluded to the potential gravity of this tragedy:

> ...almost one-fifth of America's population relies on groundwater from individual wells without the benefit of treatment systems. Inasmuch as these wells are rarely monitored, it is impossible to determine when human health is threatened by polluted water until after illness develops.[14]

Some states are almost totally dependent on ground water to supply water supplies. In Hawaii, 94 percent of the population is served by groundwater, 90 percent in Florida and 75 percent or more in New Mexico, Idaho, Iowa, Arizona and Mississippi.[15]

A major source of ground water contamination is industrial waste. EPA estimates that every day, some 50 billion gallons of newly generated liquid wastes are stored in surface water impoundments (pits, ponds and lagoons). As of mid 1980, EPA had identified some 180,000 active waste impoundments around the country, 25,749 of which were classified as industrial. Of those industrial sites studied, 72 percent were found to be unlined, and only 699 were known to be monitored.[16] In addition, the agency estimates that there are some 175,000 active and abandoned landfills that have been identified, and that "75 percent of all...disposal sites leak contaminants into the ground and ground water."[17]

Poisoning America's Wells

There are almost an endless number of horror stories that can be related about the poisoning of ground and well water supplies throughout the nation.

On Long Island, New York, where the entire population of two million people is dependent on ground water, "all three of the Island's principal aquifers are seriously and dangerously contaminated," according to EPA. In the New York metropolitan area, over 600 wells had to be closed by mid-1981 because of chemical pollution, and government officials warned that thousands of others were endangered. Residents who had used well water for drinking complained of such ailments as bronchial coughs, rashes, stomach aches, sore throats, diarrhea and other pains and illness which often disappeared when they left the area on vacation, but recurred when they returned home.[18]

The major source of Atlantic City's water supply has been found to be threatened by toxic wastes that are seeping in from a nearby landfill. On 21 September 1981, the U.S. Department of Justice filed a lawsuit charging that Proctor & Gamble, Honeywell, Union Carbide and other industrial giants were responsible for the deadly chemicals dumped at the waste site. Justice Department attorney Charles J. Walsh warned that "by the fall of 1982, the pollutants will begin seeping into Atlantic City's water supplies. The government has designated this case the most severe environmental problem in the country."[19] Numerous carcinogens have been found to be present in the contamination, including benzene at levels 11,000 times above what is considered "acceptable." The *attempted* cleanup costs are estimated at $20 to $40 million, to be borne, of course, by the taxpayers.[19b]

In nearby Farmington, New Jersey, the water supply has already been ruined by the pollution. Residents are forced to truck in water for drinking, but not enough can be stored for most residents to take baths. One resident, Dorothy Johnson, told of how she could not do laundry at home because, in the local water, clothes turned yellow. Spigot water smelled like rotten eggs, and her hot water heater had the bottom eaten out of it. Until the landfill was closed in 1976, her daughter Karen could not play outside because the odor made her sick; it was so bad that the family had to sleep with their windows closed, even during the summer. They blame the pollution for the fact that her husband, Melvin, had a kidney removed in 1977, and died of cancer in the summer of 1981. Costs to clean up the dump and provide a new water supply are estimated at $20 million.[20]

In the lovely Pine Barrens area of New Jersey, over 100 wells have been polluted and closed down by chemicals from a nearby dump used by Jackson Township. In 1971, when the community learned of plans to open a dump there, over 300 people signed a petition opposing it, on the grounds that the site contained their water supply. Nevertheless, the dump was approved and operated for years until the state closed it down. Throughout this period, township residents suffered from a variety of health disorders, including kidney ailments, miscarriages, dizziness and skin rashes.

One local resident, James McCarthy, who drunk the well water for a decade, had to have a kidney removed, and his other one was diagnosed as not functioning properly. Three of his neighbors had kidney ailments, and one of them — a 16 year old boy — lost one of his to cancer.[21] McCarthy's 9-year-old daughter, Tara, also died of cancer after a lengthy period of intense pain, including being paralyzed from the chest down.

In 1976, a study team, surveying groundwater in New Jersey to try and determine the cause of that state's extraordinarily high cancer rates, found that all 163 wells tested contained detectable levels of toxic chemicals.[22]

Every day, some 800 industrial plants along the Niagara River in New York and Ontario dump over 500 million gallons of toxic wastes into the River, endangering the health of the almost 400,000 Americans and Canadians who use its water.[23] In Connecticut, one 1979 state survey disclosed that 87 percent of the wells tested were contaminated.[24]

According to EPA, in at least 44 Massachusetts communities, water supplies are known to have become "severely contaminated," and the state estimates that "at least one-third of the Commonwealth's communities have been affected to some degree by chemical contamination." The aquifer in the Minneapolis/St. Paul metropolitan area has also been seriously polluted.[25]

From 1954 until 1977, some 400,000 cubic yards of hazardous chemical wastes from the Hooker Chemical and Plastics Corporation's plant at Montague, Michigan were dumped in leaking drums, pits, and lagoons. These chemicals ended up contaminating local wells and polluting the underlying aquifer containing some 2 billion gallons of groundwater. In addition, some 875 pounds a day of the chemicals have been flowing into White Lake, a major recreation area.

The chemicals involved include several known to cause cancer, birth defects and genetic, reproductive, behavioral, and nervous system damage, such as PCE, TCE, dioxin (TCDD), chloroform, mirex, asbestos, benzene and hexachlorocyclopentadiene (C-56). The June 1980 report on this incident issued by the Senate Committee on Environment and Public Works states:

> Persons drinking water from contaminated wells or eating contaminated fish from White Lake have been exposed to a wide range of chlorinated hydrocarbons. ... Residents have reported headaches and nausea. The contaminants include many known or suspected carcinogens, mutagens, and teratogens, and were at levels that could cause damage to human health and the environment...
>
> The ultra-toxic dioxin (TCDD) has been found on the site in concentrations of 240 parts per trillion. TCDD has produced significant toxic effects in animals at levels of 1 part per trillion...[26]

These chemicals were also dumped at various disposal sites elsewhere in Michigan and Ohio. The report described the dangers of other chemicals found to be present as follows:

> Chloroform, which was found in one well at a concentration of...6-½ times the EPA standard for drinking water, is a known carcinogen, is fetotoxic, and produces toxic effects on the central nervous system, the liver, and the kidneys.
>
> The carcinogenicity of carbon tetrachloride...has been known since 1941, and has been confirmed in repeated tests. It is also suspected of being mutagenic, and is toxic to the liver and kidneys... Mirex, which has been found in fish and the groundwater...is carcinogenic.[27]

Because of the inadequacy of present laws, especially those dealing with liability, few of the damages caused by Hooker are expected to be compensated. Hooker has agreed to undertake a cleaning program and to pay the state a fine of $1 million, but it is estimated that total costs of such an operation could total $15 million, and it is considered unlikely that much of the contamination can ever be removed.[28]

These few examples of ground water contamination are a representative sample of what is taking place throughout the country; what we are yet to learn may dwarf these incidents by comparison. As Eckhardt Beck of EPA points out, "The remaining states cannot be presumed to be untouched; we simply do not have data on them."[29]

After holding a series of hearings on the problem, the House Subcommittee on Environment, Energy, and Natural Resources, chaired by Congressman Toby Moffett (D-Conn), in late 1980 issued a report that reached the following conclusions:

- "The health of millions of Americans is threatened by government and industry's past failure to properly protect our ground water."
- "In many areas, cases of contamination may have resulted in irreversible damage to ground water resources or rendered them unuseable for decades or perhaps even for geologic time."
- "The destruction of our nation's ground water will continue unless we move immediately to locate all potential sources of ground water contamination and take action to block the further flow of toxic substances into the ground."[30]

The Prevalence of the Carcinogens TCE and PCE

As more and more wells and water tables throughout the country are examined, the more serious the problem of contamination is shown to be. Recent studies of water supplies across the country have revealed that a large number are polluted with common industrial solvents that cause cancer.

As of March 1980, when EPA's Office of Drinking Water completed an analysis of tests run by state agencies on well waters, one-third of the samples tested showed the presence of trichloroethylene (TCE). This carcinogenic chemical is used as an industrial solvent, to remove grease, and to clean septic tanks,[31] applications which probably account for most of the pollution it has caused. It has also been widely used for dry cleaning clothing and even in removing the caffeine from coffee, leaving residues behind in the process.[32] (In 1976, after the National Cancer Institute determined that TCE could cause cancer, General Foods stopped using it to decaffeinate its Sanka brand coffee. But it was replaced with a similar industrial solvent, methylene chloride, which tests showed to be even stronger as a gene altering agent and possible carcinogen.)[33]

Because TCE compounds are plentiful and cheap, they were usually discarded after use instead of being recycled, thus causing massive pollution problems.

In June 1980, California's State Department of Health reported that wells supplying over half of the water to the Santa Clara Valley contained two cancer-causing chemicals, TCE and another industrial solvent perchloroethylene (PCE). These chemicals have also been found in wells and water supplies in California's San Gabriel Valley and wells serving Los Angeles, Pasadena, Santa Monica, Burbank and other cities. One well field in the Valley which supplied drinking water to 400,000 people had to be closed because of contamination by TCE.[34]

Meanwhile, on the other side of the continent, it was found that Long Island's major water source — the underground Magothy Aquifer — had also been fouled, causing the closing of some 60 public wells and hundreds of private ones. Residents of a neighborhood near Jamesburg, New Jersey wonder what to do as an underground pool of water laced with TCE, PCE, and other hazardous chemicals migrates towards them at the rate of one-and-a-half feet per day. In fact, TCE has been found to be widely spread throughout the country: in Rockaway Township, New Jersey; Pueblo, Colorado; Des Moines, Iowa, Indianapolis, Indiana; Montague, Michigan and elsewhere in Massachusetts, Connecticut and other states.[35]

In May 1978, wells in Bedford, Massachusetts providing 80 percent of the town's drinking water were closed after being found to be heavily polluted with toxic chemicals, including TCE at levels as high as 500 parts per billion (EPA has stated that there is a risk of cancer at just 4 or 5 ppb.) A study by the State Department of Public Health revealed that in the four years preceding the closure, there was a sharp rise in deaths among the local women due to cancers of the breasts and the ovaries.[36] In other Massachusetts communities, TCE discovered in drinking water at levels of 100 parts per billion was thought to be the result of using vinyl-coated asbestos cement in pipes. Some schools were forced to use bottled water, and one placed a sign over a faucet stating that drinking from it "may be hazardous to one's health."[37]

In Tacoma, Washington, where some 250,000 tons of chlorinated chemical wastes were dumped in its bay from 1950 to 1972, large amounts of TCE, PCE and other toxic chemicals and metals were found to be present at the site of a closed Hooker Chemical Company solvent plant. The wastes were seeping into the bay, and up to one-third of the English sole examined in one area had liver tumors and degenerative diseases.[38]

The Dow Chemical Company and the Ethyl Corporation, two of the firms that manufacture TCE, claim that it is safe at the low levels at which it occurs in drinking water. But the evidence contradicts these assurances. In 1976, the National Cancer Institute found that TCE caused liver cancer in laboratory animals, and other animal studies have shown that it can damage the liver and cause paralysis, nausea, dizziness, fatigue, and even psychotic behavior. Because the chemical binds to DNA, the body's mechanism for controlling all growth and reproduction, scientists also suspect that it may cause gene mutations and birth defects.[39]

EPA estimates that the risk of cancer is increased by exposure to levels of just 4 or 5 parts per billion of these chemicals although the agency permits up to 40 ppb. of TCE in drinking water.[40]

A March 1980 Senate Environment and Public Works Commitee report describes the toxic qualities of these chemicals as follows:

> TCE is a known carcinogen, suspected of causing cancer in the human liver, and is mutagenic. Human subjected to chronic exposure to TCE have exhibited liver, kidney, and lung damage, nerve degeneration, and psychic disturbances.
>
> PCE is carcinogenic, neurotoxic, and is toxic to the liver.[41]

These findings about TCE's toxicity are especially alarming, since most if not all Americans have probably been exposed to the chemical, and much of the population regularly ingests it.

California: "Sterilizing" The Water — And The People

In recent years, incidents of massive and widespread poisoning of California's underground and drinking water supplies have prompted authorities to test water throughout the state, and the results have been shocking.

In early 1979, California health officials found dangerous levels of a widely-used, sterility and cancer-causing chemical in *half* of the irrigation and drinking wells tested in a major agricultural area, the San Joaquin Valley. Despite the fact that the pesticide dibromochloropropane (DBCP) had been banned in California two years earlier, the wells tested showed average levels five times higher than what is thought sufficient to produce cancer in a number of people drinking the water. As a result, many wells and community water supplies had to be shut down, including some new ones that had just been dug at a cost of $300,000.[42]

In just a few weeks, the state's survey found DBCP contamination in some 200 wells and water supplies serving 12 California counties and cities, such as Riverside, Anaheim and even Disneyland! Some of these areas had not been treated with DBCP for years[43] (see pages 318-19).

In some cases, this poisoning of wells and water tables has been deliberate company policy. In early 1979, a set of memos was made public from the Occidental Petroleum Corporation, whose subsidiary, Hooker Chemical Company, produces pesticides at a Lathrop, California plant. These documents indicate that the firm deliberately ignored state laws regulating the dumping of chemicals that might affect groundwater tables.[44] The chief environmental engineer at the plant repeatedly warned management that the dumping violated state law, but concluded that it would not be "wise" to disclose the information to state authorities.[45] After concluding an investigation of the matter, Congressman George Miller (D-Cal.) concluded, "Occidental purposefully ignored the law, as indicated by internal memos, and as a result polluted groundwater tables with poisonous wastes."

Since about 1953, the Occidental plant has disposed of its wastes from pesticide and fertilizer production in such a way as to allow the extremely toxic pesticides DBCP, BHC and lindane, as well as radioactive substances, to contaminate the groundwater. The Lathrop plant is located about 75 miles east of San Francisco, in California's Central Valley, one of the nation's most fertile and productive farming areas. About five tons of pesticides a year were discharged onto the ground, it is estimated.[46]

When internal company correspondence on the dumping was made public in early 1979, it clearly revealed that Occidental was well aware of the dangers posed by, and the illegality of, its actions. In a 29 April 1975 memo, plant engineer Robert Edson referred to the fact that the state Water Quality Control Board was not aware that the firm's pesticide wastes were polluting the groundwater. He also described how a dog had come in contact with some waste water that had seeped into a neighbor's field, licked himself, and died. On 25 June 1976, Edson reported that the illegal discharges had taken place "for years" and observed that the pesticides were being discharged less than 500 feet from a neighbor's drinking well. On 5 April 1977, he warned that "we have destroyed the usability of several wells in our area. If anyone should complain, we could be the party named in an action by the water quality control board . . . the basic decision is this. Do we correct the situation before we have a problem, or do we hold off until action is taken against us?"[47]

Another memo of 25 June 1977, stated that "For years, we have dumped wastewater containing pesticides and other AgChem products. Fortunately for the management of this company, no pesticide has yet been detected (in a neighbor's wells). I personally would not drink from his well." Edson's memo went on to note that "No outsiders actually know what we do and there has been no government pressure on us, so we have held back trying to find out what to do within funds we have available.[48]

When the state of California learned of the illegal dumping and began to test wells in the area in early 1979, it found extraordinarily high levels of dangerous, health-destroying chemicals. One well contained DBCP at levels 58 times above which the state set as its "action level" (at which the water is considered unfit for consumption). Other wells were found to contain high concentrations of such toxic pesticides as BHC, parathion, lindane and others.

The state of California has sued Occidental for $30 million and is seeking a $15 million cleanup effort. But private suits by affected citizens will be difficult to pursue because of the exorbitant legal costs and the burden of proving a causal link between health problems and specific chemicals. One attorney representing workers who became sterile from contact with

DBCP estimates that such a lawsuit would take five years and cost $50,000 just for expert witness fees.[49]

Although this particular type of dumping has presumably been halted or curtailed by Occidental, the pollution will continue to spread to other wells in the area for many more years: the contaminated groundwater is moving north, towards Lathrop's residential section, at a rate of 175 feet a year.[50]

Water, Water Everywhere, But Not A Drop To Drink

When the thousands of synthetic chemicals are produced each year for killing insects and other purposes, we end up drinking large amounts of them, since most chemicals introduced into the environment eventually end up in our water supplies. The drinking water of every major American city contains dozens of cancer-causing chemicals and other toxins. Douglas Costle, when Administrator of EPA, observed that Cincinnati's drinking water contains some 700 chemicals, "about 90 percent of which didn't even exist 20 years ago."[51]

The dangers of cancer-causing and other toxic chemicals in our nation's rivers and water supplies have been known and discussed for decades, but our government has yet to take effective action to address the problem.

In 1963, the former head of the Environmental Cancer Section of the National Cancer Institute, Dr. Wilhelm Hueper, and his associate, Dr. W.W. Payne, warned of these perils in an article in the *American Journal of Clinical Pathology:*

> The rapidly increasing pollution of many bodies of fresh
> and salt water with carcinogenic agents, and the inabilities
> of the presently used filtration equipment to remove
> adequately such contaminants from the drinking water
> supply, has created conditions that may result in serious
> cancer hazards to the general population. [52]

Today, with the cancer rate in America at its highest level ever, and increasing at a record rate, Drs. Hueper and Payne appear to have been proven correct.

The main sources of most of these chemical pollutants are municipal sewage and industrial discharges, and runoffs from farming areas. In addition to the hundreds of contaminants in drinking water, dozens of other chemicals, many of which are potentially hazardous, are added to drinking water at treatment plants. There are some 60 chemicals that are routinely mixed into drinking water in the process of collecting, treating, and delivering it, although some plants add less than a dozen.[53]

Chlorine, which is added to virtually all water supplies in large municipal areas, is one of the most lethal of these chemicals, and has long been known for its toxicity.

It was one of the nerve gases used by the Germans in World War I which together killed almost a hundred thousand Allied soldiers and disabled over a million more. In recent years, numerous accidents have occurred involving the release of chlorine gas being transported by truck or train (sometimes to be used in water treatment), causing the death of several people, the injury and hospitalization of hundreds (some of whom were permanently harmed) and the evacuation of thousands. The relatively low levels of chlorine added to drinking water do not appear to cause *obvious* short-term problems, but the long term effects of regularly ingesting this potent poison may prove to be extremely harmful.

Many cities use large amounts of chlorine not only to kill bacteria and prevent waterborne diseases, but also to bleach out brown coloring in waters with a high content of organic matter. Critics of present-day chlorination practices contend that the chemical is greatly over-used, and that its primary purpose is not to ensure public health but to meet arbitrary federal standards on bacteria levels.

Adding chlorine to drinking water in treatment plants is responsible for the presence of several carcinogens, since this chemical combines with others to form toxic compounds such as trihalomethanes and chloroform, which are known cancer-causing agents.

In 1975, the EPA conducted a study of six toxic chemical pollutants in the drinking water in eighty American cities and found quantities of chloroform in *all* of them. Most of these water systems also contained bromine compounds, which are mutagenic and may be carcinogenic.[54]

On 16 October 1980, Dr. Robert Harris of CEQ, revealed that several just-completed studies confirmed the link between heavily chlorinated water and malignant tumors. He said that the studies, which analyzed thousands of cancer deaths in

Louisiana, North Carolina, Illinois and Wisconsin, "add substantially to the evidence that there are measurable adverse effects from the chlorination of water; they should make believers of many of the scientists who have been disbelievers in the past."[55] One finding was that chlorinating drinking water seems to increase the chances of contracting gastrointestinal cancer by 50 to 100 percent. Dr. Harris also stated that the studies "are going to make the EPA standard look ridiculous," since they demonstrated that there is an increased risk of cancer even when chlorine is present at or below the levels allowed by the agency.[56]

The report, released in December 1980, did indeed state that the "studies have strengthened the evidence for an association between rectal, colon, and bladder cancer" and chlorinated drinking water. "These results support the hypothesis that chlorinated nonvolatile organic compounds in drinking water may be carcinogenic in humans."[57] Dr. Harris suggested that other disinfectants that kill bacteria, such as ozone, could be equally effective as chlorine in treating water.[58]

Other chemicals in our water may present equal or greater dangers. The President's Toxic Substances Strategy Committee, in reference to the 1975 EPA study, reported in June 1979,

> Only 9 to 14 percent, by weight, of the synthetic organic contaminants found in the water has been identified; some of these compounds result from chlorination during treatment and others are industrial wastes. The National Cancer Institute (NCI) has identified 23 of these chemicals commonly found in small amounts in drinking water as carcinogens or suspected carcinogens; 30 chemicals have been identified as mutagens or suspected mutagens; and 11 chemicals have been identified as promotors as well... NCI has concluded that carcinogens in drinking water pose an unacceptable risk to humans.[59]

A more in-depth analysis of ten municipal water supplies found 129 chemicals present, less than a dozen of which had been thoroughly tested for safety. Since then, some two dozen of these substances have been found to be carcinogenic, or are suspected of being so. Some 700 chemicals have been identified as contaminating municipal drinking water systems throughout the country. As of 1979, according to the Environmental

Defense Fund, only about 10 percent of these chemical pollutants found in drinking water have been identified.[60]

Twelve separate scientific studies conducted since 1974 have linked chemical levels in drinking water to cancer mortality rates. In one study of 77 cities, chloroform levels were associated with death rates from cancer of the pancreas. Studies conducted by the National Cancer Institute and the National Institute of Environmental Health Sciences indicate that chloroform residues in drinking water at levels as low as 100 parts per billion could increase cancers of the bladder by 30 percent, and cancers of the rectum and colon by 4 to 6 percent.[61]

A city with one of the highest chloroform levels is Miami, Florida. An EPA study of Miami's chloroform contamination concluded that "The level of risk...might be extrapolated to account for up to 40 percent of the observed liver cancer incidence rate," and warned that fatal diseases such as cirrhosis could also be associated with the pollution problems.[62]

The hundreds of other chemicals in drinking water, individually or in combination with each other, represent a growing threat to the American people of significant if unknown proportions. We thus find ourselves in the same dilemma as Coleridge's Ancient Mariner, who exclaimed, while dying of thirst under a blazing sun with the sea all around him,

Water, water, every where,
And All the boards did shrink;
Water, water every where,
Nor any drop to drink.

The Outlook: More Of The Same

One result of the flood of publicity over unclean drinking water in the early 1970's was the passage by Congress in 1972 of legislation — the Federal Water Pollution Control Act Amendments — enabling the government to deal with the problem. This legislation instructed EPA to compile within a year a list of poisonous chemicals, the discharging of which had to be controlled in order to protect human health and natural resources. At this time, it was estimated that some 30,000 chemical compounds were in use by industry, over 1,500 of which were suspected of causing cancer. But, incredibly, when EPA published its list, it contained *only nine chemicals!* Following that, EPA refused to issue standards limiting the discharge of these chemicals within 6 months, as the law required.[63]

As a result of EPA's failure to obey the law — which states as one of its goals "It is the national policy that the discharge of toxic pollutants in toxic amounts be prohibited," — environmentalists took EPA to court to force compliance. Eventually, EPA, as a result of these lawsuits, agreed in 1976 to carry out studies of additional chemicals and how to regulate them by 1983, as well as to set by 1977 discharge standards for 6 of the original 9 compounds.[64]

But EPA still refuses to enforce water pollution control laws. In September, 1981, the New Jersey Public Interest Research Group reported that EPA had ignored 86 percent of the Clean Water Act violations cited in the New York-New Jersey region in the last four years, pursuing only 574 of the 4,327 reported.[65]

The Reagan administration is strongly resisting efforts to have the nation's water supplies cleaned up, including trying to weaken the Clean Water Act. In January, 1982, Deputy EPA Administrator John Hernandez told a meeting of the National Association of Manufacturers that the agency would submit to Congress proposed changes to the 1977 Clean Water Act amendments that are intended to halt water pollution discharges by 1985. Calling the goal "not realistic," Hernandez invited industry representatives to let him know the "key modifications" they would like to make in the law.[66]

Opponents of laws and regulations to clean up and prevent water pollution often cite the tremendous costs involved in environmental protection. However, recent studies show that it is pollution that is costly, and that cleaning it up *saves* billions of dollars. One study, commissioned by the President's Council on Environmental Quality and released in April 1980, estimates that the removal of conventional water pollutants would, in 1983, yield "water pollution control benefits of about $6.5 billion to almost $25 billion per year," with the probable figure being about $12.3 billion annually.[67]

Most of the benefits would be attributable to improved water-based recreation, with other gains including reduced damage to crops and vegetation, improvements to human health and residential property values, and $1 billion to commercial fisheries. The author, economics professor A. Myrick Freeman III, stresses that these estimates "might understate considerably the true water pollution control benefits to be enjoyed in 1985" since they do not include preventing damage by certain toxic pollutants to such things as shellfish beds, commercial fisheries, and effects on human health by organic chemical and heavy metal contamination of drinking water. According to CEQ, "... total benefits could grow dramatically once the benefits from reducing these pollutants are included."[68]

But regardless of how the laws are implemented or changed, the harsh reality of the situation is that with over 4 billion pounds of synthetic chemicals being produced each week, these toxic compounds will continue to end up in our water supplies and be consumed by humans in ever-increasing amounts.

As the government weakens or refuses to enforce its standards for toxic chemical residues in our water, food and air, it legitimizes and perpetuates this vicious circle of lethal contamination. We have, in short, effectively created a politically and economically feasible formula for our own destruction.

What You Can Do

There are several steps that can be taken by individuals and communities to reduce the levels of toxic contaminants in their water supply. For instance, carbon water filters can be purchased and placed on home faucets, and larger versions can be used in municipal water plants to greatly reduce the presence of cancer-causing chemicals in drinking water. Using bottled water for drinking can also reduce the intake of these carcinogens, although some brands are freer of impurities than others. Some harmful substances can be removed from tap water by letting it run for a couple of minutes in the morning to flush it of lead, cadmium, and other toxic metals that build up from the plumbing. If you want to have your water tested, contact your local water authority, health department, or environmental agency for information on which laboratories perform such analysis.

Chlorine is often added to drinking water in municipal areas, but it is misused to an extraordinary extent. Often, it is added to rid water of dark coloring from organic matter. Reducing that kind of use could help eliminate many cancer-causing compounds that result from the chlorination process. If you boil your water or put it in a blender for a few minutes, some of the chlorine and other gaseous compounds will evaporate.[69]

Only in the last few years have we become aware of the alarming dimensions of this problem. We will begin moving toward a solution as soon as our government decides to implement and enforce laws designed to protect our nation's precious supplies of water.

Chapter Five

Air Pollution: 35 Million Americans In Danger

In 1969, the American Public Health Association estimated that some 10 to 20 people a day were dying from air pollution in New York City.[1] Today, we have a broader perspective on the massive mortality caused by contaminated air: we now know that air pollution is probably killing literally hundreds of thousands of Americans each year,[2] and that the health of some 35 million urban residents is in serious jeopardy.[3]

Almost two decades ago, President Lyndon Johnson issued a colorful and accurate warning:

> Very plainly, we are left with but a single choice. Either we stop poisoning our air — or we become a nation in gas masks, groping our way through dying cities — and a wilderness of ghost towns.[4]

The President's message addressed a problem that was obvious even then: our metropolitan areas were becoming unliveable. Since then, despite numerous new laws, studies and promises from politicians, our air has in many respects continued to worsen. While truly significant improvements have been made in some areas, air pollution remains a serious problem, a major aspect of being the presence of dozens of chemicals and other pollutants known to cause cancer, disability and death. The pre-

189

sence of just one of these substances — sulfates — is estimated to cause some 200,000 deaths a year among adult whites.[5]

One graphic example often cited of the damaging effects of air pollution is that of Cleopatra's Needle, the striking granite obelisk that was placed in New York's Central Park in 1883, after some 3,500 years in the Egyptian desert. During less than a century in the Park, it has lost several inches of granite and undergone greater deterioration than in the 35 centuries it spent being pounded and ground by the sand and wind, and baked by the hot Egyptian sun. This same air—which has eaten away part of this solid granite object—is breathed in and out of the lungs of each of New York's seven million residents at an average rate of 16,000 quarts a day per person.[6]

Caution: Don't Breathe The Air

People who wish to avoid or minimize their contact with cancer-causing chemicals can choose not to smoke, drink tap water or eat meat and other high-fat foods. But one cannot choose to stop breathing, and most Americans are regularly exposed to serious air pollution, which may kill hundreds of thousands of us each year.

Air pollution is hardly a new phenomenon; some cities, such as London, have experienced it for centuries. Indeed, in 1952, an atmospheric inversion there claimed some 4,000 lives. Ten years later in London, 700 people died during another period of heavy air pollution. Other lethal air pollution episodes include New York City in 1953, in which 200 people were killed; the Meuse Valley Belgium where 63 died in 1930, and Donora, Pennsylvania, where 20 were killed in 1948.[7]

Today, much of the pollution is less visible than earlier, soot-type filth in the air, and it does not kill as quickly as some versions. But, over the long term, it is just as destructive of human health, and in many cases, more so. Breathing polluted air can bring on chronic, irreversible and often fatal lung damage in even the healthiest individuals, causing such obviously-induced diseases as "black lung" from coal dust in coal miners, and "brown lung" from cotton dust in textile workers. At lower levels of pollution, the effects are less apparent and not as well understood, but nevertheless have subtler consequences such as shortened life spans, disability, cancer, increased asthma attacks and generally poor health. The

May 1980 Report to the President by the Inter-agency Toxic Substances Strategy Committee states that the lives of 35 million Americans are in danger because of dirty air:

> Of all urban dwellers, one in five — more than 35 million people — are at special risk from such illnesses as emphysema and bronchitis as a result of exposure to air pollution because of age or health. In general, industrialized, densely populated metropolitan areas have higher cancer mortality rates than rural areas, especially for lung cancer... In Great Britain, lung cancer rates declined first in areas where the clean air laws were applied.[8]

Many studies conducted in the 1950's show correlations between air pollution in English cities and bronchitis, pneumonia, lung cancer and infant deaths among the local residents. Lester Lave and Eugene Seskin's famous 1970 study of 114 Standard Metropolitan Statistical Areas found air pollution — especially particulates and sulfates — to be correlated with total death rates as well as the death rate for infants. Another study of U.S. communities found increased prevalence of chronic respiratory disease in both smokers and non-smokers in more polluted areas.[9]

The Lethal Role Of Chemicals

Carcinogens and other toxic chemicals are so widespread throughout the air that exposure to them simply cannot be avoided. This is especially true for people living in urban areas; they die of lung cancer at nearly double the rate of those in rural areas. In fact, there are strong statistical correlations between cities with high industrial pollution levels and death rates from lung cancer.[10]

Three EPA studies of air pollution in large cities found several, such as Los Angeles, Oakland, Phoenix, and others in Texas and New Jersey, with chemical levels equal to or greater than readings taken outside of homes in the Love Canal area.[11]

Several studies have been conducted in Japan and Europe on how childhood growth and development, and the incidence of cancer are affected by certain air pollutants, such as asbestos. As Dorothy Noyes Kane writes in the November 1976 issue of *Environment* magazine.

> ...the investigations indicate that air pollution from sulfur
> dioxide, nitrogen oxide, dust, arsenic, ammonia, fluorides,
> and other materials is responsible for an adverse effect on
> children's growth and development as manifested by
> retarded bone maturity, certain abnormalities in blood
> chemistry, and an excess of larger than normal tonsils and
> lymph nodes.[12]

Despite the passage of various anti-pollution laws, America's air quality is continuing to deteriorate. One reason for this is the government's inability or unwillingness to enforce the laws.

The results of EPA's inaction have been most apparent to those living in metropolitan areas. Since the 1950's, the visibility in the northeastern United States has decreased by 10 to 40 percent, and by 10 to 30 percent in the Southwest.[13] In 1977, in two of the largest urban areas, New York City and Los Angeles, air pollution levels were in what EPA defines as the "unhealthful range" on over two-thirds of the days.[14] In New York, according to EPA, "air pollution remains a serious problem ... Despite cleaner cars, carbon monoxide and ozone are still at levels typically twice those acceptable for protection of human health."[15]

In early October 1980, Los Angeles experienced almost two weeks of some of its worst smog and air pollution ever, keeping joggers off the streets; children, old, and sick people indoors and most of the residents uncomfortable, with irritated eyes, ears and respiratory systems.

Ironically, it was near the height of this pollution that presidential candidate Ronald Reagan issued his widely ridiculed statement that air pollution had been "substantially controlled;" that millions of dollars were being wasted on auto emission controls and that Mount St. Helens had released more sulfur dioxide pollutants into the air "than [have] been released in the last 10 years of automobile driving or things of that kind that people are so concerned about." (In fact, man-made emissions of sulfur dioxide total 81,000 tons a day, compared with 1,500 tons a day for the heaviest 100 days for the volcano.)[16]

The state of New Jersey contains one of the greatest concentrations of industrial polluters found anywhere in the

This plane is spraying sulfur on grapevines, but far more toxic compounds are usually used on agricultural crops. This results in the massive poisoning of farmworkers, and contamination of our food with chemicals that are known to cause cancer, birth defects, and genetic damage at extremely low doses. Cancer now strikes one American in four, and kills over 1,000 of us every day. *Photo courtesy of Gene Daniels, EPA, 1972.*

America the Poisoned by Lewis Regenstein

20 Years Later, Still...

Rachel Carson, author of the classic book *Silent Spring*, documented the immense harm that pesticides were causing to humans, wildlife, and the environment. The chemical industry tried to stop the publication of the book in 1962, suggesting that it was part of a communist plot against America. Today, 20 years later, almost all of the toxic chemicals she warned about are still in widespread use. The few that have been restricted have often been replaced by equally or more hazardous compounds. And a thousand new and largely untested chemicals are introduced each year. *Photo courtesy of Shirley Briggs/Rachel Carson Council, Inc.*

America the Poisoned by Lewis Regenstein

Poisoning Our Wildlife

Such birds of prey as falcons, ospreys, and even our national symbol, the bald eagle, have been decimated by pesticide spraying, in some cases to the point of near extinction. Pesticides can kill the birds directly, or, in lower doses, prevent them from reproducing. Since DDT and a few other pesticides have been restricted, several of these species have begun to make a comeback in some areas. *American Bald Eagle on nest: Photo courtesy of Karl Kenyon, Bureau of Sport Fisheries & Wildlife. Peregrine Falcon: Photo courtesy of Mike Smith, U.S. Fish & Wildlife Service. Osprey in flight: Photo courtesy of Luther C. Goldman, Bureau of Sport Fisheries & Wildlife.*

America the Poisoned by Lewis Regenstein

Poisoning Our Food

When "cropdusters" spray pesticides on farmland, less than one percent may hit the target insects, and only a fourth may land in the crop field, with the rest drifting miles away. Since birds, frogs, and predatory insects are killed off, the spraying of chemical insecticides is usually counter-productive. Although the use of such chemicals has increased tenfold in the last 30 years, losses to insects have doubled. We can grow more food, more cheaply without relying on the massive use of pesticides. Newly-developed biological methods of pest control emphasize such techniques as crop rotation, soil tillage, insect traps, and the release of wasps and sterile bugs. Government studies show that such systems, if used on major crops, could reduce pesticide use by 50 percent over the next 5 years, and 70–80 percent in the next decade "with no reduction in present crop yield levels." *Biplane: Photos courtesy of Charles O'Rear, EPA-Documerica/Rural America.*

It is hardly surprising that we are in the midst of a cancer epidemic. Several hundred pesticide ingredients used on our food are known to cause cancer and birth defects in animals, in some cases at the lowest doses tested. By the time the government got around to banning major uses of some the most deadly chemicals, such as DDT, dieldrin, BHC, and PCB's, these poisons were being found in the tissues of literally 99 percent of all Americans tested, as well as in our food, air, water, and almost all mothers' milk samples. In fact, breast milk has been so contaminated with banned, cancer-causing chemicals that it would be illegal to sell it in supermarkets. *Photo courtesy of USDA.*

America the Poisoned by Lewis Regenstein

Poisoning Our Forests

Herbicides are sprayed on forest land to kill off unwanted trees and other vegetation, which *supposedly* speeds the growth of commercially valuable timber (although studies do not confirm this). This provides a rationale, albeit a false one, for increased cutting of national forests by the timber industry and the U.S. Forest Service. Such spraying has caused countless miscarriages and other health problems among residents of nearby towns throughout the west, apparently including cancer and birth defects. These chemicals often wipe out the local wildlife, as well as pets and livestock. *Photo courtesy of Wes Guderian, The Oregonian/Rural America.*

America the Poisoned by Lewis Regenstein

Poisoning our Environment

The amount of toxic wastes produced in the U.S. each year amounts to over 600 pounds for every man, woman, and child in America, or perhaps even double that. 90 percent of these wastes are improperly disposed of, simply because there is no safe place to discard such deadly substances. There are literally thousands of hazardous dump sites across the country that are as dangerous as Love Canal or more so. As a result of such contamination, water supplies throughout the nation are becoming polluted with cancer-causing chemicals. *Photos courtesy of Ruffin Harris/Environmental Defense Fund. "Danger" sign: Photo courtesy of EPA-Documerica.*

America the Poisoned by Lewis Regenstein

It is common to see discarded pesticide containers scattered along roads, yards, farms, and in cities. These empty and partly empty barrels of endrin and methyl parathion were left near Arkansas City, Kansas, by a group of aerial sprayers who used an airstrip without permission. Some of the cans still contained chemicals, and the smell of pesticides continued to be strong several months after they were discarded. The air field was used by young people for drag racing on weekends. No measures were taken to keep children or animals away, and several pet dogs died after walking through the field. As little as a teaspoonful of methyl parathion absorbed through the skin can kill a person. *Photo courtesy of Larry Miller/Caldwell, Kansas, 1976.*

America the Poisoned by Lewis Regenstein

Poisoning Ourselves

This 1961 USDA photo shows a helicopter spraying the deadly herbicide 2,4-D over "sagebrush infested rangeland" in Utah's Fishlake National
Forest "to improve forage production and summer grazing for livestock."
This chemical, the nation's most widely used weedkiller, not only kills and
deforms fish and wildlife, but can also contaminate meat from livestock
grazed on sprayed areas. After 2,4-D was sprayed along forests and roadsides, 9 out of 10 pregnant women in one area of Montana had miscarriages;
in Ashford, Washington, 10 out of 12 pregnancies ended in miscarriages. *Barrels: Photo courtesy of Larry Miller/Caldwell, Kansas. Helicopter: Photo courtesy Ma*
Robinson, USDA.

America the Poisoned by Lewis Regenstein

ACROPOLIS BOOKS LTD.
Washington, D.C. 20009

world; it also enjoys the dubious distinction of having the highest cancer rate of any state in the union. As the Environmental Defense Fund notes in its authoritative book on chemical pollution, *Malignant Neglect* (1979).

> There are over 1,200 chemical plants in the state, including an area adjacent to Newark Airport that may be the biggest complex of petrochemical plants and oil refineries anywhere. In New Jersey, each year 24,000 new cases of cancer are diagnosed and 14,000 people die of the disease. For American males, the state rates first in the incidence of cancer of the bladder, second for rectum and large intestinal cancers, and third for cancers of the lung and stomach. A state senate commission studying the problem has stated that "...air contamination by carcinogens is a primary cause of many types of cancer."[17]

Another study of men living in a highly polluted industrial area of Los Angeles County, containing chemical plants and oil refineries, found that they died of lung cancer at a rate over one-third higher than other men living elsewhere in the County. The greatest number of deaths from lung cancer occurred in the vicinity of an air pollution measuring device in the middle of the industrial area, where the recorded levels of benzo(a)pyrene were highest.[18] This kind of air pollution may be killing or destroying the health of millions of Americans.

Ironically, indoor air pollution can be equally or more hazardous, and may cause over 10,000 cancer deaths a year. Vapors from such products as urea formaldehyde insulation, household oven cleansers and grease dissolvers, synthetic fibers and household pesticides can be deadly. Decaying radon gas, seeping out of the ground, brick walls and tap water can cause 10,000 to 70,000 fatal lung cancers a year depending on air circulation, according to an EPA working document.[19] Attempts by environmentalists and health officials to restrict the use of formaldehyde — which is used in numerous consumer products, permanent press fabrics, carpets and as insulation in some 400,000 to 600,000 homes — have been stymied by the Reagan administration, which even tried to fire a government scientist who warned of its dangers.[20] (see p. 382).

Sulfates: 187,000 Deaths A Year

The tremendous and largely undetected impact of air pollution on human health is documented in a 1978 study which found that up to 212,000 adult white Americans may be killed each year just by the presence of sulfates in the air. The study, by Professors Robert Mendelsohn of the University of Washington in Seattle, and Guy Orcutt, of Yale University, estimates that the annual number of deaths in adult whites from exposure to airborne sulfates is between 163,587 and 211,781, with the expected level at 187,686. Sulfur dioxide causes between 7,730 and 39,782 deaths each year, with the probable level at 23,756. Equally lethal is carbon monoxide, killing between 7,416 and 35,389 white adults, or probably 21,403.[21]

The authors conclude that "there appears to be a definite association between pollution and deaths from heart and circulatory failure."

Their research also found that the most polluted air was present in the northeastern and north central areas of the country, and that residents of those areas have about twice as great a chance of dying from air pollution than people in the rest of the nation. The worst single area identified was in eastern Ohio, which had the greatest concentration of sulfates, probably from the presence of many major power plants. The cleanest air was found in states west of the Great Plains.

Mendelsohn and Orcutt estimate that each year a total of about 142,000 adult white deaths, or 9 percent of all mortality, are caused by air pollution. Because the presence of some substances, such as ozone and nitrogen dioxide, was negatively correlated with mortality, or had no apparent effect, the study puts the total number of adult white deaths associated with air pollution at between 87,570 and 197,260, or between 5.4 and 12.3 percent of all deaths, as opposed to the projected total number for sulfates.[22]

Numerous other studies also demonstrate that sulfates cause breathing difficulty and various adverse health effects. A number of research papers published by EPA on this subject document an association between sulfate levels in the air and such effects as increases in mortality, heart and lung disease symptoms, chronic bronchitis, asthma attacks, lung disability and susceptibility to viral disease of the respiratory tract. Lave and Seskin's 1970 study of 114 metropolitan areas found a strong correlation between sulfates and the overall death rate, including infant mortality and deaths from cancer and heart disease.[23]

In February 1981, the results of a government study were released showing that sulfates and other air pollutants, just from coal-fired power plants, may help cause the deaths of 8,000 people a year in the Ohio Valley. It revealed that 163,000 people there could die of heart and lung disease caused by pollution in the next 25 years if current trends continue, but stricter controls could reduce the figure by one-third. Weak enforcement, on the other hand, could increase the death toll to 200,000. These conclusions were the result of a five year study conducted by some 100 researchers at 10 universities (the Ohio River Basin Energy Study).[24]

And in 1975 Dr. Leonard Hamilton, a scientist at the Brookhaven National Laboratory who studies the health effects of pollutants, estimated that "acid sulfates from fossil fuel... emissions are responsible for 7500 to 120,000 deaths a year."[24b]

Acid Rain

Sulfates are also a major factor in the recently recognized phenomenon called "acid rain," a critical problem of world-wide dimensions which threatens to damage the earth's capacity to produce food.

It is only in the last few years that the dangers of acid rain, one of the most serious forms of air pollution, have been appreciated. Acid precipitation in the form of rain, snow, hail, dew and frost is occurring throughout large areas of the United States, Canada, Europe and Japan. In the northeastern United States, during the past 25 years, the acidity of rainfall has increased some 50-fold! It is now up to 100 times more acid than normal rainfall, and some typical summer rains have the acidity of lemon juice, according to an EPA study released in mid-1980. As a result, government officials and scientists predict that in just 15 years, over 50,000 lakes in the United States and Canada, or even several times that number, could become "dead" — devoid of fish and plant life.[25]

The origin of acid rain can be found in smoke emissions from a variety of sources (tall stacks at power plants, smelters, car exhausts) containing oxides of sulfur and nitrogen, which are transformed in the atmosphere into sulfuric and nitric acids. Because of the high smokestacks, these pollutants are transported hundreds or thousands of miles before falling to earth as acid rain.[26] Thus, pollution from the huge concentration of coal-fired power plants and industries in the Ohio River Valley — which are responsible for one-fifth of all sulfur dioxide emissions across the United States — ends up killing trout in New York state. EPA estimates that some 28 million tons of sulfur dioxide emissions are spewed out across America each year, mainly by about 200 coal-fired plants built years ago and thus exempt from the rules governing new plants.[27]

At the present time, there are no reliable estimates on the losses to cultivated crops and forest productivity, but the damage to agricultural areas and natural eco-systems is enormous. Studies that have been conducted show that acid rain causes drastic reduction in the growth of radishes, soybeans, kidney beans and other legumes.[28]

Killing Thousands of Lakes

It is also thought that acid rain is responsible for wiping out the fish populations in some 170 to 264 high-altitude lakes, encompassing 11,500 acres, in the Adirondack Mountains of upper New York state. Another 256 Adirondack lakes are considered endangered. A March 1980 report issued by the Senate Environment and Public Works Committee states, "Ninety percent of the Adirondack lakes above 2,000 feet elevation are barren because of acid rain contamination." (Other studies place the percentage at 50 percent.) Officials estimate that 1,000 lakes or more in Wisconsin are at the point of disaster, and 10,000 in the northern part of the state are threatened. Although the biggest problems have been in the East, acid conditions are also turning up in states as far away as Colorado and California.[29]

In the state of Pennsylvania, the rainfall is among the most acidic in the world (100 times the normal level). The head of the Pennsylvania Fish Commission wrote in the June 1980 issue of *The Pennsylvania Angler* magazine:

Pennsylvania's cold-water streams are faced with their greatest threat to survival in history. Greater than the deforestation which caused the loss of much of our original brook trout habitat by the turn of the century, greater than strip mining, industrial and municipal pollution and channelization, each of which has caused the loss of thousands of miles of fishing waters. The threat comes in the invisible form known as acid precipitation and is a greater threat than floods, droughts, diversions, or over-exploitation.[30]

As fish and amphibians such as frogs and salamanders die off, birds and mammals that feed on them are deprived of an important food source, and they, in turn, disappear. Creatures that are especially susceptible to acid rain include river otters, minks, muskrats, loons, herons and various other species of ducks and waterfowl.

In recent years, it has become apparent that acid rain is a worldwide problem. Winds from Great Britain kill salmon and other fish in Scandinavia. Acid rain falls on Bermuda, an island in the middle of the Atlantic Ocean with no appreciable industry, as well as on Hawaii's Mauna Loa volcano.[30b] In one well-known incident in 1974, a storm in Scotland produced rain that was the acidic equivalent of vinegar.[31]

In Greece, Rome, and elsewhere in Europe, the Acropolis and other priceless and historic buildings, statues, and monuments, built centuries ago, are disintegrating because of acid rain and other air pollution. From the Egyptian temples at Karnak to the U.S. Capitol and the Statue of Liberty, irreplaceable cultural treasures are dissolving before our eyes.

In Sudbury, Ontario, International Nickel (Inco Ltd.) operates a coal-fired smelter with a 1,250 foot stack that has been responsible for one percent of all the world's sulfur dioxide (SO_2) emissions, and is blamed for harming or rendering lifeless tens of thousands of Canadian lakes. It produces 2,500 tons a day of SO_2, a two-thirds reduction from earlier years of some 7,000 tons. A study by the Ontario Ministry of Environment reports that some 140 lakes are already showing serious damage from acid rain, and another 48,000 lakes could be endangered in the next two decades. Salmon populations in Canada are also threatened, and they have already disappeared from several rivers where they were once abundant.[32]

Although this threat has only recently become apparent in North America, Europe has experienced the problem for years. In southern Norway and Sweden, salmon fishing is all but wiped out; thousands of lakes and streams have lost their fish populations and have become practically lifeless, and Scandinavian scientists estimate that acid rain has caused a 15 percent reduction in timber growth.[32b]

In Sweden alone, some 15,000 lakes are without fish because of acid precipitation caused by industries in England and northern Europe,[33] especially West Germany. Sweden has estimated the 1979 loss to its commercial and sport fishing industry at $50 to $100 million.[34] A four percent annual decline in overall forest growth in southern Sweden has also been attributed to acid rain.[35]

High acidity also causes the water and the fish to absorb higher levels of toxic metals, such as lead and mercury, and it erodes drinking water pipes, depositing metals into the water. In one area of Sweden, the inhabitants' hair turned red from drinking water carried by copper pipes eaten away by acidity.[36]

Some of the most damaging effects of acid rain are so insidious they are not noticed, but this makes the damage no less severe. Marble buildings, monuments and statues in the eastern United States are slowly being dissolved — literally — by acid rains, including the famous Cleopatra's Needle in New York's Central Park.

Acid rain hits the pocketbook of every American in the form of increased inflation; higher taxes, health and insurance costs, and food and lumber prices; damage to private property (such as automobile paint corrosion) and loss of tourism to affected areas.[37]

The actual estimates for the economic impacts of acid rain are staggering. In joint Senate hearings held on 23 September 1980 by the Select Committee on Small Businesses, and the Environment and Public Works Committee, Dr. Tom Crocker of the University of Wyoming cited $5 billion as the 1978 figure for damages just in the eastern United States. This included $2 billion in materials, $1.75 billion in forests, $1 billion in agriculture, $250 million in lakes, rivers and other aquatic ecosystems, and $100 million in health, water supplies and other areas.[38]

Senator Gaylord Nelson (D-Wis), chairing the hearings, observed that if the 100,000 now-threatened lakes in northern

Wisconsin were significantly harmed, the state's $5 billion a year tourist industry would be "severely impacted." Jay Heffern, of the Minnesota Pollution Control Agency, noted that because agriculture accounts for a third of the state's employed work force, a reduction in agricultural productivity of only 2 to 3 percent could cost the state half a million dollars a year. With 2,300 lakes there considered to be endangered by acid rain, the "tourism and fishing economics could suffer an annual loss of $78 million," according to Heffern. In New York, the Adirondack Park Agency estimates that the state loses at least $1-½ million annually just from decreased sport fishing.[39]

On 18 June 1980, Douglas Costle, then Administrator of EPA, stated that "the damage being done may be irreversible; in fact, is probably irreversible." Thus, every day that passes without steps being taken to address the problem means more permanent damage to the environment, the economy and our system of food production.

Ironically, this problem was identified, though not acted on, over a century ago. In 1852, Robert Angus Smith, a British chemist, noticed what appeared to be a correlation between the sooty air of heavily-industralized Manchester, England and "acid rain," as he termed it two decades later in a 600 page book.[39b] Now, over one hundred years later, he has been proven correct; but we still seem unwilling to address this critical problem.

The Outlook For Acid Rain: More Of It

The acid rain problem is likely to worsen, especially with the projected increase in the use of coal. In December 1977, a report by a Presidential panel of scientists stated that one result of burning more coal will be that "agricultural and forest production may decrease in large regions of the country, and fish populations will disappear or be seriously reduced in certain lakes." Human health will also feel the impact in the form of increased incidence of serious respiratory ailments:

> We can anticipate an increase in the number of asthma attacks in susceptible individuals, increased incidence of exacerbation of acute and chronic respiratory disorders.[40]

The best way to reduce acid rain is, of course, to reduce the burning of coal. But coal-fired plants can also be made much cleaner by adding "scrubbers" to reduce emissions before they leave the stacks. Existing scrubber technology can remove up to 90 percent of sulfur emissions.[41] Washing the coal before burning can also remove much of the sulfur, making the coal not only cleaner but often more economical to burn.[42] It is estimated, for example, that if all Ohio electric power utilities were required to wash coal before burning it, sulfur dioxide emissions in the state could be reduced by 25 to 30 percent at plants where no other control measures are required.[43] The National Commission on Air Quality has estimated that sulfur dioxide emissions could be reduced 40 percent at a cost of raising utility rates in the East by less than 2 percent.[44]

In Japan, between 1968 and 1975, strict sulfur oxide controls, the use of low sulfur fuels and the removal of sulfur cut emissions in half while energy use doubled. Some 1,200 scrubbers have been installed in plants there, six times as many as in the U.S.[44b]

The Carter Administration's plan to boost coal use has also elicited cries of outrage from abroad. On 18 April 1980, Canada's Minister of the Environment, John Roberts, protested the Carter Administration's $10 billion program to get utilities to switch from oil to coal, stating, "acid rain is literally a rain of death." Roberts estimates that 70 percent of the acid rain falling on eastern Canada comes from the U.S. and causes up to $4 billion a year in damages, mainly to forests and fisheries.[45] Douglas Costle estimated that the resulting conversion would increase sulfur emissions by 16 percent, producing a 10-15 percent increase in acid rain, much of which will fall in Canada.[46] The increased burning of coal, he told a Senate subcommittee on April 1980, could put an additional 330,000 tons of sulfur dioxide (SO_2) and 200,000 tons of nitrogen oxides (NO) into the northeastern U.S.' air each year.[47] (In 1980, the U.S. spewed over 26 million tons of SO_2 and 22 million tons of nitrogen oxides into the air; the amount of SO_2 produced by the U.S., Canada and Europe combined is estimated at some 100 million tons.[47b])

On 5 August 1980, the United States and Canada signed an agreement to work toward a treaty controlling the spread of acid rain, but the document contained no requirements for substantive steps to be taken toward realizing this goal. The agreement contains no new controls on emissions, and commits each nation to appoint members to study groups which will — you guessed it — issue a report on the problem![48]

In March 1981, Canadian officials charged that the U.S. had "largely ignored" the bilateral agreement. They complained that while Canada had strengthened its Clean Air Act and placed additional controls on some plants and programs, the U.S. had not responded in kind. As the *Toronto Globe and Mail* reported on 4 March 1981:

> When the memorandum was signed, a U.S. official at the ceremony remarked: "Of course, this all goes to hell if Ronald Reagan gets in power." Canadian officials now fear he was right.

The U.S.' inaction on acid rain has become the single greatest diplomatic problem between these two countries. On 6 October 1981, Raymond Robinson, a Canadian environmental official, told a U.S. congressional hearing, "Actions that would focus exclusively on further research would allow the situation to grow substantially worse and might frustrate any hope of controlling it in the foreseeable future." But while Canada complains — justifiably — about acid rain caused by U.S. sources, it still permits motor vehicles to emit three times as much nitrogen oxide as an American vehicle. Canada also continues to allow the operation of the single biggest source of acid rain — International Nickel's Sudbury, Ontario plant, which spews sulfur dioxide throughout Canada and the eastern U.S.[49]

Some of the most pessimistic projections about the future effect of acid rain come from the U.S. government itself. Some agencies have called attention to the seriousness of the problem, but to little avail. Indeed, the *Global 2000* study, issued in July 1980 by CEQ and the State Department, predicts that such adverse effects in water, soil and plants "are likely to become increasingly severe." According to the study, this accelerated damage will take place mainly because a projected 58 percent increase in the combustion of oil, a 13 percent increase in that of coal and a 43 percent increase in burning of natural gas by 1990 will greatly increase sulfur oxide and nitrogen oxides and thus exacerbate the acid problem.[50]

Other chemicals in the air will also damage the world's ability to grow food; as the government's *Global 2000* study points out, the situation continues to worsen:

> Air pollution already causes significant damage to agricultural crops. In the United States, air pollution damage to crops in Southern California alone cost farmers $14.8 million per year during the 1972-76 period... Celery, potatoes, and tomatoes were especially hard hit. Over the next two decades, the impact of air pollutants on agriculture can be expected to increase, especially for farmers downwind of industrial centers.[51]

In light of the present political realities, there appears to be little hope that any effective action will be taken to address the acid rain problem adequately. Secretary of Interior James Watt, who serves as Chairman of the Cabinet-level Council on Energy and the Environment, has characterized acid rain as "another scare tactic" by environmentalists.[52] And EPA Administrator Anne M. Gorsuch is a former Denver legislator with an anti-environmental, anti-regulation background. At the time her nomination was announced, she stated that the causes and solutions for acid rain were "largely speculative," and could not be linked for certain to power plant discharges. She also suggested that the required use of scrubbers to remove sulfur and other pollutants was not rational.[53] Under Gorsuch, EPA has continued to resist taking action, and the agency has even proposed to double permissible nitrogen oxide emissions from cars.

In addition, the policies of the Reagan administration, stressing less regulation of polluters and the burning of more fossil fuels, are sure to exacerbate this already critical problem.

None of this should come as a surprise to anyone familiar with the views and policies of President Reagan. In October 1980, he said that trees and other vegetation were responsible for 93 percent of the nation's nitrogen oxide pollution, and that "some doctors" were investigating the possibility that nitrogen oxides "might be beneficial to tubercular patients." He also promised that, as president, he would "turn to your industries for help in coming up with reasonable rules," and that he would "see to it that EPA has leaders who know and care about the coal industry."[54]

Melting The Polar Ice Caps

Another serious potential impact of the increased burning of fossil fuels such as coal and oil could be a change in the earth's climate and the permanent flooding of coastal cities. Because atmospheric carbon dioxide allows sunlight to pass through to the earth but absorbs and traps part of the heat re-emitted and given off by the earth, it tends to raise the surface temperature of the globe in what is called the "greenhouse effect."

Testifying before the Senate Energy Committee on 3 April 1980, Gus Speth, then Chairman of the President's Council on Environmental Quality (CEQ), warned that there was "broad consensus that use of fossil fuels at historic growth rates could lead to a doubling of atmospheric carbon dioxide (CO_2) within the next few decades and consequently to significant climate changes..."[55]

Speth listed the "profound" socio-economic effects that could occur, including the permanent flooding of coastal cities such as New York and Los Angeles:

> [f]arming regions might warm up, dry out, and become less productive: dust bowls could be created. These consequences could render human settlement patterns and capital infrastructures obsolete. The sea level could rise due to melting of polar glacial ice sheets. The resulting coastal inundation could force eventual evacuation of lands now considered to be among the world's most desirable.

He also stressed that unless action were taken immediately to address the problem, it could quickly overwhelm our ability to mitigate it:

> The insidious nature of the CO_2 problem is that if a response is postponed until significant and harmful climate changes are actually observed or until scientific uncertainties are largely resolved, it may be too late to avoid even more severe climate changes. Once the effects of increased CO_2 concentrations are visible enough to arouse concern throughout the world, they may be virtually irreversible for centuries... It would probably take several hundred years to regain a climate similar to our present climate... Indeed, it might be impossible to return to a state approaching present conditions.[56]

Already, there are alarming indications of a trend toward global warming. Recent studies demonstrate that the typical summer ice pack has been reduced since the early 1970's by over one-third. The Antarctic ice pack has shrunk by almost a million square miles because of the accelerated melting of sea ice, and the edge of the summer ice pack has retreated some 140 miles.[57]

In addition to the burning of fossil fuels, several other environmental factors are at work, each of which simultaneously helps to generate the greenhouse effect. We are also removing several key factors which absorb and break down carbon dioxide. The on-going devegetation of much of the earth's surface, especially the cutting of the tropical rain forests, not only releases the carbon stored in the biomass, but also eliminates sources of photosynthesis, the process by which plants absorb carbon and release oxygen.[58]

The rate at which the world's forests are being cut is almost inconceivable. In the last 30 years, half have been destroyed. Each year, between 27 million and 50 million acres of tropical forests are lost, at a rate of one or two acres a second.[59] (In 1967, an American industrialist, Daniel K. Ludwig, bought up and began developing an area of the Brazilian Amazon rain forest equivalent to the size of state of Connecticut![60]) Even in the United States, recent changes in federal timber policy ordered by Presidents Carter and Reagan are permitting extensive cutting of the national forests far in excess of sustained yield management. Much of the timber being taken out of American forests, especially from the western states and Alaska, is not used domestically but is shipped overseas to Japan.

This massive worldwide deforestation, combined with pollution, will be a major factor in wiping out hundreds of thousands, perhaps millions, of species of animals and plants in the next two decades, most of the latter unstudied for their potential value as food or medicine. (Half of our prescription drugs and most of our food come from plants.) As the *Global 2000* report observes,

> ...between half a million and 2 million species — 15 to 20 percent of all species on earth — could be extinguished by 2000, mainly because of loss of wild habitat but also in part because of pollution.[61]

The increase in carbon dioxide resulting from worldwide deforestation could have profound and irreversible consequences that could threaten the survival of another unique species — human beings. A report issued by CEQ in 1978 points out that "once in the atmosphere, excess carbon dioxide will remain there for many centuries,"[62] and warns that "the climatic and ecological consequences of a temperature rise of 3°C. or more . . . could be catastrophic":

> . . . energy policy decisions made today can . . . radically
> affect human well-being 50 to 100 years from now, and
> then continue to exert influences for a milleneum after
> that. Every ton of coal burned adds 3 tons of carbon
> dioxide to the atmosphere, and burning the world's known
> coal reserves would produce an amount of carbon dioxide
> eight times that currently in the atmosphere.[63]

What all this means is that unless the western world, especially the United States, quickly changes its energy policies, people living in cities on the Atlantic, Pacific and Gulf coasts may someday find their homes underwater. An enormous calamity, considering the fact that over half the U.S. population lives in counties bordering the oceans, the Gulf or the Great Lakes; 70 percent live within 50 miles of a coastline.[64]

Cleaner Air, Lower Costs

Although reducing and preventing pollution can be expensive, it is far cheaper in the long run — both in terms of saving money and lives.

In April 1980, CEQ issued a report estimating that 14,000 lives and billions of dollars were saved in the United States in 1978 alone as a result of improvements in air quality since 1970, the year the Clean Air Act was passed.

The report, prepared for CEQ by economics professor A. Myrick Freeman III, concludes that "national benefits which have been realized from reductions in air pollution since 1970 lie in the range from roughly $5 billion to $51 billion per year," with the best estimate of air quality improvement benefits enjoyed in 1978 being $21.4 billion. Such savings include reduced damage to human health, crops, forests, vegetation buildings, and other property.

If the "best estimate" for 1978 benefits is used — $21.4 billion — this represents almost $5 billion more than the costs of complying with the Clean Air Act, and more than $2 billion above *all* spending on air pollution control for that year.

These figures are considered to be quite conservative since they do not include actual benefits from preventing further degradation of air quality, or benefits accruing to Canada and Mexico. In addition, Freeman estimates that "between 2,780 and 27,800 lives are saved per year as a result of air pollution control. The most likely number is 13,900."[65]

An earlier study by Lester Lave and Eugene Seskin, *Air Pollution and Human Health*, calculates the cost benefits of regulating air pollution, and comes up with similar figures. They conclude that controls on stationary pollution sources (such as power plants) yield over $16 billion in savings on health costs (not including property damage) at a cost of only $9.5 billion, producing a net benefit in dollars of $6.5 billion, to say nothing of the savings in terms of deaths and illness prevented. They estimate that if all health benefits from improved air quality (such as increased earnings) are included, the combined savings to society amount to "some $23 billion a year (or) considerably higher," in addition to an increase in life expectancy of almost one year for the average person.

A study recently published by The New Jersey Economic Policy Council estimates that government regulations requiring reductions in the levels of sulfur oxides and particulates were saving 2,000 lives a year, and $166 million to $2.1 billion in costs in that state alone.[66]

Unfortunately, EPA's lethargic enforcement of the Clean Air Act may soon become a moot issue. The Act came up for renewal by Congress in 1981, and industry geared up for a major fight to cripple the law. As Douglas Costle pointed out in June 1980, business and special interest groups were expected to spend some $55 million in contributions to the 1980 elections in an attempt to elect a Congress that would be receptive to their attempts to "gut" the Act. "Probably at no time since the late 1800's — the era of trusts, of watered stock, of anything-goes profiteering — have private interests had such influence in Congress as they have today," said Costle.[67]

In fact, corporate PAC's (political action committees) gave $19 to $30 million to congressional and presidential campaigns, with over a third coming from the oil and gas industries. Contributions from the oil industry more than doubled between the 1978 and the 1980 elections, rising to over $4.5 million for 1979-80.

In addition, seven major industries working to weaken the Clean Air Act gave over $1 million to the 1980 campaigns of key House and Senate members whose congressional committees had jurisdiction over the Act. Almost half of this money, and 98 percent of that given by the steel industry, was contributed by firms that EPA had cited for violating the Act. The largest donations from corporate PAC's went to two Republican members of the Senate Environment and Public Works Committee: Steven Symms (Idaho) and James Abdnor (S.D.), each of whom received some $170,000. In 1981 and '82, both Senators worked to weaken the Act and to push for industry-supported amendments.[68]

Interestingly, all five candidates who received over $100,000 from the oil and gas interests were conservative Republicans running on anti-environmental platforms who succeeded in defeating pro-environment incumbent Democrats. These Republicans are: Senators Charles Grassley (Iowa), Steven Symms (Idaho), James Abdnor (S.D.), Dan Quayle (Ind.), and Congressman Jack Fields (Tex).[69]

Ranking third in amounts given by all corporate PAC's were those set up by the Dow Chemical Company, which donated $304,138 to campaigns for federal offices.[70]

Industry's efforts to buy a new Congress exceeded its wildest expectations. The Republicans gained a majority in, and took control of, the Senate, with Ronald Reagan being elected on a platform of weakening environmental laws. He once said of the Clean Air Act, "I'd invite the coal and steel industries in to re-write the Clean Air regulations."[71] Sure enough, EPA's initial 1981 proposal to amend the Act called for crippling revisions of major portions of it, prompting Congressman Henry A. Waxman (D-Cal.), Chairman of the House Subcommittee on Health and the Environment, to describe the draft as "a blueprint for the destruction of our clean air laws." Fortunately, two other key Republicans, Senators Robert Stafford (Vt.), Chairman of the Environment and Public Works Committee, and John Chafee (R.I.), Chairman of the Subcommittee on

Environmental Pollution, fought to resist a major gutting of the Act, with the fight continuing into 1982.

Doubling Energy Supplies Through Conservation

It should be noted, as environmentalists have stressed for years, that it is not necessary to poison the air, kill millions of people and destroy agricultural productivity in order to provide America with adequate energy as oil reserves dwindle and foreign sources are threatened. The alternative — emphasizing conservation and solar energy — is not only cheaper, but will enhance America's national security and strengthen our economy while maintaining our present standard of living.

Indeed, an impressive argument can be made that cracking down on the emission of hazardous pollutants could force a change in technology that could not only clean up the air but help solve our energy problems and reduce taxes and inflation.

The simple, uncontested fact central to our energy problem is that America wastes about half of the energy it consumes. This means that realistic energy conservation measures could effectively double our energy supply. Such a course of action could be undertaken without Americans changing their basic lifestyles. In fact, other countries with comparable standards of living use and waste far less energy. America, with some 5.8 percent of the world's population, accounts for 33 percent of the world's energy consumption.[72]

As an environmental task force put it in *The Unfinished Agenda* (1977), each baby born in America is equal to several dozen born in underdeveloped nations in terms of energy used and pollution generated:

> ...each new American consumes...60 times as much energy as the average South Asian. Thus, the 1.3 million Americans added each year through natural increase (not counting immigration) constitute at least as much of a burden on the global resource base as the annual increase of 12 million in India.[73]

Put another way, the average American used 57 times as much commercial energy in 1974 as the average citizen in India, and 122 times more than the average Nigerian, according to the United Nations *Statistical Yearbook* for 1975.[74]

In 1979, the Harvard Business School issued a report, *Energy Future,* concluding that over the long run, the United States can use up to 40 percent less energy with "virtually no penalty for the way Americans live":

> If the United States were to make a serious commitment
> to conservation, it might well consume 30 to 40 percent
> less energy than it now does, and still enjoy the same or
> an even higher standard of living. That savings would not
> hinge on a major technological breakthrough, and it would
> only require modest adjustments in the way people live.[75]

The results of such a policy, according to the report, would include less damage to the environment and public health, a stronger dollar, less dependence on OPEC oil and a general reduction of tension in western society. Energy savings could be achieved by implementing such measures as stricter fuel efficiency standards for cars and trucks, and tax incentives for conservation investments in the industrial, residential and commercial sectors.

CEQ's 1979 tenth annual report makes essentially the same points, stressing that "our year 2,000 energy consumption ... need not increase by more than about 10 to 15 percent; at the same time, total real GNP could increase sharply to nearly double today's levels":

> technology is now available to increase U.S. energy pro-
> ductivity in an economical manner. More efficient
> buildings can save the home owner several dollars for
> every dollar judiciously invested in energy-conserving
> features. By selecting vehicles with more efficient designs,
> automobile users can economically reduce fuel costs by 50
> percent or more. A wide variety of conservation tech-
> nologies available to industry today can often provide a 30
> to 50 percent per year return on investment.[76]

A recent Department of Energy study concludes that tighter efficiency standards for appliances could save one-and-a-half million barrels of oil a day, and stricter standards for home insulation could save almost 6 million. Almost 3 million barrels a day could be saved by raising automobile gas mileage to 55 miles per gallon. By the year 2,000, 25 to 45 percent of the fuel needed to operate the nation's transportation system could be derived from alcohol-based fuels made from such sources as wood or grain.[77]

According to former Deputy Undersecretary of Energy Jackson S. Gouraud, if each of the 50 million Americans who still drive to work by themselves carried one passenger, 400,000 barrels of oil a day would be saved. This is the amount that could be made by eight pollution-producing synthetic fuel plants costing $15 to $30 billion. Moreover, writes Gouraud, conservation is essential if America's financial resources are not to be drained by foreign oil producers:

> ... the only solution to our energy problem is to cut our demand for oil through conservation and solar conversion ... In 1970, our imported oil cost about $3 billion; in 1979, it cost $45 billion; this year (1980), it will cost about $90 billion for about the same amount of oil, more than the net assets of I.B.M., General Electric, General Motors, and Ford ... Our dependence on imported oil is now costing us $10 million an hour, every hour, all year long.[78]

In what could be "curtains for OPEC," 165 million barrels of oil a year could be saved if American homeowners simply installed and used windowshades, according to a mid-1980 study by North Carolina State University. The study found that shades could save the average home $168 a year in fuel and utility costs, and $314 a year in areas with extreme temperatures. In the summer, it was estimated, lowering the shades before sunlight pours in could reduce summer utility bills by $100.[79]

Some energy authorities also favor a 50 cent a gallon tax on gasoline, which they believe would reduce consumption by as much as 700,000 to one million barrels of oil per day. (The United States was importing 6-½ million barrels of oil per day as of 1980.)[80]

National legislation requiring a deposit on soft drink bottles in order to encourage their re-cycling would save an estimated 100,000 barrels of oil a day, in addition to reducing litter.

There are numerous other measures that could easily be taken to conserve energy. Some of the billions of dollars spent each year by the federal government on building and improving highways could be used instead on mass transit. (The National

Transportation Policy Study Commission estimates that $900 *billion* will have to be spent in the next 20 years — or $45 billion a year — to build and maintain highways just to keep the system in the condition it was in 1975.)[81] Building more highways, of course, encourages the consumption of more gasoline and causes more pollution.

Indeed, the path of conservation is not only a cleaner but also a much cheaper one to follow, since it would cut down on the burning of increasingly scarce and expensive fossil fuels. The price of coal and oil are artifically held at unrealistically cheap levels which do not take into consideration the immense costs involved in extracting, transporting, spilling and burning these fuels, or in disposing of their wastes. If the social and environmental costs, such as destruction of the land and human health, were included in their price, these fuels would be much more expensive. But even at current prices, conservation costs less. As James Murphy of the Natural Resources Defense Council points out,

> An investment in residential weatherization comparable to that which the Federal Government plans to spend on coal conversion would save about four times as much imported oil — 1.6 million barrels per day as compared to only 400,000 barrels per day projected for utility, coal conversion.[82]

In fact, the cooling, heating, and lighting of the world's buildings uses a quarter of the world's energy, but simple architectural changes could reduce such consumption substantially. These requirements can be supplied or supplemented through "passive solar design," which utilizes sunlight, shade, and natural ventilation, such as by placing windows and overhangs on the south side to catch the sun in winter and shade in the summer. Such elementary innovations in design could cut fuel consumption in buildings by three quarters, and save the United States millions of barrels of oil a day.[83]

But the few conservation measures put in effect in recent years are being reviewed or rescinded by the Reagan administration. For example, on 17 February 1981, the President cancelled a rule requiring that all public buildings keep to a maximum temperature of 65 degrees in winter, and no less than

78 in the summer. The Department of Energy had estimated that this regulation had saved $4.5 billion, or 123 million barrels of imported oil.[84]

Solar Energy: Clean, Cheap and Renewable

Another cheap, renewable and environmentally-benign way of meeting America's and the world's energy needs is through solar energy. In its 1978 report, *Solar Energy: Progress and Promise,* CEQ concluded that "the nation could meet as much as one-quarter of its energy demands from solar sources by the end of the century and perhaps as much as one-half by the year 2020 ... For the period 2020 and beyond, it is possible to speak of the United States' becoming a solar society — one in which solar is not a supplementary source but rather a primary source of energy."[85]

Unlike the United States, many other countries already make extensive use of solar energy. Denis Hayes, when head of the federal Solar Energy Research Institute, pointed out:

> About one-fifth of all energy used around the world now comes from solar resources: wind power, water power, biomass, and direct sunlight. By the year 2000, such renewable energy sources could provide 40 percent of the global energy budget; by 2025, humanity could obtain 75 percent of its energy from solar resources.[86]

The development of solar energy continues to be hampered by federal programs that subsidize non-solar sources, such as federal price controls (which have recently been modified but not eliminated). As CEQ reported in 1979, "... the introduction of renewable sources is presently hindered by widespread subsidies to conventional energy sources at both the federal and state levels."[87] For these and other reasons, including the influence of the oil industry, solar and other renewable energy sources received less than one percent of U.S. research and development funding from World War II to 1978.[88]

The outlook is not encouraging. President Reagan, upon assuming office, quickly ordered drastic cutbacks in conservation and solar energy programs. The administration's first budget, released in February 1981, contained cuts in the Energy Department's solar development program totalling $1.9 billion by the end of 1986, and slashed its energy conservation budget by $2.4 billion. The Department of Housing and Urban Development's $125 million a year solar energy program was eliminated entirely. Such action was not unexpected. On 25 November 1980, Ronald Reagan summed up his philosophy on such matters by stating that "conservation means we'll all be too hot in the summer and too cold in the winter."[89]

Reagan's 1982 budget cut funding for solar energy projects by some 43 percent to $303 million, while boosting subsidies to foster nuclear energy to $1.6 billion. And EPA's initial 1983 budget proposal contained additional deep cuts, including a 50 percent reduction in funds to enforce air pollution controls for power plants and other stationary sources of pollution.[90] For the Energy Department's 1983 budget, the Office of Management and Budget in December 1981 proposed to cut conservation programs by 98 percent, and solar by 86 percent.[91]

There is no question that one day conservation and solar energy will be the foundation of our nation's energy policy, if for no other reason than the increasingly scarce and expensive supply of oil. The only question is, will we make the inevitable switch to these benign, cheap and dependable energy sources before or after we have destroyed our health; disrupted and polluted the environment, perhaps beyond recovery; and spent ourselves into bankruptcy?

What You Can Do

Preventing and mitigating the damaging effects of air pollution involve more than just staying indoors on days of heavy air pollution. A few changes in our energy policies could result in a dramatic drop in pollution levels without disrupting the standard of living of the average American.

As has been stressed in this chapter, installing scrubbers in power plants, burning low-sulfur coal, curtailing the cutting of forests, and, most of all, emphasizing energy conservation and solar power could yield tremendous dividends in terms of lives saved, a healthier populace, less dependence on foreign oil sources, and a more productive economy. This could all be achieved at potentially great savings in terms of reduced taxes, health costs, and damages to crops and property.

The methodology to accomplish these objectives is already largely available. All we are lacking are the political will to pursue this course of action, and the leadership to take us down the road of benign and dependable energy sources.

Chapter Six

Poisoning The Planet: Our Export Trade In Death

One of President Reagan's first acts upon entering office was to rescind an executive order designed to protect American consumers and foreign nations from particularly hazardous products that cause death and disability and are ostensibly banned for sale in the U.S. As a result, U.S. as well as foreign nationals will continue to be exposed to extremely dangerous chemicals and other products that cannot legally be sold in the U.S.

There is no shortage of examples of deadly U.S. products being shipped abroad. In 1977 and 1978, after the U.S. banned the sale of children's nightwear treated with a cancer-causing chemical (tris) that may cause hundreds of thousands of cancer cases among the children who were exposed, several million of these garments were exported for sale overseas (see pp. 281-92). Also in 1977, after the Consumer Product Safety Commission (CPSC) moved to ban the sale of infant pacifiers which could be swallowed and caused a number of deaths by choking, several U.S. firms exported some half a million of the devices abroad. And after an intra-uterine birth control device called the Dalkon Shield, first marketed in 1971, was implicated in the death of at least 17 women and serious uterine infections in over 200,000 others, the manufacturer — A.H. Robins Co. — withdrew it from the domestic market. It was then sold overseas and distributed by the U.S. Agency for International Development (AID). In 1975, AID had to recall those that it had sent abroad, but A.H. Robins continued to market the device, and it is still in widespread use in underdeveloped countries.[1]

But the really massive trade in banned products is carried on by the U.S. chemical industry. Each year, hundreds of millions of pounds of pesticides banned for use in the U.S. are shipped abroad, where they poison humans by the hundreds of thousands, as well as livestock, fish and wildlife and the environment.

These toxins also end up poisoning American consumers when they reenter the country on imported fruits and vegetables. Less than 1 percent of imported food shipments are checked by the Food and Drug Administration (FDA) for pesticide residues, and Congress' General Accounting Office (GAO) recently reported that "half of the imported food that FDA found to be adulterated ... was marketed without penalty to importers and consumed by an unsuspecting American public."[2]

The outlook is for greatly increased use of hazardous pesticides in underdeveloped countries. This not only threatens the world's ability to grow uncontaminated food, but over the long term places in jeopardy the earth's entire system of food production.

Caution: Made In America

When the chemical industry learns that one of its products is too dangerously toxic to be sold in the U.S., it has found that leftover stocks can be sold, and production often continued, by shipping the product overseas. This trade in unregistered chemicals, and in those that have been banned in the U.S., is resulting in the death and disability of countless thousands of people in other countries.

There are numerous documented cases of foreign deaths and serious illnesses caused by U.S.-exported pesticides. In Iraq in 1971-72, between 450 and 1,000 people were killed, and 6,000 to 60,000 gravely poisoned by U.S. grain that had been treated with a methyl mercury pesticide that had been made in the U.S. but was banned for use domestically and in other developed countries.[3] In Bahia, Brazil in 1975, the death of 13 children by poisoning was attributed to some aldrin put out to kill stray dogs.[4] In Pakistan in 1976, at least 5 people were killed and 2,900 others poisoned by the commonly-used pesticide malathion. In Colombia in 1975, 70 miscarriages are believed to have been caused by 2,4,5,-T. Three years later, massive aerial spraying of crops is reported to have killed thousands of animals, caused large-scale illness among local residents and even forced many peasant farmers to migrate from the area involved.[5]

In Turkey in the late 1950's, over 3,000 Turkish children became horribly disfigured after eating bread contaminated with the fungicide hexachlorobenzene. For years after the incident, the children were still suffering from stunted growth, darkened skin, sensitivity to light and a deformity manifested by grotesque clumps of hair which grew in thick tufts all over their foreheads.[6]

In the Shiekdom of Qatar in June and July 1967, some 700 people had to be hospitalized, and 24 died, in the Persian Gulf city of Doha, after they experienced convulsions, severe abdominal cramps, vomiting and confusion. The source of the disaster turned out to be bread that had been made with flour laced with the pesticide endrin, both products having arrived from Houston on a ship aboard which the contamination occurred.[7]

Across the U.S. border, in Tijuana, Mexico in September 1967, 600 children were poisoned and 17 died after eating pastries containing sugar that had been contaminated with parathion. Two months later, a similar incident occurred in Chiquinquira, Colombia, when 600 people were sickened and 80 died from consuming bread made with flour containing the same pesticide.[8]

None of these tragedies was unusual; in fact, such incidents are inevitable in light of the lack of restraint on the part of the chemical industry in dumping such toxic chemicals on under-developed lands. There are numerous examples that could be cited of U.S. firms taking action that practically guarantees to expose residents of foreign nations to the dangers of lethal pesticides, many of which are so dangerous that they are banned in the country of origin.

In September 1976, the Hooker Chemical and Plastics Corporation voluntarily agreed to the cancellation of the U.S. registration of benzene hexachloride (BHC), of which Hooker was the primary U.S. producer. BHC had been found to be extremely potent in causing cancerous tumors and birth defects in laboratory animals. The chemical, a remarkably persistent chlorinated hydrocarbon, had been found in the tissues of 99 percent of Americans tested. But in announcing the halt in U.S. sales, Hooker stated that it would dispose of its existing stocks abroad and would continue to sell it to foreign buyers.[9] In 1976, 456,000 pounds — a third of that year's production — was exported.[10]

This meant, of course, not only that foreigners would still be exposed to Hooker's admittedly dangerous product, but also that Americans would continue to ingest it as a contaminant on imported foodstuffs. Chocolate lovers will be unhappy to learn that BHC is heavily used in cocoa-producing nations to control insects that infest cocoa beans. In the U.S., tolerances have been established for BHC residues found on the raw beans and in finished chocolate products.[11]

BHC is but one of a multitude of banned chemicals that America is selling abroad, contaminating not only foreign lands but also food grown for shipment to the U.S.

Poisoning U.S. Consumers

According to the World Health Organization (WHO), the "vast majority" of the 800 million or so pounds of pesticides that less developed countries use each year are sprayed on crops that are exported to the U.S. and other western countries. In this way, Americans continue to be exposed to chemicals that are banned in this country.

Moreover, there is over a score of potentially carcinogenic pesticides that the FDA cannot detect with its food testing program, and often the FDA cannot identify the chemicals that are found. For instance, the six major coffee producing countries in Latin America use some 94 different kinds of pesticides; of these, 64 cannot be detected by FDA tests, and thus cannot be prevented from contaminating U.S. coffee.[12] In a 1977 study by FDA, 45 percent of the green (unroasted) coffee beans tested contained illegal residues of banned or restricted pesticides.[13] Among the chemicals FDA has detected on green coffee beans imported from South America and other countries in Africa and Asia are such banned, cancer-causing pesticides as BHC, dieldrin, DDT, DDE and heptachlor, in addition to toxic chemicals that have been restricted in the U.S. or are undergoing cancellation review, including malathion, diazinon and lindane. In 1976, some 2.75 billion pounds of coffee were imported into the U.S.[14]

(In March 1981 a study was released showing a statistical link between drinking coffee and cancer of the pancreas. Scientists at the Harvard School of Public Health reported that "coffee use might account for a substantial proportion of the cases of this disease in the U.S.," which is the fourth most common cause of cancer deaths among Americans. The scientists estimated that coffee might cause over half of the nation's pancreatic cancer causes, which account for some 20,000 deaths a year)[15]

In addition to coffee, cacao from Ecuador and sugar and tea from India come into the U.S. with aldrin, dieldrin, chlordane and heptachlor — all largely banned in the U.S. but exported abroad. Bananas are repeatedly sprayed with such banned pesticides as kepone, dieldrin and DDT.[16]

After several years of investigation, the General Accounting Office on 22 June 1979, issued a report to Congress on the export of pesticides abroad, and how this was leading to contamination of imported foods. GAO's investigation found that U.S. pesticide export policies have created major hazards not only abroad, but for U.S. consumers as well. The report determined:

> In some foreign countries, pesticides known or suspected of (sic) causing cancer, birth defects, and gene mutations are carelessly or excessively used . . . Pesticide use patterns in foreign countries clearly indicate that a large portion of food imported into the U.S. may in fact contain unsafe pesticide residues.[17]

As is shown in the following GAO table, pesticides largely banned in the U.S. for use on food crops (many of which cause cancer in laboratory animals at the lowest doses tested) are routinely used on crops in foreign countries for import into the U.S.:

**Food on Which Foreign Countries Allow Use
of Suspended and Canceled Pesticides[18]**

Pesticide	Country			
	Ecuador	Guatemala	Costa Rica	India
Aldrin	cacao coffee	coffee sugar	coffee	sugar tea
Dieldrin	coffee	bananas sugar coffee	bananas coffee cacao	
Heptachlor		sugar	sugar cacao	sugar
Chlordane	cacao		coffee	
DDT		bananas		
Kepone		bananas		

GAO found that FDA's processes for testing imported food-stuffs for contamination were woefully inadequate and ineffective. GAO identified 130 different pesticides, without U.S. tolerances and which are used on just ten imported commodities, which FDA cannot detect on its two most commonly used tests. GAO found a total of 178 pesticides for which tolerances had been established in the U.S., and 90 other pesticides used abroad but not allowed in U.S. food, which FDA's usual techniques could not detect. Of the 268 pesticides with U.S. tolerances, FDA can normally detect approximately 90.[19]

Moreover, less than one percent of the hundreds of thousands of shipments into the U.S. are tested at all.[20]

But even when FDA detects illegal residues of pesticides on imported foods, GAO found, "the food will probably be marketed and consumed rather than detained or destroyed." This is because FDA allows the distribution of imported foods, especially those that are perishable, *before* analysis of them is completed. Thus, by the time contamination is discovered, the product "often has already been marketed and consumed."[21]

Nor does FDA generally penalize importers for distributing adulterated food, and "even importers with histories of repeated violation are not penalized and their imports are seldom detained pending analysis." Typical examples of FDA's enforcement methods reported by GAO revealed gross incompetence and indifference on the agency's part. In the Dallas, Texas district, a shipment of imported cabbage with a "pronounced insecticide-like smell" was allowed to be marketed by the importer, who had a history of shipping adulterated products. Complete analysis of the cabbage (by which time it could not be recalled) revealed illegal residues of BHC, the extremely potent cancer-causing substance that had been cancelled for use in the U.S. in 1976. Of 47 shipments of Mexican peppers from importers with histories of adulterated imports, only 18 were denied entry. One pepper from a marketed shipment that was later analyzed was found to have 29 times the allowed level of pesticide residues.[22]

The result of this mismanagement, summarizes GAO, is that "half of the imported food that FDA found to be adulterated . . . was marketed without penalty to importers and consumed by an unsuspecting American public."[23]

The GAO report was severely critical of EPA and FDA for not cancelling some 300 tolerances that had been established on different food crops for 14 pesticides that had been suspended, cancelled, or greatly restricted. These tolerances were still in effect up to six years after the use of these 14 pesticides had been banned or curtailed in the U.S. because of the hazards they posed to human health. Such a policy, said the report, is dangerous and misleading because it "could condone foreign use of these pesticides by giving the appearance that the U.S. approves of their use;" thwarts "Federal efforts to protect the consumer from pesticides" that have been shown to be especially hazardous and "places American farmers at a competitive disadvantage" to foreign ones since U.S. farmers cannot use such pesticides, which are often cheaper and more persistent. The unnecessary maintenance of such tolerance and action levels, concluded GAO, "to a large extent defeats the purpose of EPA's pesticide registration program — to eliminate consumer exposure to pesticides posing unreasonable, adverse effects.[24]

500,000 Reported Poisonings A Year

There is no legal prohibition against exporting unregistered pesticides, as long as the label notes this and the importers and the foreign government are so notified. In addition, the U.S.

exports each year tens of millions of pounds of registered but highly toxic pesticides which require strict regulations for the training and certification of people who use them in this country. But when these extremely dangerous poisons are shipped abroad, they often end up in countries where there is little knowledge of these hazards and where such training and warning labels are largely ignored. In Malaysia, for example, the deadly herbicide 2,4,5-T is sold in bottles, and is often found in stores for sale alongside similar bottles of food sauce.[25]

The World Health Organization (WHO) considers pesticide poisoning of farmworkers a major health problem in underdeveloped countries. It estimates that pesticides cause some 500,000 human poisonings each year, 5,000 of which are fatal.[26]

However, these officially-cited figures are thought to represent only a fraction of the actual poisonings that occur each year, most of which are never reported. For example, a nurse at the Social Security clinic in Tiquisate, Guatemala told a reporter that during the cotton spraying season, "we treat 30 or 40 people a day for pesticide poisoning: the farmers often tell the peasants to give another reason for their sickness, but you can smell the pesticide in their clothes."[27] [28] In Costa Rica, authorities claim to have found about 300 cases of pesticide poisoning a year, but an independent investigation turned up 700 cases in three months at just two hospitals.[29]

It would, in fact, be impossible to count the real number of pesticide poisonings even if all cases were reported, since it cannot be determined how many people will develop delayed responses or will contract cancer decades later from such exposure.

The most direct victims of the chemical industry's "dumping" policies are, of course, the peasants and Indians of South and Central America who actually apply the poisons to the crops and later harvest them, usually barefoot. Most of these farmworkers are illiterate and thus cannot understand the warning labels on the containers. One 1977 study by a Central American institute found that over 75 percent of the pesticide users observed were illiterate.[30] This lack of literacy results in excessive amounts of chemicals being applied to the crops, and recommended precautions to exposure are ignored. A study group from the University of California which visited Pakistan in 1974 reported, for example, that "one customer, lacking a suitable container, unwrapped his turban, poured a granular pesticide therein, and replaced it on his head for transport."[31] Indeed, a 1976 report by WHO states that an average of up to 40 percent of the spray applicators in some countries incurred pesticide poisoning during the spraying season.[32]

One of the worst situations prevails on Guatemala's Pacific Coast, where there are vast cotton growing operations, yielding enormous profits for the plantation owners each year. Here, pesticides are sprayed more intensely than perhaps anywhere in the world, and little thought is given to the effects of these poisons on the local people.

On an August 1977 technical aid mission to Guatemala and Nicaragua sponsored by the Department of Labor, the U.S. delegation learned that children were being used as "flagmen" to mark the location of fields to be sprayed. According to Sheldon Samuels, the A.F.L.-C.I.O. representative on the trip, cotton fields were being sprayed in this way 40 to 50 times a year, and "the death of children from acute pesticide poisoning was not considered unusual."[33]

In the lowland coastal area where most of the cotton is grown, a familiar and vicious cycle has been established: the spraying and clearing of forest lands to establish cotton fields has destroyed the natural ecological balance and eliminated the birds and predatory insects that preyed on boll weevils and other insects harmful to cotton. This has caused a population explosion of cotton pests and an intensification of spraying, with larger amounts of more toxic pesticides having to be used. Some cotton fields are sprayed as much as 50 times in the plants' three months growing cycle.[34]

About a million people in just this one area are exposed each year to this massive chemical war on nature. Alan Riding, a reporter for *The New York Times*, gives an account of what life is like for the local people. He describes one villager rushing her five children into her home — a wooden hut — as she hears the spray planes overhead: "Within minutes, dead fish (from the stream) surfaced; within hours, four of her ducklings had died." But her misfortune, he points out, is but one among hundreds of thousands:

Some 370,000 people live permanently in the cotton-growing lowlands, but in the picking season from

November to March, as many as 600,000 Indians come down from the highlands to supplement their incomes working for stretches of 30 to 90 days in the fields. They earn $1.25 for a 12-hour shift.

The Indians are ignorant of the dangers of pesticide poisoning. They eat sweet-tasting cotton buds that have already been sprayed, they enter the fields too soon after fumigation, and they take water for washing and drinking from polluted irrigation canals.[35]

One local clinic treats up to 40 people a day for vomiting, dizziness, extreme weakness, liver damage and other manifestations of pesticide poisoning.

Because of the political power of the plantation owners and the lack of it among their victims, pesticide spraying is not a major political issue in the country. But the extent of discontent and frustration among a large part of the population became apparent in 1976, when a guerrilla group — the Guatemalan Army of the Poor — staged a midnight raid at La Flora and destroyed 22 pesticide spray planes. However, the destroyed crop dusters were soon replaced by new ones, and the spraying resumed as before.[36]

While the conditions in Guatemala are among the worst in the world, they are not qualitatively different from those which prevail throughout much of Latin America. The typical situation has been described by Dr. Louis Falcon, an entomologist from the University of California who often travels to Latin America:

> The people who work in the fields of Central America are treated like half-humans. When an airplane flies over to spray, they can leave if they want, but they won't be paid their seven cents a day, or whatever. They often live in huts in the middle of the fields. Their homes, their children, and their food get contaminated.[37]

The average DDT level in the blood of Guatemalans and Nicaraguans is reportedly more than 30 times that of Americans. One study showed that levels of the carcinogenic pesticide aldrin (largely banned in the U.S. since 1975) on cabbage in Central America was almost 2,000 times the allowable level in the U.S.[38]

Since epidemiological studies are rarely conducted on the peasants of underdeveloped nations, it is impossible to know how many thousands or millions are being condemned to a premature death or ruined health each year.

AID: Financing Deadly Exports

Each year, it is estimated, about 4 billion pounds of pesticides are produced worldwide — almost a pound for every man, woman and child on earth![39]

Of the over one-and-a-half billion pounds of pesticides produced each year in the U.S., some 30 to 40 percent are purchased for export to foreign countries. Many of the hazardous products shipped abroad are not only permitted, but are paid for, by the U.S. government as part of its foreign aid program.

Since 1957, the U.S. Agency for International Development (AID) has used taxpayers' money to pay for the export abroad of $500 million worth of pesticides.[40] From 1969 to 1974, AID paid for the export abroad of some $17 million worth of pesticides *each year,* many of which were banned or severely restricted in the U.S., such as DDT, 2,4,5-T, 2,4-D, aldrin, endrin and heptachlor. (Such AID-financed exports, while substantial, accounted for only 12 percent of the pesticides sent by the U.S. to developing countries.)[41]

According to Congress' General Accounting Office (GAO), U.S. firms exported more than 550 million pounds of pesticides in 1976, almost a third of which — over 160 million pounds — were not registered for use in this country, including many whose registration had been suspended, cancelled, or denied.[42] Others had never undergone review or testing by the government.

In the early 1970's, AID vigorously promoted the use of pesticides in underdeveloped nations and routinely shipped and distributed abroad pesticides that were not approved for use in the U.S. Only after being taken to court in 1975 by the Natural Resources Defense Council and other environmental groups did AID reexamine and modify this policy. As a result, AID announced in 1977 that it would be more careful and discriminating in financing the export of pesticides, and would place greater stress on nonchemical insect control techniques. In fiscal 1979, AID's financing of pesticides for developing countries had dropped to $8.3 million, about half that spent in previous years, and most of which went to malaria control programs.[43]

The U.S. government's foreign aid program, designed in part to improve the nation's image abroad and to help underdeveloped countries, has often had just the opposite effect in dealing with pesticides, as Jacob Scherr of the Natural Resources Defense Council points out. On three occasions — in 1977, 1978, and 1980 — the United Nations Environment Program issued resolutions requesting that nations not allow the export of hazardous substances without notifying and obtaining the permission of the importing countries. Dr. J.G. Kiano, head of the Kenyan delegation to the 1977 meeting, warned that third world nations would no longer tolerate being used as "guinea pigs" and "dumping grounds for products that had not been adequately tested." On 17 December 1979, a resolution was adopted by the U.N. General Assembly urging member states to "discourage the exportation of such products to other countries."[44]

But despite some improvements in AID's policies, it is still deeply involved in subsidizing the U.S. chemical industry and paying for much of the poisoning of the Third World.

Even DDT is still produced for sale abroad and is exported at a rate of 44 *million* pounds a year, mainly to underdeveloped countries. In fact, the 1972 U.S. ban on most uses was a temporary bonanza for foreign customers since it made more of the chemical available at cheaper prices.[45]

Much of the trade in this largely-banned chemical is financed by U.S. taxpayers through AID, which buys it for use by third world nations in anti-malarial spraying programs. However, many species of mosquitoes have developed a certain amount of tolerance or even immunity to DDT and other pesticides, in part because of repeated exposure to these chemicals from the spraying of cotton fields and other crops. As *Science* magazine reported in June 1979, "Thought to be virtually wiped out in the 1960's by house-to-house spraying of DDT, malaria is now resurging in more than a dozen countries because so many species of the *Anopheles* mosquito — a malaria carrier — have become resistant to pesticides."[46]

Often, the response to the problem of increasing insect resistance has been to spray greater amounts more frequently, and/or to use more toxic chemicals. In areas of Guatemala and Nicaragua, the number of times fields are sprayed has increased from about eight per growing season in the 1940's to 40 or 50 in 1977, thus continuing the vicious cycle.[47]

Commenting on AID's pesticide program, the A.F.L.-C.I.O. wrote in December 1977:

It has been estimated that the release of about $10 million worth of chemical pesticides into the cotton fields of Guatemala each year provides a return of $60 million or more per year in revenues derived from cotton exports. The yearly costs to society and the environment for this exchange is about $200 million when reasonable costs are assigned to the economic, social, and environmental problems generated by the use of pesticides.[48]

Often, the use of insecticides by untrained people results in the replacement of one pest with a far more destructive one. DDT has been blamed for causing an outbreak of hemorrhagic fever (or "black typhus") in which over 300 people died in the Bolivian village of San Joaquin in 1963. A team of U.S. health officials who were sent in to halt the epidemic determined that the source was a species of mouse-like rodent called *lauchas;* they had rarely been seen in the village, but an invasion of them preceded the outbreak of the disease. This had, in turn, been caused by the poisoning off of almost all of the village cats — which had formerly preyed on the rodents — by DDT, which had been used to kill off malarial mosquitoes.[49]

Destruction Of Wildlife

The routine pesticide poisoning of people in foreign lands, as widespread and tragic as it is, hardly compares to the enormous devastation of wildlife that takes place, especially in agricultural areas.

In Kenya, Lake Nakura and its millions of spectacular pink flamingoes used to be a major tourist attraction. Now the birds — and the tourists — have virtually disappeared, due in large part to the presence of DDT and dieldrin that has contaminated the water from nearby farms.

In Zambia and Zimbabwe (formerly Rhodesia), where DDT is used extensively, the eggs of fish eagles have begun to show alarming signs of thinning, which indicates a threat to their ability to reproduce. High levels of DDT are also showing up there in breast milk, dairy products, corn and beef.[50]

Even wildlife preserves, where the animals are supposedly protected from the intrusion of humans, are not immune from the destructive effects of pesticides. Losses of wildlife have become so routine that they seldom warrant the attention of governments or the press, but a 1973 incident was so severe that it could not be ignored:

In the fall of that year, tens of thousands of birds, including many representing rare and endangered species, were wiped out at one of Europe's most important refuges. In the lagoon located at Coto de Donana, Spain, a major nesting area and stopover point for hundreds of thousands of migrating birds, pesticide pollution decimated populations of flamingos, storks, herons, geese, ducks, egrets, spoonbills, sandpipers, herring and laughing gulls, mallards, snipes, and many others. One of the most tragic losses at Donana was that of at least 150 of the 500 extremely rare spoonbills, one of only three remaining colonies then left in Europe. At least 35,000 birds are estimated to have perished, but that figure may greatly understate the number lost.

The deaths were attributed to pesticides that contaminated the reserve's waters from surrounding rice plantations, where such chemicals are heavily used. The 82,000 acre refuge had been called by its head zoologist "one of the few places left in Europe where nature has been undisturbed...a symbol of mankind's ability to live in harmony with nature." But in late September 1973, press reports of the disaster referred to the park as "a scene of desolation," a "death stop" for Europe's migrating birds.[51]

Another example of the ruinous impact insecticides can have on the environment when improperly used was described on 25 July 1979 by Craig Van Note, head of the Monitor Consortium in Washington, D.C., a coalition of national conservation and animal protection groups. According to Van Note, a shipment of the pesticide aldrin, produced by Shell Chemical in western Europe, was used to poison off the elephants and other wildlife throughout a huge area of the African nation of Zaire (formerly the Belgian Congo).

In April and May 1978, thousands of elephants were killed in the vast forest west of Kisangani in northeastern Zaire. "While the death toll was staggering for the elephants, it was also virtually complete for all other animals over a region of hundreds of square miles," Van Note testified. The pesticides were obtained from corrupt government officials, who diverted 20 metric tons that were to have been used for spraying nearby coffee plantations and delivered the chemicals to gangs of poachers:

> The pesticide was poured into the waterholes in the marshy areas where the elephants came to drink. Within days, thousands of elephants were poisoned. It took up to a week for them to die. The sickened beasts staggered through the forest in agony. The poachers, eager to get the ivory first, frequently hastened death by running up to the weakened elephants from the rear and disemboweling them with machetes.

> Now the forests are silent for miles and miles along the Congo River, report recent travellers. The poison wiped out the birds, the monkeys, the antelope and all the other creatures of the forest. The killing continues in other areas of Zaire on a smaller scale, with bullets and poison. The elephants will die as long as the rich nations create the market and allow the free trade in ivory.[52]

He could have added that the wildlife of Africa and other under-developed countries will continue to be poisoned off, accidentally and intentionally, as long as western nations allow their toxic chemicals to be dumped on the Third World.

Marketing Techniques For Toxics

Often, it is American companies with overseas operations that import and use large quantities of chemicals that are banned or restricted in the U.S. In Costa Rica, a subsidiary of Standard Oil of California — the Ortho division of Chevron Chemical Company — is that country's biggest importer for seven carcinogenic pesticides that have been largely or completely banned for use in the United States: aldrin, dieldrin, endrin, chlordane, heptachlor, DDT and BHC. Other U.S. and multi-national firms that sell and/or import deadly chemicals in the third world include some of the world's industrial giants: American Cyanamid, Allied Chemical, Bayer, Dow, Ciba-Geigy, Hercules, Hoechst, ICI, Dupont, Hooker, Monsanto, Pfizer, Shell, Stauffer, Rohm & Haas, Union Carbide, Velsicol and others.[53]

But it's not just greedy American businessmen who have an interest in dumping poisons on Third World nations. Western European nations, mainly England, France, Germany, the Netherlands and Switzerland, exported five times more pesticides in 1974 than did the U.S.[54] (It should also be noted that the Communist nations — the Soviet Union, its Eastern European satellites and Cuba — have been no more conscientious in their pesticide policies than the western world. But because of their unworkable economic systems, they are often not able to compete in terms of production and export of chemicals.)

The internal corruption so rife in underdeveloped nations contributes significantly to the pesticide problem. Louis Falcon cites the example of a former minister of agriculture in Nicaragua who also owned interests in a firm that was importing pesticides. In that country, he notes, "pesticide importing has been a $20 million a year business in a country with an entire operating budget of only $90 million. The important people are simply not interested in pesticides becoming a controversy."[55]

Dr. Harold Hubbard, of the World Health Organization (WHO), points out:

> One of the reasons governments of less developed
> countries don't do anything about pesticide problems is
> that the people who use pesticides, the people who import
> pesticides, and the people who "regulate" pesticides are the
> same people. It's a tight little group in each developing
> country. [56]

Describing a 1978 trip to Africa, EPA Deputy Administrator Barbara Blum described how her agency's counterpart in Nigeria consisted of one person, who did not have any real ability to ban pesticide imports if he wanted to:

> I was discussing the pesticide problem with the Minister of
> Environment of Nigeria. The whole EPA in Nigeria is one
> person. The Minister has one professional and one secre-

tary. He said that the person you notify in this country is me. There are just three of us, and I really can't cope with it.[57]

Kepone: How Allied Chemical Contaminated The World's Largest Estuary

There are grave dangers to American citizens and to our environment in allowing the production for sale abroad of pesticides that are too toxic to be used in the U.S.

One of the most disastrous cases of pesticide pollution occurred in the mid-1970's, when large parts of the James River and the surrounding Chesapeake Bay area — among the world's most productive fishing areas — had to be closed to fishing. This happened after kepone, a cancer-causing chemical, was illegally dumped into the river by Life Sciences Inc., in Hopewell, Virginia, a spin-off firm of the giant Allied Chemical Corporation.[58, 59, 60]

Production of kepone was halted in July 1975, after Virginia state health officials found at least 70 to 75 workers at the plant, plus ten spouses and children, had been poisoned by the pesticide. At least 29 of these victims had to be hospitalized with such untreatable ailments as brain and liver damage, sterility, tremors, blurred vision, skin discoloration, joint and chest pains, stuttering, anxiety, involuntary movements of the eyeball, loss of memory, twitching eyes and slurred speech. The Medical College of Virginia reported that 14 of the former employees were probably sterile.[61]

The health-damaging effects of kepone were not slow to show up, but rather appeared in workers within three weeks of the plant's beginning operation in April 1974. Several of them became so sick with tremors, dizziness and nervousness that they sought medical help, and some were even hospitalized. Usually, the patients were told that the ailments were caused by tension, overwork and anxiety, and they were given tranquilizers; at least one worker was referred to a psychiatrist.[62,63]

In 1976, the year after production ceased, a National Cancer Institute study was released reporting that kepone causes cancer and liver damage in laboratory animals.[64] But this was hardly the first indication that the chemical was toxic; indeed, this had been known for almost two decades. According to a Library of Congress/Congressional Research Service study,

"Corporate Crime," published in May 1980 by the House Sub-committee on Crime, Allied Chemical knew of the chemical's health hazards well over a decade earlier:

> The carcinogenicity of this pesticide, as well as its toxic effects on the reproductive and central nervous systems, were reportedly discovered in the early 1960's through studies sponsored by the manufacturer, Allied Chemical Corporation.[65]

An extremely stable, persistent, and bio-accumulative chemical, kepone is steadily spreading outward and into the Chesapeake Bay, the world's largest and most productive estuary. The James River is one of the most important and richest fishing grounds on the East Coast, and is the "seedbed" for an estimated one-quarter of the nation's oysters.[66] The bay, which is 195 miles long and has 7,000 miles of shoreline, supports a seafood industry worth an estimated $87 million a year in 1977. In terms of catch, it produces some 90 percent of America's soft-shell crabs, 40 percent of all commercial oysters and 15 percent of the softshell clams, in addition to enormous numbers of fish caught by sport and commercial fishermen. A task force formed to study this problem has warned that the river and bay may remain contaminated with kepone for another two centuries.[67]

As a result of the contamination, over 100 miles of the James River and its tributaries, including the lower Chesapeake Bay, were closed to commercial fishing, destroying a multi-million dollar industry and eliminating the livelihood for an estimated 3,500 families.[68] The ban on fishing remained in effect for five years until pressure from the fishing industry caused it to be prematurely lifted, even though kepone levels remained dangerously high.

A Library of Congress study estimates the monetary damages of this pollution to be $2 billion just to private citizens.[69] EPA has put the price tag for cleaning up the river at over $2 billion, noting that if action had been taken in 1975 when the Hopewell plant was discovered to be polluting the river, the clean-up could have cost a "mere" $250,000.[70] Allied's fine amounted to $13 million.

All of this led Senator Patrick Leahy (D-Vt.) to observe at a December 1977 hearing that he was chairing on pesticides and worker safety,

> Allied Chemical took a position that makes Pontius Pilate look like Mother Theresa of Calcutta. That is giving them the benefit of the doubt. And with all this buck-passing, the workers ended up getting clobbered in the end. I think it is a great human tragedy.[71]

Incredibly, EPA did not notify foreign nations — for which the chemical was being produced — of the dangers of kepone until 1978, four years after the problem was known. The agency belatedly issued a notification of the pesticide's 1976 cancella tion only after receiving a letter of inquiry from Congress' General Accounting Office, which was conducting an investigation of this and related hazardous export problems.[72] This failure to alert other governments may have harmed not just foreign nationals but U.S. consumers as well, since kepone is known to have been used on Guatemalan bananas and other products destined for export to the United States.[73]

Leptophos: Dead Water Buffaloes, Paralyzed Workers

Leptophos, also called Phosvel, was produced in the early 1970's by Velsicol Chemical Company (a division of Northwest Industries of Chicago) in Bayport, Texas and eight other plants in the U.S. for export to foreign countries. Production was finally halted in the U.S. in 1976 after workers producing it incurred severe neurological damage. But by this time, countless farmworkers and consumers in over 50 foreign countries to which it had been sent had been exposed to the poison, as well as millions of U.S. citizens who consumed vegetables imported from Mexico that had been contaminated with leptophos.[74]

The workers at Velsicol's Bayport, Texas plant suffered partial paralysis, blurred vision, dizziness and other severe neurological disorders, including spastic paralysis of the lower extremities. The disaster that befall Velsicol's workers was not discovered and publicized by the company or even by government health officials, but rather by Peter Milius, a *Washington Post* reporter who was tipped off to the situation by Sheldon Samuels, an AFL-CIO occupational health official who had learned of it.[75] Milius told of such horror stories as seeing a worker "eating a baloney sandwich in the lunch room with Phosvel in his moustache." The federal Occupational Safety and Health Administration (OSHA) later charged Velsicol with 44 safety violations for its handling of the chemical.[76][77]

One worker, John Wright, a 33 year-old former paratrooper, was almost totally disabled by an ailment known as "spastic paralysis of the lower extremities;" he could barely walk, and then only with a cane. His weight dropped from 140 pounds to 110, and he began to talk with a stutter. As he described, in an embarrassed manner, his condition,

> My spine is deteriorating. It's dissolving. My legs started acting up. I couldn't control my balance. I couldn't even walk. When I'd get up in the morning, I'd fall...My little girl, she's only five years old...she calls me Crooked Legs.[78]

When Wright returned to work after one period of being in and out of hospitals and doctors' offices, he found he had been demoted and his salary cut. He was later told he should quit. When interviewed, he was confined to his home and living on Social Security disability checks. His children, he said, did not understand what was wrong with him, and sometimes the boys "still try to get me out there to play with them: I just tell them I ain't got time."[79]

Another former employee, whose legs were partially paralyzed and had to use leg braces to walk, drowned in the summer of 1975 after falling out of a boat.

Since leptophos can produce symptoms similar to those of multiple sclerosis — vision impairment, depression, amnesia, tremors, bladder dysfunction, loss of sensation and muscular coordination and rolling of the eyeballs — some of the workers were diagnosed by their doctors as having this disease. One worker, Esequiel de la Torre, was committed to an asylum because he was "disoriented, unable to walk without staggering, having auditory hallucinations and wandering around aimlessly."[80]

The Coverup Of The Dangers By Velsicol And EPA

There is, in fact, strong evidence that Velsicol suspected or knew about the health hazards of leptophos, and tried to conceal them. Years before nerve damage showed up in plant workers, similar symptoms appeared in laboratory animals dosed with the pesticide. As early as 1969, two years before Velsicol began producing leptophos, the firm had a study done on chickens to determine the chemical's toxicity. The pesticide caused the chickens to suffer such nerve disorders as staggering, weak legs, breathing difficulties, droopiness, comb turning blue; five of the eighteen hens tested died. [81, 82]

In its December 1976 report on the incident, the Senate Subcommittee on Administrative Practice and Procedure, chaired by Senator Edward M. Kennedy, noted that "EPA received repeated warnings from scientists, within and outside the agency, that leptophos was neurotoxic in mammals and fowl."[83]

After citing this and other detailed evidence of leptophos' ability to cause nerve damage and central nervous system disorders, the Subcommittee report concludes that EPA was aware of this information:

> Yet, EPA chose to discount all of this evidence, including its own study conducted in 1973, which seriously undermined Velsicol's claims that leptophos was safe. Instead, EPA chose to credit and rely upon reports developed for and submitted by Velsicol — all of which concluded that leptophos was safe. And on May 31, 1974, EPA granted tolerances for leptophos in and on tomatoes and lettuce.[84]

Not only was this not an isolated case, the Senate investigators determined, but "the superficiality of EPA's scientific review in granting tolerances for leptophos is quite typical of its entire program."[85]

Velsicol also attempted to stymie a direct investigation of its worker health problems by the federal government. When inspectors from the National Institute for Occupational Safety and Health (NIOSH) visited the Bayport plant in February 1976, the company withheld vital information on the health problems of its workers. [86, 87, 88]

Interestingly, some of the first safety tests on leptophos were performed by Industrial Bio-Test Laboratories (IBT), which reported no evidence that it caused nerve damage, even though the test animals did show such neurotoxic effects as loss of control of their legs.[89] Both IBT and Velsicol have been indicted on charges of manipulating and concealing evidence that certain other chemicals caused cancer and birth defects. (See pp. 121-23, 367).

Exporting Leptophos Abroad

Despite the mounting and irrefutable evidence of the chemical's hazardous nature, leptophos continued to be produced and exported to Brazil, Colombia, Egypt, India, Indonesia, Israel, Pakistan, South Vietnam and some 40 other less developed countries under an export license from the U.S. Department of Agriculture. This continued long after leptophos' threat to the human nervous system had been identified, and numerous cases of human and livestock poisoning had occurred. In Egypt, for example, between 1971-76 it caused the death-by-paralysis of some 1,300 water buffalo working in flooded crop fields; an unknown number of humans are reported to have died as well.[90]

Yet Velsicol continued to export the chemical for use on vegetable and grain crops, all the while proclaiming its safety. In Indonesia, according to one official there, the chemical was sold "alongside the potatoes and rice . . . people just collect it in sugar sacks, milk cartons, and coke bottles."[91] Between 1971-76, some 14 million pounds were exported to 50 countries, much of it by the U.S. Agency for International Development (AID) at a cost to the taxpayer of several million dollars.[92]

Even after the chemical was banned, the knowledge of leptophos' hazards did not prevent Velsicol from selling off its leftover supplies of the chemical. The month after revocation of its tolerance, Peter Milius and Dan Morgan reported on 26 December 1976 in *The Washington Post:*

> Velsicol still has about 1.5 million pounds of Phosvel
> stored in two warehouses in Houston. It remains free
> under the law to export that, and is doing so as it can. Last
> week, federal inspectors reported that a relatively small
> amount was being prepared for shipment in liquid form.

"The labels were in Arabic," one inspector said, "it's going to Syria."[93]

According to Velsicol, it had sold leptophos to 50 countries around the world since 1971, 29 of them in 1976. As Senator Kennedy's subcommittee wrote in its report, this "raises the possibility that the U.S. may still be importing leptophos-contaminated foodstuffs from other countries."[94]

In 1976, a total of 1,771,167 pounds and 83,570 gallons of leptophos was sold abroad. And in 1977, the year following the disclosures of the chemical's hazards, over a million and a half pounds and 22,157 gallons of it were shipped overseas.[95]

U.S. Consumers Exposed to Leptophos

U.S. consumers were also exposed to leptophos. American imports of beans, tomatoes, peas, cantaloupes, eggplants, peppers, squash and cucumbers from Mexico were found by inspectors from the Food and Drug Administration (FDA) to contain residues of leptophos from 1972-1976, even though EPA had only established a leptophos tolerance for lettuce and tomatoes in May 1974. In fact, in 1975, residues of the chemical were found by FDA in almost half the truckloads sampled of tomatoes from Mexico, which supplies about a quarter of all tomatoes sold in the U.S. [96, 97]

These tolerances were opposed by some scientists from EPA and other agencies who were alarmed by evidence not only that leptophos caused nerve damage, paralysis, and death in test animals, but also that it bioaccumulates, making even tiny doses potentially dangerous over time. But Velsicol pressured EPA to retain the tolerances, and they remained in effect until November 1976,[98] almost a year after production of the chemical had been halted.

"Incredible as it sounds," wrote the *Houston Post's* Harold Scarlett on 16 January, 1977, "Phosvel almost became a staple — a kind of 'secret sauce' — in the diets of millions of Americans."

Reagan Scuttles Export Restrictions

In 1978, an inter-agency working group, under White House consumer advisor Esther Peterson, was formed to consider what U.S. policy should be on the export of hazardous products which are banned or significantly restricted in this country. The working group's 1980 report concluded that the present situation could result in great damage to U.S. national interests:

> Sale abroad of banned products tends to undermine foreign confidence in American-made products. Among the potential consequences are losses in export trade and thus negative effects on our balance of payments, and possible adverse long-term effects on foreign markets. Uncontrolled export of hazardous substances also tends to damage our relations with foreign countries...to these countries, a U.S. government policy that tolerates unrestrained export of banned products could appear callous or hostile and thus be detrimental to U.S. foreign policy interests...[99]

The report warned that unless the U.S. placed controls on dangerous exports, its foreign policy objectives could be seriously harmed:

> If the United States does not exercise special vigilance over the export of some banned or significantly restricted products which are particularly hazardous, our economic and diplomatic ties with other countries could be jeopardized. Citizens and governments of foreign countries may develop increasingly hostile attitudes towards this country and its products.[100]

But even this modest, repeatedly watered-down proposal drew heavy fire from industry and its supporters in Congress and the White House. The Chemical Manufacturers Association (CMA) attacked the document in a massive, 49-page legal brief drafted by the influential Washington, D.C. law firm of Covington and Burling. With the help of the National Agricultural Chemicals Association, the FMC Corporation, and other chemical and pharmaceutical firms, CMA was able to enlist the help of several key legislators, such as Congressmen Thomas S. Foley (D-Wash.), Harley O. Staggers (D-W.Va.), and James T. Broyhill (R-N.C.), all of whom protested the move.

Within the Carter administration, opposition to the proposed export curbs came from Commerce Secretary Philip M.

Klutznick, U.S. Trade Representative Rueben Askew and his deputy, Robert D. Hormats.[101] (On 3 March 1981, Hormats was appointed by President Reagan to be Assistant Secretary of State for Economic Affairs.) While this opposition was not able to stop the proposal from being temporarily implemented, it was able to weaken it substantially.

And outside of Peterson's office, there was widespread opposition or indifference to the proposal among White House aides, despite President Carter's emphasis on the "human rights" issue in foreign policy. As one of the President's aides told me confidentially, "The business community cared a lot about this issue, but it wasn't high on anyone else's list of priorities — at least no one important was interested."

On 15 January 1981, after two-and-a-half years of study and debate, President Carter, in the closing days of his administration, issued an executive order limiting the export of particularly hazardous products. This action, taken despite the opposition of the chemical industry, the Commerce Department and the Reagan transition team, essentially strengthened the labeling and notification procedures for such exports.

In addition, certain "especially hazardous substances" *could* be banned from export if the Departments of State and Commerce determined that they constituted a "substantial threat to human health or safety or to the environment," or would cause "clear and significant harm to the foreign policy of the U.S." Yet, even such dangerous products could still obtain export licenses in "exceptional cases." The order did not actually ban any substance, no matter how dangerous, from being exported.

Nevertheless, President Reagan, having been in office for less than a month, rescinded the executive order on 17 February 1981, thereby guaranteeing that people in foreign lands, as well as U.S. consumers, will continue to be exposed to chemicals so toxic that they cannot be sold in the U.S.[102]

And in September 1981, it was revealed that the Reagan administration had drafted plans to expedite the ability of U.S. firms to export banned products. Complaining that restrictive rules had "placed U.S. exports at a competitive disadvantage," the Departments of State and Commerce proposed to do away with most remaining regulations requiring that, before an extremely dangerous product is sent abroad, the government involved be notified.[103]

The Future: More Pesticides, Less Food

The outlook worldwide is for increased use of pesticides, especially in the lesser developed countries (LDC's). The *Global 2000 Report to the President,* drafted by the President's Council on Environmental Quality (CEQ) and the State Department, states that world use of pesticides, herbicides and chemical fertilizers will more than double between 1975-2000, and in the underdeveloped world the increase will be between fourfold and sevenfold, or more.[104]

Defenders of the use of pesticides and other toxic chemicals often claim, but rarely with reliable documentation, that the use of these poisons is necessary if the world is to grow enough food to satisfy its increasingly large and hungry population. In fact, just the opposite prevails: the increasing contamination resulting from pesticide spraying and other pollutants in the air and water is a serious threat to the world's food production system over the long term, and could help bring about or exacerbate massive illness and starvation in future years. Moreover, most of the polluting pesticides exported to underdeveloped countries are used on cash crops destined not for internal consumption but export to the rich, developed nations.[105]

This increased pesticide use will severely damage or destroy one of the world's most promising schemes for producing high-protein food — the "farming" of fresh-water fish, which can yield as much protein per acre as rice farming, and could account for as much as half the amount of fish now being taken out of the oceans.[106]

The use of chlorinated hydrocarbon pesticides can have even more devastating impacts on the wider marine ecosystem by disrupting the entire natural food chain. Even low concentrations of these chemicals are known to kill or damage phytoplankton communities — vital estuarine organisms that are on or near the bottom of the food chain and upon which higher life forms depend. Thus phytoplankton-consuming fish and shellfish populations found in the rich coastal waters, which serve as hatcheries and nurseries for many species, can be seriously affected.

The *Global 2000* report also predicts that in many ways, the use of chemical pesticides will be destructive and self-defeating, with the results including:

(1) biological amplification and concentration of persistent pesticides in the tissues of higher-order predators, including humans.

(2) development of increased resistance to pesticides by numerous insect pests, and hence possible declines in yields through increased vulnerability to pests.

(3) destruction of natural pest controls such as insect-eating birds and predatory insects, and hence further increases in the cost of — and decreases in the effectiveness of — preventing crop losses caused by pests.

(4) emergence of new pests previously not troublesome.

(5) increased poisonings of farm workers and families from nonpersistent pesticides.[107]

Thus, in the next decade, it can be expected that the chemical industries of the U.S. and the developed countries will continue to increase their efforts to drench the underdeveloped world in toxic chemicals now known to be poisonous to humans and the environment. But if this seems unreasonable, one must ask why the industry should treat the rest of the world any better than it has treated us.

What You Can Do

The virtually unrestricted export of dangerous chemicals could be stopped by executive order of the President (a version of which President Reagan rescinded in 1981) or by legislation. Congressman Michael Barnes (D-Md.) has introduced a bill in the House of Representatives, H.R. 2439, which would "restrict the export of goods which have been found to be hazardous to public health." Senators and Congressmen should be encouraged to support this legislation.

At the same time, the U.S. Agency for International Development should stop exporting such restricted chemicals as DDT. Instead, it should intensify its efforts to promote IPM techniques abroad and to reduce the dependence of farmers in foreign countries on toxic chemicals. And the FDA and EPA should act to effectively prohibit the import into the U.S. of food contaminated with residues of cancer-causing pesticides banned for use in this country. Such an action, which could be accomplished by removing existing food tolerances for such compounds, would immediately stop the use of such pesticides on vast amounts of food grown abroad for export to the U.S.

In the meantime, in order to minimize consumption of domestically-banned chemicals found on imported food, you should peel or thoroughly scrub fruits and vegetables. This includes banana peels and the outsides of other produce that are handled, since they are sprayed with extremely toxic compounds and may well contain significant chemical residues.

Part Two

The Impending Disaster

In 1976, the spraying of such deadly pesticides as endrin, toxaphene, and parathion over millions of acres of Kansas and Oklahoma caused what the *Kansas City Star* called "the worst pesticide application disaster the nation has ever known." Wildlife was wiped out by the millions, mainly fish birds, and frogs, but pets and livestock were also poisoned and killed. People, towns, homes and cars were repeatedly sprayed, and a yellow school bus full of children was even doused by a cropduster. Many residents were sickened, and damage was estimated in the millions of dollars. When Caldwell, Kansas science teacher Larry Miller and his mother were investigating pesticide spraying in Oklahoma in 1979, two spray planes twice buzzed their truck. One plane then landed and blocked the public road on which they were driving. The pilot jumped out and ran at the Millers, threatening to "get" or "kill" them. As Miller drove away at high speed, the second plane chased him. Although he photographed and reported this incident, and others in which he was directly sprayed, to the local authorities and the Federal Aviation Administration, no action was ever taken against the pilots. "There does not seem to be a law against spraying people," Miller notes. *Photos courtesy of Larry Miller/Caldwell, Kansas.*

Chapter Seven

Our Chemical Heritage: Cancer, Sterility, and Deformed Children

...some of the pesticides... are so long-lasting and so pervasive in the environment that virtually the entire human population of the Nation, and indeed the world, carries some body burden of one or several of them.

Library of Congress study accompanying Surgeon General's report on toxic pollution, August, 1980[1]

Man-made toxic chemicals are a significant source of death and disease in the U.S. today.
Gus Speth, Chairman, President's Council on Environmental Quality (CEQ), 29 June, 1980[2]

...70 to 90 percent of all cancers are caused by environmental influences and are hence theoretically preventable.

CEQ, 1978[3]

I t is now apparent that a major potential threat to the life and health of every single American exists in the pervasive presence of poisonous chemicals in the environment. These toxic substances, many of which cause cancer, birth defects, nerve and brain damage, and gene mutations, are so prevalent

245

in our daily lives that they are simply impossible to avoid. Deadly synthetic chemicals are now present in our food, air, water, and even our own bodies—including mothers' milk—where they remain and accumulate.

By the time many of the most dangerous pesticides were banned or restricted as powerful cancer-causing agents, they had contaminated virtually every American and most of our food supply. Dieldrin was being found in the flesh of 99.5 percent of all human tissue samples tested, as well as in 96 percent of all meat, fish and poultry.[4] BHC had been detected in 99 percent of all Americans tested,[5] and heptachlor in 95 percent, as well as in 70 percent of meat, poultry, fish and dairy products.[6] DDT and PCB's were turning up in almost all human tissue samples, fresh water fish, meat and dairy products[7] (see chapters on dieldrin, DDT, and PCB's).

Now that we know that almost every American carries detectable traces of several generally banned carcinogens, we must await the inevitable effects. How long can our bodies continue to accumulate such lethal poisons before succumbing to cancer? What will be the effects upon our genes, upon future generations? The answer, of course, is that we do not know, because never before in human history have such deadly substances existed, much less entered and lodged in our bodies.

Cancers usually take 20 to 30 years to appear after a person has been exposed to a carcinogen, and government studies are now detecting a regular increase in the cancer rate. Here is thus the first harvest of disease following the seeds sown in the 1950's and '60's, when chemicals first began to proliferate throughout the environment.

But the worst appears to lie ahead. Each year, some one thousand new chemicals are introduced to pollute our air, water and food. Most cancers — the estimates run between 60 and 90 percent — are attributed to environmental factors, including the widespread use since World War II of toxic chemicals. A 1978 report issued by the President's Council on Environmental Quality (CEQ) unequivocally states that "most researchers agree that 70 to 90 percent of all cancers are caused by environmental influences and are hence theoretically preventable."[8]

The Current Cancer Epidemic

With exposure to toxic chemicals beginning at the moment of conception and continuing throughout life, one in four Americans now contracts cancer.[9] This means that some 60 million Americans now living can expect to get the disease.[9b] Two-thirds of them will die of it, which comes to more than 400,000 deaths a year — over a thousand a day!* To put that number in perspective, consider that fewer Americans died in the battles of World War II plus the Korean and Vietnamese conflicts combined. Each year, cancer kills about 8 times more Americans than do automobile accidents.[10]

For Americans of all ages, cancer is the second biggest killer, after heart disease; and in 1975 cancer caused 20 percent of all deaths. In 1900, cancer ranked between 8th and 10th as a cause of death, and was responsible for only about 3 percent of all mortality. Cancer is now the leading cause of death for women between 30 and 40, and for children aged 1 to 10. Thus, the burgeoning cancer rate is not caused by people living longer; it has now become a disease of the young as well as the old. It is 2nd only to accidents as the chief cause of death for Americans under 35.[11] It is no exaggeration to say that we are in the midst of a cancer epidemic, and that this may be just the tip of an oncoming iceberg: the cumulative effect of the thousands of new chemicals in the food chain and the environment may not become fully apparent for another decade or so.

While the chemical industry loudly complains about the alleged costs of health and environmental laws, the truth is that the monetary costs of pollution, and the inflation and higher taxes it causes, are many times greater, to say nothing of the lost lives and ruined health. Over a million people are now under treatment for cancer, and the Department of Health and Human Services estimates that the costs to the nation from cancer, including medical expenses and lost productivity and income, may total $30 billion a year.[12]

---

*The American Cancer Society in 1981 projected that 420,000 Americans would die of cancer that year, as compared to 412,000 in 1980, and 404,000 in 1979. Some 120,000 of these deaths, mainly lung cancers, are thought to be caused by cigarette smoking.

In order to determine scientifically the effects of toxic chemicals on our health, CEQ in 1977 formed a research team of 18 federal agencies called the Toxic Substances Strategy Committee (TSSC). After three years of study, the committee in 1980 issued a report clearly demonstrating that perhaps millions of Americans are killed or disabled by carcinogenic synthetic chemicals. TSSC confirmed that cancer incidence rates were on the increase — "about 10 percent between 1970 and 1976" — while this period saw all other causes of death decline.[13, 14, 15] It pointed out that "Cancer is the only major cause of death that rose continuously from 1900 to 1978. Recent figures show that both incidence (new cases) and mortality (deaths) rates are increasing."[16]

The report stated clearly that almost all cancers are caused by substances present in the environment:

> It is generally believed that most cancer cases, possibly 90 percent, are related to at least one environmental factor — that is, external non-genetic factors like air pollution, smoking, chemical and other contaminants in the workplace, and dietary components.[17]

In announcing the findings, Gus Speth, Chairman of CEQ, emphasized that "man-made toxic chemicals are a significant source of death and disease in the United States today."[18] And it is no coincidence, CEQ member Dr. Robert Harris pointed out, that the production of chemicals greatly increased between 1950 and 1960, and the dramatic increase in the cancer rate showed up 20 to 25 years later, "the lag time one might expect."[19]

As Dr. Joseph Highland, Chairman of the Environmental Defense Fund's Toxic Chemicals Program, testified before a Congressional committee, "Through our current inaction, we have jeopardized our futures," and the worst is yet to come:

> As cigarette consumption rose, so did lung cancer rates, but only after a characteristic "lag time" or "latent period" of approximately 30 years. The lung cancer incidence we see today clearly reflects the smoking hazards of 30 years ago. Consequently, if a pattern similar to that of smoking occurs in that a 30-year lag time exists between exposure to these chemicals and the manifestation of disease, the

effects of general exposure may not be seen until the mid-1980's.[20]

On 27 August 1980, the Surgeon General of the United States, Julius Richmond, released an assessment of the threat to public health caused by toxic chemicals. Just as the Surgeon General's 1964 report was instrumental in convincing the public of the hazards of cigarette smoking, this 1980 report was equally forthright in declaring that the threat posed by hazardous chemicals is severe:

> ...it is clear that it is a major and growing public health problem. We believe that toxic chemicals are adding to the disease burden of the United States in a significant, although as yet ill-defined way. In addition, we believe that this problem will become more manifest in the years ahead...We believe that the magnitude of the public health risk associated with toxic chemicals currently is increasing and will continue to do so until we are successful in identifying chemicals which are highly toxic and controlling the introduction of these chemicals into our environment.

The Surgeon General also reported that while "full implementation of recent environmental control legislation" could reduce the public exposure to such chemicals, nevertheless "through this decade we will have to confront a series of environmental emergencies and, therefore, are developing strategies to respond to their public health implications."[21]

An accompanying report prepared by the Library of Congress makes it clear that this problem is not confined to the U.S., but is of worldwide dimensions:

> In the case of chemicals such as some of the pesticides (aldrin, dieldrin, DDT, etc.), PBB's and PCB's...these are so long-lasting and so pervasive in the environment that virtually the entire human population of the Nation, and indeed the world, carries some body burden of one or several of them...a significant proportion of the world's population, perhaps all of it, is exposed to the cumulative if unknown effects of a plethora of man-made pollutants from a number of diverse and often distant sources.[22]

A more recent study on cancer is even more ominous. A June 1981 report by the National Cancer Institute states that the average American now stands almost a one in three chance of contracting cancer: "The cumulative incidence rate for all races, both sexes, all areas combined . . ., which can be interpreted as the probability of developing cancer from birth to age 74, is approximately 31 percent."[23]

Mother's Milk: "Keep Out Of Reach Of Children"

Cancer-causing chemicals commonly found in the fat tissue of Americans include aldrin, chlordane, DDT, dieldrin, dioxin, benzene hexachloride (BHC), endrin, heptachlor, mirex and PCB's, all of which have been banned for all or major uses. In 1980, a chemist doing research for the federal government on dioxin levels in Vietnam veterans found that his samples were so contaminated with DDT (largely banned in 1972) and PCB's that the dioxin was initally obscured.[24]

Like silent time bombs ticking away, these toxic chemicals accumulate and build up until, at a time and in a manner still not understood, they set off the runaway growth process in the cells that we call cancer.

These particular carcinogens are members of a chemical family known as chlorinated hydrocarbons, which are known not only for their toxicity, but also for their endurance and ability to remain for long periods in the environment. When ingested by fish, wildlife or humans — either through eating or breathing them, or by absorption through the skin — these substances accumulate in the body and are stored in fat tissue. There they can stay indefinitely unless they are expelled through loss of weight or lactation

Because toxic chemicals are stored in human fat tissue, they become highly concentrated in mother's milk. Its tragic contamination has been dramatized by The Ecology Center in Berkeley, California which issued a poster of a nude and pregnant woman with a label on her breasts reading, "Caution: keep out of reach of children." There is more truth than humor in this display, for we have succeeded in poisoning this vital source of infant nourishment.

When a newborn baby begins to breast-feed, it may directly ingest extraordinarily large and concentrated doses of stored chemical carcinogens from the body fat used to manufacture mothers' milk. As Stephanie Harris of the Environmental Defense Fund has described the process,

> During lactation, the body fat reservoir is called upon to provide the energy to create milk and milk fat. The chlorinated hydrocarbons are mobilized through the bloodstream to the mammary gland where they are incorporated in the milk fat and then excreted. These residues are then transferred to the nursing infant.[25]

A 1976 EPA study of cancer-causing chemicals in the milk of nursing mothers across the U.S. found that virtually all samples contained DDT and PCB's, 81 percent showed levels of dieldrin, and 87 percent had BHC. In some areas of the country, locally-used carcinogens were regularly detected, such as dioxin, kepone or mirex.[26]

The fact that alarming levels of cancer-causing pesticides continue to be present in mother's milk and human tissue samples long after they have been banned or severely restricted shows that the problems caused by these extremely stable and persistent chemicals will trouble us for the foreseeable future.

Dangers To The Infant

Since mothers' milk is often an infant's only source of food, the baby ingests a much larger portion of these and other chlorinated hydrocarbon pesticides on a body-weight basis than does an adult, and the effects could thus be much more serious. The World Health Organization (WHO) has estimated the levels at which chemicals may be consumed without causing serious harm. These Acceptable Daily Intake (ADI) levels are regularly equalled or exceeded by nursing infants in the United States.

Indeed, it would be illegal to sell most mothers' milk in interstate commerce. Because the amounts of chemicals such as PCB's and DDT exceed the maximum daily permissible intake levels, much of the milk would be subject to confiscation and destruction by the Food and Drug Administration (FDA). As the Environmental Defense Fund has pointed out, the average nursing infant receives one hundred times more PCB's on a body-weight basis than does an adult, and the amount ingested is ten times that set by FDA as allowable. This is true for other pesticides as well. EPA's 1976 survey found that the carcinogenic pesticide dieldrin was being ingested by the average infant at 9 times the permissible levels, and in one case the levels exceeded the FDA standards 700 fold![27]

All of this does not mean that mothers should avoid nursing their infants. Human breast milk is nutritionally far superior to cow's milk and formula, which often are also contaminated with these and other chemicals, as is the tap water used to mix with formula. And mothers' milk contains various unique substances which confer immunity to sickness and disease, and is — in general — the ideal infant food.[28]

Alas, the conscientious mother has no way to completely protect her child from contact with carcinogens; exposure can only be minimized. In fact, the poisoning of the fetus begins while it is still in the womb. Pregnant women have an increased metabolic rate, and this hastens the transfer of toxic chemicals from the blood of the mother to that of the child. Moreover, the baby is much more defenseless and sensitive to these poisons, since its immature liver cannot detoxify the chemicals as well as that of an adult. The fact that developing fetuses and infants appear to be more susceptible to contracting cancer from exposure to carcinogens has been demonstrated in laboratory tests on animals.[29, 30, 31]

In a June 1980 "Position Document," EPA stressed that young children are definitely at greater risk from such chemicals, and that even a single exposure can be dangerous.

EPA also notes that "the potentially higher exposure per unit of body weight for young children and the possibility of underdeveloped detoxification systems in very young children would also appear to pose significant concerns... Although feto-toxic effects are considered a chronic human risk, it should be pointed out that fetotoxicity can result from a single exposure."[32]

Thus, the exposure of humans to carcinogens begins literally at the moment of conception. When a woman becomes pregnant, many of these chemicals are released and "migrate" toward the fetus. The post-World War II generations of Americans are the first in history to spend their entire lives — from conception on — surrounded by toxic chemicals.[33]

Dead And Deformed Babies

Damage to unborn generations is another price we pay for our chemical saturated society. Many toxic chemicals in widespread use have been shown to cause birth defects in test animals, and presumably can and do cause them in humans as well.

No one knows how many birth defects occur in America, but the number is substantial, and far higher than is generally realized or reported. According to the March of Dimes Birth Defects Foundation, each year some 233,000 infants are born with birth defects, with another 53,700 threatened by extremely low birthweight (4 lbs, 6 oz. or less). In addition, the Foundation estimates that 500,000 stillbirths and spontaneous miscarriages occur annually, mainly because of abnormal fetal development.[34] (A 1968 study estimated that of 5.7 million conceptions that occur in the U.S. each year, almost a third of the fetuses, "nearly 1.7 million, are aborted spontaneously, are miscarried, or are born dead.")[35]

These estimates for the number of impaired births may be far too low. A January 1981 report by the Council on Environmental Quality (CEQ), "Chemical Hazards of Human Reproduction," estimates the percentage of newborns with birth defects that can be ascertained by early childhood at "about 16 percent, twice that recognized in the first year."[36]

More than a million people are hospitalized each year for treatment of birth defects, and over 60,000 deaths occur because of such damage, it is estimated.[37]

The number of these defects caused by toxic chemicals is not known, but could be substantial. The May 1980 "Report to the President," by the inter-agency Toxic Substances Strategy Committee, observes that very little "is known about the origins of human birth defects, the causes of more than 80 percent have not been identified."[38]

Writing in the 15 May 1980 *American Journal of Obstetrics and Gynecology*, Dr. Lawrence D. Longo warns of the as-yet-unstudied and unknown dangers of pesticides to infant brain development:

Despite literally mountains of evidence that these substances produce manifold problems, little work is being
carried out on the less obvious effects on the developing
organism. For instance, even a single exposure to some
common organophosphate pesticides (such as malathion,
parathion, Mipofax) can alter brain electrical activity for
years... Some preliminary studies suggest altered learning
patterns and behavior.[39]

Some scientists fear that toxic chemicals — many of which are
mutagenic in test animals — may be affecting humans in the
same way, causing genetic changes that will affect future generations in ways impossible to foresee or prevent.

According to Dr. Jean Priest, a genetic researcher at Emory
University in Atlanta, scientists cannot yet tell what the effects
will be of the thousands upon thousands of chemicals to which
Americans are being exposed: "The effects of mutations take at
least one generation and maybe many generations (to show up).
It's only in our generation that there has been a huge increase in
chemicals that are not found naturally."[40] Dr. George
Streisinger, of Oregon's Institute of Molecular Biology, spe-
cializes in studying chemicals that cause mutations. He believes
that this process involves damage to or interference with the
DNA molecules, which carry the "blue print" or "computer
code" for the makeup of the body:

> ...a defect in that blue print causes a hereditary disease.
> These effects would, of course, persist for untold genera-
> tions in the future.[41]

We are thus in danger of irreparably harming the human gene
pool, which the National Academy of Sciences describes as
"...the primary resource of mankind, today and tomorrow,
...the culmination of three billion years or evolution and
natural selection."[42] As John Gofman and Arthur Tamplin have
observed,

> Changes in the chromosomes of sperm or our precursor
> cells may be transmitted to all future generations of
> humans. The heredity of man, his greatest treasure, is
> thereby at stake. Once irreversibly injured, the chromo-
> somes cannot be repaired by any process known to man.[43]

Ironically, in order to preclude the possibility of having a defective child and avoid losing their jobs, some women workers are voluntarily undergoing permanent sterilization in order to continue employment in jobs that expose them to high levels of toxic chemicals.

After an American Cyanamid plant in Willow Island, West Virginia restricted females capable of bearing children from certain jobs in 1978, five women employees had themselves surgically sterilized. In 1977, an Allied Chemical plant in Danville, Illinois laid off five women employees because of concern that a fluorocarbon could cause fetal damage. After two of the women underwent sterilization and reclaimed their jobs, it was decided that the product was not teratogenic after all.[44]

In 1979, the Occupational Safety and Health Administration (OSHA) unsuccessfully brought charges against American Cyanamid for having "adopted and implemented a policy which required women employees to be sterilized in order to be eligible to work in those areas of the plant where they would be exposed to certain toxic substances." And in September 1980, OSHA cited the Bunker Hill Company of Kellogg, Idaho, for a similar exclusion policy in its lead smelting operation, charging that it required some women to be sterilized in order to work in areas where they would be exposed to lead.[45]

Thus, the trend in U.S. industry is to remove from the workplace not toxic chemicals that can harm a fetus, but rather the possibility that there will be a fetus. Commenting on this development, labor leader Anthony Mazzocchi points out that the logical extension of such a policy would affect men as well:

> Ultimately, it will be quite clear that women and men alike suffer from exposure to lead and other toxic chemicals. When that happens, the industry initiative may be to have men sterilized. We will then enter the stage of the neutered worker.[46]

Indeed, recent studies indicate that a high number of birth defects — perhaps a substantial majority — may be related to factors affecting the male instead of the female. Toxic chemicals, including those especially damaging to the sperm, tend to concentrate and collect in the male reproductive tract for future elimination. According to the CEQ report "Chemical Hazards of Human Reproduction,"

> Semen can serve as a route of excretion of chemical agents ... Chemicals in the semen can be transmitted to the female during intercourse and absorbed through the vaginal mucosa ... thus, impairments generally associated with a stage in reproduction that occurs in the female may actually be the result of exposure in the male.[47]

A study by the University of Southern California Medical School, released in the summer of 1981, showed a link between the exposure to chemicals in the work place and brain tumors in the children of workers. The association was especially striking among workers in the aircraft industry, and those exposed to paint fumes and who worked around chemical solvents. It is theorized that the tumors are caused by genetic damage to the father's reproductive system, or by chemicals brought home on his clothes.[48]

Damage to the developing fetus and nursing infant, through contaminants that enter human bodies, is more than a theoretical possibility; epidemiological studies in several countries (e.g., Vietnam; Alsea, Oregon; Seveso, Italy; Kyushu and Minamata, Japan; all referred to earlier in this book) have confirmed that maternal contact with toxic chemicals has indeed killed and deformed babies. It is now generally recognized that no drug has been proven safe for pregnant and nursing mothers, and they are warned to avoid *all* drugs, even alcohol, aspirin, and caffeine in coffee and tea. Yet, ironically, the developing fetus and the nursing infant are regularly exposed to dozens of cancer-causing and otherwise toxic pesticides and chemicals.

The long-term impact of repeated low doses of this combination of poisons, which continue throughout life, remains to be seen, but the results to date suggest that our toxic environment may have already set the stage for an epidemic of future fetal abnormalities.

The Threat Of Sterility

In several widely-publicized occupational health disasters, it was discovered that workers had been rendered sterile by exposure to pesticides they were producing, such as DBCP, leptophos and kepone (all of which are discussed elsewhere in this book). But the general population has also been, and in some cases continues to be, widely exposed to these and/or other such pesticides and chemicals that leave residues on domestically-grown and imported fruits, vegetables and meat, as well as to chemicals such as PCB's that are known to reduce and damage human sperm.

Exposure to the pervasive contaminant lead — which is present in humans at levels several hundred times what our prehistoric ancestors carried — appears to decrease, and cause malformations in, male sperm. Lead is present in food, air and water; contact with it thus cannot be avoided.[49]

Several studies indicate that the average sperm count of American men has dropped by one-third in the last 30 years, at least in part because of the effects of PCB's and other chemicals (see page 298), especially chlorinated hydrocarbon pesticides such as DDT.

Studies conducted in 1980 by Dr. Ralph Dougherty of Florida State University (FSU) found that 23 percent of the students there were functionally sterile. Testing in Houston, Texas turned up an identical rate. Other recent studies have shown an estimated sterility rate of 10 to 20 percent for the total male population, as compared to a reported rate of just one-half of one percent in 1938. Dr. Dougherty attributes a major share of the blame for this drastic reduction in fertility to 27 chlorinated hydrocarbon chemicals that have become widespread in the food chain and the environment.[50] Dr. Dougherty's research indicates that the average sperm count in American men has dropped 40 percent in the last 30 years — from 100 million to 60 million sperm per milliliter of ejaculate.[51] Just in the last 15 years, the average number of sperm cells per cubic centimeter of semen has dropped from 50 to 100 million sperm cells to less than half that — 20 to 40 million, according to some reports.[52]

An analysis of the effects of toxic chemicals on human health, by Eric Jansson of Friends of the Earth, estimates that eliminating just 20 chemicals from the food chain and the environment could decrease male sterility and birth defects by 30 percent, and sharply reduce the cancer rate as well.[53]

While it is still too early to draw firm conclusions, such data suggest that we are reducing — or eliminating — our ability to reproduce ourselves. If and when the final proof arrives, it may be too late to remedy the problem.

Affecting How We Think, Feel, And Perform

Toxic substances in the environment may be affecting how we think, feel, and act to a degree that may be impossible to fully assess. In addition to physical illness, toxic chemicals and metals such as lead, mercury, and cadmium appear to be causing mental, emotional, and behavioral problems among a large segment of the population. Chemicals, food additives, and drugs are not generally tested for such effects; but since subtle behavioral changes are often difficult to detect, they may be extremely widespread throughout the population.

A 1975 NIOSH survey of 485 painters exposed to the fumes from paints, solvents, and thinners found that over 70 percent suffered intoxication and disorientation from the use of these chemicals.[53b] With some substances, such as lead, their presence in the environment is so prevasive that the entire population is contaminated with significant levels, and an uncontaminated control group would be virtually impossible to set up.[53c]

Many toxic substances interfere with brain development in unborn and very young children, and cause mental problems in older children. It is known, for example, that children with high levels of lead in their blood have significantly more difficulty with classroom behavior and learning ability.

As pointed out by the Center for Science in the Public Interest, which sponsors a project and newsletter on "Environment & Behavior," such effects caused by toxins can range from subtle changes to "severe and permanent mental disorders." These include lowered IQ's, mental illness, irritability, inability to concentrate or pay attention, hyperactivity, confusion, and tremors.

Industry's attitude towards this situation has changed little since a British publication in 1902 described a rubber plant in which contact with the chemical carbon disulfide was causing workers to become so psychotic, paranoid, or disoriented that they were leaping out of the windows. Instead of having the chemical removed from the workplace, the plant simply had the windows barred.[53d]

The Tip Of The Iceberg

Today, we are being exposed to more toxic chemicals than we were yesterday, and there will be even more tomorrow. The production of new, untested chemicals is a fabulously profitable "growth" industry. It has overwhelmed the government regulatory agencies, as well as our bodies' defense mechanisms.

The manufacture of synthetic organic (petroleum-based) chemicals in the U.S. has grown from less than a billion total pounds in 1941 for all such chemicals, to 172 billion pounds in 1978 for just the 50 major ones. In that same year, world trade in these chemicals rose by 18 percent.[54] Production of the top 50 inorganic chemicals totalled 350 billion pounds[55]—almost 100 pounds for each person on earth!

The American Chemical Society's registry of chemicals reported in the literature since 1965 contains over 4 million compounds, with some 6,000 a week being added in the 1970's.[56] There may be as many as 70,000 of these chemicals in commercial production in the U.S.; and according to CEQ, "The environmental and health effects of most of these substances have not been adequately studied."[57] Since the second World War, some 50,000 new chemicals — most of them never tested for safety — have been introduced into the environment, with about 1,000 new ones being added each year.[58]

In the Report to the President from the Toxic Substances Strategy Committee (TSSC), it was concluded that "The magnitude of the toxic substances control problem...is staggering."[59] This is hardly an exaggeration.

In its inventory of hazardous chemicals sufficiently dangerous to be subject to regulation by the Toxic Substances Control Act, EPA had listed over 55,000 chemicals by late 1980, and the agency stated that it expected to add hundreds of compounds each year.[60]

As of 1980, only one percent of all the chemicals in commerce had been shown to be safe or their use restricted.[61] Only 1500 or so chemicals had been tested for their ability to cause cancer.[62]

According to EPA, a third of the 1,500 active ingredients used in pesticides are toxic and a fourth cause cancer.[63]

These chemicals are, for the most part, substances that human cells have never before come into contact with in the entire history of human evolution, and with which our bodies do not know how to cope — to expel or digest — when we are exposed to them.

Scientists have long contended that there is no "safe" level of exposure to carcinogens. Since many carcinogens are stored and build up in our bodily tissues, especially the fat, repeated "insignificant" exposures over a period of time can accumulate to levels sufficient to trigger the process that causes cancer and other ailments.

Despite the often-expressed belief that a person has to consume enormous quantities of a carcinogen to be affected, the 1980 TSSC report emphasized that even slight exposure to a mild carcinogen can cause cancer:

> Methods do not now exist for determining a "safe" threshold level of exposure to carcinogens ... prudence requires that no safe threshold levels be assumed to exist.[64]

Significantly, a substance that is only mildly carcinogenic and produces tumors in only 10 percent of the test animals, could, if it had a similar effect on the American population, mean over 20 million cancers over the lifetime of a generation.[65]

Moreover, non-cancer-causing chemicals can damage or modify cells in such a way as to enhance the effect of carcinogens on them.

While some chemicals thought to be potentially unsafe are being tested individually, the only truly valid test of the dangers of the chemicals in our environment would be one virtually impossible to conduct: a study combining the hundreds or thousands we are daily exposed to in order to determine their synergistic effect, how they react in us when mixed together. While a chemical may not be carcinogenic by itself, or may be a weak one, such as saccharine, it may help promote the carcinogenicity of other chemicals through processes that are still not clearly understood.

Promoters, rather than "initiating" cancers, appear to increase the effect of a carcinogen, and can speed up the cancer process and increase the number of tumors.[66] According to the President's Council on Environmental Quality, "many cancers probably result from the combined effects of long-term exposures to several 'weak' carcinogens acting together (in an additive or synergistic way) rather than from sporadic exposure to a few 'potent' carcinogens. If so, most cancers may be the result of several causative agents, some of which are potentially preventable."[67]

By the time the deadly results of exposure show up decades later, it is usually too late to help the numerous people affected, or even to know the precise cause of the cancer. Already, it is impossible to determine a cause-and-effect relationship for most of the cases of cancer that are contracted by one out of four Americans.

As Dr. Liebe F. Cavalieri, a professor of biochemistry at Cornell Medical College and a member of the Sloan-Kettering Institute for Cancer Research, recently wrote:

> the implication that low levels of exposure (to carcinogens) are somehow unimportant is grossly misleading. The fact is that low-level effects at the molecular level can be additive or even multiplicative. Moreover, many non-carcinogenic chemicals enhance the cancer-causing effects of carcinogens ... It is a terrible mistake to consider carcinogens individually; an integrated analysis of the multiplicity of chemicals Americans are being exposed to and a plan of attack based on their cumulative effects are desperately needed.[68]

While it has become fashionable in some quarters to adopt the position that *everything* causes cancer if given to test animals in high enough doses, this is simply untrue.

By 1977, some 1,500 chemicals suspected of being potentially carcinogenic had been extensively tested for safety, with only some 600 to 800 showing "substantial, positive evidence of carcinogenicity." In one series of animal studies using high doses of 120 pesticides and industrial chemicals, only 10 percent of these chemicals proved to be clearly cancer-causing, with an additional 10 percent showing indications of carcinogenicity. According to CEQ, "the large majority of chemicals tested at high doeses in animals to determine their cancer-causing potential have (sic) not been found to cause the disease."[69]

But even so, if 10 percent of the tens of thousands of chemicals in widespread use prove to be carcinogens, that will amount to several thousand substances.

PBB: The Poisoning Of Michigan

Some Michigan farmers are said to be selling for hamburger animals that are referred to as "leaners" — that is, animals that can stand up only when they are packed in tightly enough to lean on one another.

**Jane E. Brody, The New York Times,
12 August, 1976**[70]

In 1976 alone, Michigan residents ate on estimated five million to seven million pounds of hamburger contaminated with PBB ... More than 90 percent of the breast milk samples showed PBB and expanded breast milk testing since then has confirmed those finding.

Detroit Free Press, 14 and 15 March, 1977[71]

PBB's are persistent and can be passed on for generations. PBB's are stored in the body fat, where they can remain indefinitely: during pregnancy they can cross the placenta to the developing fetus ... Scientists at Harvard University and the National Cancer Institute have found that PBB's contain two suspected carcinogens, napthlene and furon ... Animal experiments have shown that PBB is ... capable of producing physical defects in offspring in utero.

**Subcommittee on Crime, U.S. House of
Representatives, May, 1980**[72]

Perhaps the nation's greatest agricultural disaster, and one of the worst incidents of mass poisonings of a population, occurred in Michigan in 1973 and 1974. It affected almost every resident of the state and demonstrated how rapidly even small amounts of a toxic chemical, improperly handled, can be dispersed throughout the environment and the food chain, poisoning millions of people.

In mid-June 1981, a federal government study determined that PBB — a fire-retardant present in the bodies of 90 percent or more of the residents of Michigan — caused cancer of the liver in animals and/or shortened their lifespans.[73]

This was ominous news for the people of that state, almost all of whom ingested PBB in 1973-74, when it was accidentally mixed with livestock feed and dispersed throughout the state. Cows, chickens and other livestock that became sick were sold directly for meat or for slaughter.

The response of state and federal officials to the incident amounted to a cover-up, and had the effect of magnifying the problem rather than containing it. As a result of the accident, the health of hundreds of people was ruined; millions of chickens and tens of thousands of other farm animals had to be destroyed; and nearly every person in the state of Michigan — some nine million people — ended up with this cancer-causing chemical in his or her body. Since over 90 percent of the breast milk samples tested showed that PBB was present, future generations are also at risk.[74]

Unfortunately, the PBB contamination appears to be spreading. A $2.25 million study released in December 1981 indicated that 97 percent of the state's residents had traces of the chemical in their tissues.[74b]

While a few hundred pounds of PBB caused a major health disaster in Michigan, millions upon millions of pounds of this chemical have been produced and sold throughout the country. In 1974, the last year that Michigan Chemical Corporation produced PBB, 4.8 million pounds of it were made.[75] An estimated 12 million pounds of Firemaster — a product made with PBB that was responsible for the Michigan tragedy — was manufactured and distributed to locations now unknown.[76] The disposal of PBB products in dumps, landfills, and elsewhere will spread this deadly chemical throughout the environment and perhaps allow it to enter drinking water sources and the food chain on a nationwide scale, making the Michigan experience with PBB a national one.

Although it is impossible to know how many other PBB-like incidents have occurred but gone undetected, this tragedy should serve as a warning to us. Similar problems can be expected from the marketing and disposal of the thousands of persistent and toxic chemicals now in use, as has already happened with PCB's. Yet we continue to produce and market tens of thousands of enormously hazardous chemicals on a massive scale, oblivious to the possibly disastrous consequences of our thoughtless action.

Cancer In The Workplace

While all Americans experience repeated low-level exposure to toxic chemicals, a large percentage of U.S. workers are especially in danger because they are regularly subjected to dangerously heavy exposure to many types of cancer-causing chemicals.

One of the major findings of the TSSC report was based on 1978 studies by the National Cancer Institute, the National Institute for Occupational Safety and Health (NIOSH) and the National Institute for Environmental Health Sciences demonstrating that 20 to 38 percent of all cancers are associated with occupational exposure. According to the report, one in four workers today — approximately 20 million Americans — is exposed on the job to toxic chemicals so dangerous that they are regulated by the Occupational Safety and Health Administration (OSHA).

As NIOSH puts it, "one of four U.S. citizens is exposed in the workplace to some substance capable of causing death or disease."[77] In fact, worker exposure to just six chemicals — arsenic, asbestos, benzene, chromium, nickel and petroleum distillates — are thought to account for 20 to 38 percent, or at least 120,000 cases, of the over 600,000 cases of cancer that Americans contract each year.[78]

Cancers incurred because of carcinogens in the workplace provide some of the clearest examples of how such chemicals can directly affect those exposed to them. The first generally acknowledged case in which an environmental agent was pinpointed as the source of human cancer occurred in 1775, when a London physician, Sir Percival Pott, discovered that soot was the cause of the rare cancer of the scrotum so common among chimney sweeps of that day.[79]

Since then, scientists have discovered various other "occupational diseases:" cancer of the bladder among workers in the dye industry, leukemia in benzene workers, lung cancer in asbestos workers.

Even painters are at risk from toxic chemicals found in solvents and pigments. Studies by the National Cancer Institute have shown that artists die from leukemia and bladder cancer at a rate several times higher than the national average. Painters in New York were found to have a life expectancy of 11 years less than the average American.

Early symptoms of poisoning among such groups of artists as painters, silk screeners and stained glass and jewelry makers include headaches, blurred vision, slurred speech, dizziness and exhilaration, progressing to more severe ailments such as paralysis, hallucinations and central nervous system disorders. A study of paint hazards by John Hopkins University found "minimally, over 300 toxic materials and 150 carcinogens potentially present in paints." One painter complained that when the work force showed up on its job each morning, "we had to haul dead rats out of the rooms we had painted the night before."[80]

Unfortunately, measures to improve the hazardous conditions prevailing for painters and many other workers have now been delayed or cancelled by the Reagan administration. At the behest of the Chemical Manufacturers Association, the U.S. Department of Labor in early 1981 froze a proposal to require that the hundreds of thousands of products containing toxic chemicals be given labels with more adequate warnings of their hazards and how to use them.[81]

An extremely rare and incurable form of liver cancer — angiosarcoma — often shows up in workers exposed to vinyl chloride. Thousands of coke oven workers are forced to inhale soots and tars chemically similar to those of the London chimney sweeps, and die of lung cancer at 10 times the rate of their fellow steel workers. In 1973, half of the retired workers from one benzidine plant were found to have bladder cancer. Many groups of workers who come in contact with inorganic arsenic die of lung and lymphatic cancers at 2 to 8 times the normal rate.[82] Shoe factory workers contract cancer of the sinuses and nasal cavity at seven times the national average, and leukemia at double the average rate; printing pressmen get cancer of the mouth 125 percent more often than the average person; and coal miners contract stomach cancer at a 40 percent higher rate.[83]

In 1980, what OSHA calls the "largest single series of presumably occupation-related brain cancers" in medical annals was discovered at a Union Carbide petrochemical plant in Texas City, Texas, where workers were dying of brain cancers at 4 to 5 times the national average.[84]

Recently, researchers discovered an alarming pattern among such workers: employees of chemical and oil processing plants, where numerous toxic chemicals are present, die of brain cancer and brain tumors at double the normal rate. Plants surveyed included a Union Carbide petrochemical plant, a Dow Chemical plant, and refineries operated by Texaco, Gulf and Mobil oil companies.[85] A high rate of broken chromosomes was also discovered among workers at this same Dow plant in the early 1970's, but Dow admits that it has taken "no additional safeguards" on its production line despite the findings.[86]

It is not just plant workers who are affected by industrial chemicals, but also people who live in the vicinity of such industries. Children who lived around copper smelters have been found to have abnormally high levels of arsenic in their bodies, which shows up in the hair and urine. Indeed, much or all of the highly-industrialized state of New Jersey — with over 1,200 chemical plants — appears to be contaminated with airborne carcinogens, and has the highest cancer rate of any state in the union.[87]

Tragically, U.S. workers are still being denied vital information on dangers to their health. In August 1981, press reports revealed that the U.S. government had long been aware of hundreds of thousands of workers who had been exposed to carcinogens where they worked but had not tried to notify them of the known perils to their health. Nor has an attempt been made to contact and inform the 21 million workers — one in every four — known to have been exposed to hazardous substances regulated by OSHA. In 1977, NIOSH representatives testified at a congressional hearing that finding and contacting all of these 21 million workers could cost as much as $40 million, and medical surveillance (not treatment) could cost $54 billion.[88]

Instead of contacting the workers, NIOSH has used its lists for research, literally waiting for the workers to die so that the data on causes of death can be compiled and analyzed. Ironically, this information is then supposedly used to set regulations for the control of hazardous substances to protect workers' health.[89]

What is at stake here is not just the lives of U.S. workers, but the lives of all of us. Because of the huge numbers of cancer-causing chemicals generated by our highly industrialized society, all Americans are in a sense "occupationally exposed" to these deadly compounds. As the Council on Environmental Quality put it in a recent publication, "factory workers . . . have often unwittingly served as an early warning system for society."[90]

Asbestos: 60,000 Deaths A Year

A particularly graphic — and tragic — example of an occupa-tionally-induced cancer epidemic is the case of the affliction of asbestos workers, a situation that may foreshadow the effects of toxic chemicals 20 to 40 years after exposure.

Asbestos is a fibrous mineral, one of the most potent cancer-causing substances known to man. According to the Depart-ment of Health and Human Services, "asbestos is so widely used that our entire population is exposed to some level."[91] Asbestos fibers or bodies are found in the lungs of almost the entire American population, with particular concentrations in urban dwellers. A University of California study found asbestos in 96 percent of the residents of urban areas who were tested, with airborne fibers being the probable source.[92]

The main sources of the asbestos that is present in the air of virtually all municipal areas are the linings of automobile brakes and clutch plates, and the demolition of buildings in which asbestos has been used extensively in fireproofing, insulating walls and ceilings, as a spray-on wall covering (which has been widely used in schools), and in cement, shingles and caulking compounds.

Since 1941, an estimated 8 to 11 million American workers have been subjected to direct and repeated occupational expo-sure to asbestos, some 4 million of whom were heavily exposed.[93] By the end of the century, it is estimated, between 500,000 and several million Americans will have died from such exposure.[94] In 1978, the Department of Health, Education and Welfare predicted that from then well into the next century, some 60,000 deaths *a year* would occur in the U.S. because of asbestos.[95] Some of this mortality will be among the family members of workers who were exposed to asbestos dust brought home on workers' clothes. Some studies indicate that just one exposure to these fibers can cause cancer decades later.[96] (In December, 1981, part of a study done for the Labor Department by Dr. Irving Selikoff was made public, stating that up to 12,000 asbestos-related deaths would occur each year among such workers, totalling over 200,000 excess deaths by the end of the century.[96b])

Ironically, the dangers of asbestos were known to industry and the government in the early 1900's, and in 1918 the U.S. Bureau of Labor Statistics reported that insurance companies considered asbestos workers to be poor risks because of their extraordinarily high mortality rates. But these known hazards were covered up by the industry, as is demonstrated in a 1 October 1935 letter from the president of a major firm, Raybestos-Manhattan, to the attorney for the Johns-Mansville Corporation, the country's largest asbestos producer. The letter expressed relief that articles published in England on asbestos hazards were not attracting attention in the U.S., and stated, "I think the less said about asbestos, the better off we are." The attorney responded, "I quite agree with you that our interests are best served by having asbestosis [a disabling and incurable crippling of the lungs] receive the minimum of publicity..."[97]

After a lengthy Congressional investigation of this situation, Congressman George Miller (D-Cal.) observed: "This cover-up, which we have now traced from the 1930's to the 1950's, continued into the 1970's," with the industry persisting in downplaying and denying the dangers of its products and avoiding adequate warning labels. "The failure to notify workers, customers, and the general public about health risks associated with asbestos," said Miller, "contributed substantially to the continued usage of the material, and the resultant exposure of millions of people to severe health disorders."[98] Today, one asbestos worker in two dies of cancer.[99]

U.S. firms are now shifting their operations to foreign countries to evade U.S. health regulations that somewhat limit worker exposure to asbestos. And the Reagan administration's anti-regulatory policies have delayed and placed in doubt various proposed health regulations dealing with this and other toxic substances.

Meanwhile, in foreign nations where few such laws exist, asbestos production continues much as it always has. In March 1980, a former employee of a British-owned asbestos plant in Bombay, India described conditions in the air there as being so heavy with the fibers that it resembled "a dust storm." The products were not labelled to indicate any health hazard, and the workers were given no protection against, or even information on, the dangers: "They are completely covered in dust and look like they work in a flour mill, white from head to toe."[100]

Failing To Enforce The Law

In the 1970's, as the problem of pervasive toxic chemical contamination grew to mammoth proportions, public concern mounted, as did demands for a law to protect the American people from deadly chemicals. This resulted in the passage in 1976, after a long struggle, of a law designed to address the problem: The Toxic Substances Control Act (TSCA). Unfortunately, non-enforcement of the law's major provisions by EPA has rendered the legislation largely ineffective.

The passage of TSCA after six years of efforts by environmentalists and other citizen groups was a potentially valuable step in the fight to control poisonous chemicals. But this law, hailed as a far-reaching and comprehensive program for action, was considerably weakened by industry lobbyists before its passage.

For example, many categories of the most toxic substances are not subject to TSCA, such as pesticides, drugs, tobacco products, nuclear materials, cosmetics, food and food additives. And instead of requiring that all new chemicals be proven safe before entering the marketplace, Congress essentially placed the burden on EPA to consider new chemicals safe until proven dangerous, thus making it impossible to screen adequately the thousand or so new chemicals introduced each year.

EPA's timid implementation of the law has greatly eroded whatever effectiveness it could have had. In an article on EPA's administration of TSCA, *New York Times* reporter Richard Severo wrote on 6 May 1980:

> More than three years after passage of the landmark Toxic Substances Control Act, the federal agency responsible for administering the law has yet to order testing of any of the 50,000 chemicals on the market.
>
> The EPA has missed almost every important deadline established by the act on January 1, 1977.[101]

The law requires that industry notify EPA before production is planned for new chemicals. It also gives EPA authority to screen new chemicals and to take action against new or existing substances found to be an unreasonable threat to public health or the environment. But the agency has not used its power to act effectively in this regard. As of the end of 1981, only a handful of chemical substances had been regulated under TSCA.

In October 1980, Congress' General Accounting Office issued a report sharply criticizing the agency's non-enforcement of the 1976 law, saying that "...almost 4 years later, neither the public nor the environment are (sic) much better protected."[102] The GAO report goes into great detail in describing the painfully slow progress in controlling toxic chemicals. For example, "Although actions have been taken to control three chemicals, no chemicals have been tested and basic data is lacking on most of the other 55,000 chemicals now in use ... At present, EPA has allotted enough staff and money to initiate only two or three control actions on existing chemicals annually."[103]

The report describes at length how the major provisions of the law giving EPA "broad authority to protect the public and the environment from the harmful effects of chemicals" have not been implemented.

The law requires EPA to take prompt action on high risk chemicals identified for priority testing. However, this provision has also not been adhered to:

> EPA is required to initiate rulemaking procedures within one year to require the testing or to specify its reason for not doing so. EPA has not required the testing of any of the 38 chemicals or categories recommended thus far and has not issued any final testing standards ...

Moreover, the incredibly slow pace at which chemicals are being regulated is expected to continue:

> EPA estimates show that it will take at least 18 months to issue an Advanced Notice of Proposed Rulemaking and about 5 years to issue a final test rule ... 9 or more years may elapse before a potentially harmful chemical is regulated.[104]

Not only has EPA not taken action to control toxic chemicals, it has also failed even to gather data on those chemicals which may require regulation:

When TSCA was passed, little or no information existed
on the number of chemicals in commerce, how they were
being used, and which ones were toxic. Despite almost 4
years of activity, EPA has not collected this information on
existing chemicals...EPA has requested and obtained
health and safety studies on only 10 chemical
substances.[105]

In response to this devastating indictment of EPA, the agency's
Assistant Administrator for Planning and Management,
William Drayton, Jr., essentially admitted that GAO's criticisms
were valid. In an 8 September 1980 letter to GAO, Drayton
wrote: "In general, we found the report to be a fair summary of
EPA's experience in implementing TSCA," and claimed that the
agency was taking action "to correct most of the problems."[106]
However, there is no reason to believe that EPA will do any
better over the next four years than it did in the previous four.

TSCA, which was never effectively implemented under the
Carter administration, is being effectively dismantled by the
Reagan administration. As the *Chemical Regulation Reporter*
observed in its 15 January, 1982 issue,

...no testing has been conducted under the act.
Monitoring or regulating existing chemicals that might be
hazardous has been unsuccessful, and the agency has no
access to chemical manufacturers' health and safety studies
or records. Outside of the barebones parameters set down
by Congress, TSCA remains a law virtually undeveloped
despite five years of attempted implementation.

The newsletter also reported that in response to a proposal
from the Chemical Manufacturers Association, "EPA officials
have endorsed...exemptions that would eliminate 75 to 80
percent of the chemicals that now must undergo pre-manufac-
ture review."

Describing the anti-regulatory policy of the new administration, John A. Todhunter, EPA's Assistant Administrator for
pesticides and toxic substances, indicated that the major change
in the agency's implementation of TSCA would be the elimination of its "adversarial" relationship with industry.[106b] What

this means, according to a confidential EPA source working on TSCA, is that "you either acquiesce and do nothing, or you become a proponent of industry. This is what we've been doing across the board."

Meat, Cancer And Vegetarianism

The prime source of toxic pesticides and other chemicals for most Americans is in the consumption of foods high in fat content, such as meat and dairy products. A vegetarian diet, or one that minimizes the consumption of animal products, can substantially reduce one's exposure to most of these cancer-causing chemicals.

It has long been known that toxic chemicals in our food, especially in meat, represent a serious threat to human health. In June 1980, the President's Toxic Substances Strategy Committee's report on the relationship between chemicals and cancer stated,

> like smoking, diet is a very important contributor to cancer. According to one estimate, diet is a factor in approximately 25 percent of cancer in males and 45 percent in females in the United States; another study places the estimate at 50 percent for both sexes.[107]

The dangers of lethal chemicals in the American diet were dramatized on 5 April 1973, when the FDA announced that it was banning as a suspected carcinogen the use of Violet No. 1. This artificial food coloring, which was shown to cause cancer in laboratory animals, had been used by the Agriculture Department to stamp meat with the grades "Choice," "Prime" and "U.S. No. 1 USDA." The food coloring had been used for 22 years in a wide variety of foods, drugs and cosmetics (including candy, ice cream, pet food and bakery products). But almost a third of all the Violet No. 1 was used by Agriculture's federal meat inspection program to assure the consumers of the quality of the product they were buying, thereby poisoning them in the process.[108]

There is an extensive collection of literature available on the presence in our food of cancer-causing chemicals deliberately added to preserve, sweeten, color, flavor or otherwise enhance the appeal of such products. While the potential dangers of these additives are substantial, the most widely-documented hazards stem from pesticides and other chemicals which unintentionally contaminate food, and which concentrate and build up in the flesh — especially the fat tissue — of animals and fish.

While pesticide residues are commonly found on virtually all types of food, many of these toxic substances accumulate in the fatty tissues of animals in much greater concentrations than are found on fruits and vegetables. Meat contains approximately fourteen times more pesticides than do plant foods; dairy products 5½ times more.[109]

Thus, by eating foods of animal origin, one ingests greatly concentrated amounts of hazardous chemicals. Analysis of various foods by the FDA shows that meat, poultry, fish, cheese and other dairy products contain levels of these pesticides more often and in greater amounts than other foods. A 1976 survey by the U.S. Department of Agriculture found that beef, chicken, turkey and veal were most likely to have such residues. High-fat dairy products, such as cream, butter, high-fat cheese and ice cream are also frequently heavily contaminated.[110]

In its 1975 annual report, the Council on Environmental Quality wrote: "most of the DDT that people ingest comes from dairy and meat products; in 1973, these two food groups accounted for 95 percent of the daily intake of DDT in the United States."[111] Although DDT levels appear to be declining in many areas, a similar pattern of food contamination can be assumed to exist for many other pesticides closely related to DDT that are still in widespread use. One FDA study, published in 1958, found that the breast milk of mothers who ate meat had double the level of DDT found in comparable non-meat eaters. Another study, published in 1979, found that the level of these pesticides in the breast milk of vegetarian mothers was significantly lower than that in mothers who had conventional diets.

This latter study concluded that,

A vegetarian diet which includes no meat, fish, or poultry
is an important factor in reducing organochlorine
(chlorinated hydrocarbon) contamination of breast milk.
Among vegetarians, those consuming dairy products less
frequently had lower levels of contaminants.[112]

There are dozens of known potential carcinogens and other harmful chemicals present in meat. A report issued in January 1979, by Congress' General Accounting Office (GAO) showed that fourteen percent of the dressed raw meat and poultry available in supermarkets could contain illegal toxic chemical residues:

> Of the 143 drugs and pesticides GAO identified as likely to leave residues in raw poultry and meat, 42 are suspected of causing cancer, 20 of causing birth defects, and 6 of causing mutations.[113]

The report went on to point out that products found to be tainted cannot usually be recalled since slaughterhouses cut up and ship out their meat within 24 hours, and it requires 6 to 25 days to chemically analyze meat samples. This makes it impossible to test the output from slaughterhouses or even to protect the public when contaminated meat is located, the report noted: "Neither the U.S.D.A. nor the F.D.A. can locate and remove from the market raw meat and poultry found to contain illegal residues."[114]

The consumption of animal fat has for many years been strongly associated with high cancer rates. For example, a study released in 1965 by Dr. Ernst Wynder, head of the American Health Foundation, showed a remarkable correlation throughout the world between diets high in animal fat and cancer of the colon. The Japanese, who ate only one-fourth as much fat as Americans, had less than one-fifth the colon cancer rate: 2.5 per 100,000 verses 14 per 100,000 in the United States. Similar patterns prevailed for western countries with a high fat diet, such as Belgium, Canada, Denmark, England, France, the Netherlands and West Germany. Moreover, Wynder noted that when Japanese moved to the United States, the amount of fat in their diets increased, as did their rate of colon and breast cancer. Similarly, the rise in the cancer rate in Japan has paralleled the introduction of hamburgers and other high-fat western food.[115]

Most fish contain numerous toxic synthetic chemicals. When the National Water Quality Laboratory tested fish from Lake Ontario in 1976 for the presence of toxins, it found the following chlorinated organic chemicals present in just one alewife: mirex, DDT, hexachlorobenzene, chlordane and PCB's — all of which are known carcinogens — in addition to DDD, DDE, heptachlorostyrene, nonachlor, octachloronapthalene, octachlorostyrene, pentachloroaniline, pentachlorobenzene, tetrachloroanthracene, tetrachlorobenzene and trichlorobenzene. Had the alewife been tested for other compounds, this list could have been twice as long.[116]

These chemicals enter the water in the form of industrial emissions and discharges, agricultural runoffs, and contaminated dust. Such toxic chemicals as DDT, PCB's and dieldrin accumulate in fish at concentrations thousands of times higher than in the waters the fish inhabit.

From a health standpoint, the least dangerous fish to eat are smaller, deep-ocean fish that do not live or spawn near the coast, such as cod, halibut and pollack, or freshwater fish from high-altitude streams or lakes that are not contaminated from industrial or agricultural runoffs or dumping.[117]

But even these fish will carry some pollutants. Unfortunately, uncontaminated fish and other animal products may simply no longer exist.

The Impending Tragedy

This, then, is the foremost health and environmental crisis of our times. It is perhaps, as EPA has described it, the most serious such problem our country has ever faced: the pervasive presence of dozens upon dozens of toxic chemicals which, at minute doses, cause cancer and/or birth defects in laboratory animals, and which presumably will affect many humans in a similar way. These toxins have been so widely distributed throughout our environment and food chain that most Americans carry detectable residues of them in their bodies. Even if all toxic chemicals were immediately banned, their effects would continue to show up for 20 or 30 years or more, and the environment would remain saturated with many of them for decades to come, in some cases for centuries. Philip Shabecoff, who has covered the environment for *The New York Times* for a decade and is not given to overstatement, says that "unless we can find a way to solve this toxic chemicals problem, our future is going to be very much in doubt."[118]

If any one or a combination of these poisons has a serious health impact on only a small percent of the humans exposed — and it would be astonishing if *many* of these chemicals did not cause such effects — this will have implications for the health of tens of millions of Americans, as well as for unborn generations. In the long run, what may be at stake — and we will not know for sure until it is too late — is the very integrity of the human race.

When the final answers are in, we may well have sealed the fate of a generation of Americans — and perhaps of future ones as well. What we know so far indicates that only the magnitude of the impending tragedy, not whether there will be one, remains to be seen.

What You Can Do

With our knowledge of the causes of cancer, birth defects, miscarriages, and sterility still in a relatively primitive stage, it is difficult to know precisely how an individual can avoid these health disorders.

There are several things a person can do to minimize exposure to these dangers. They include not smoking, not getting X-rayed unless absolutely necessary, drinking bottled or filtered water and maintaining a sensible diet, one that minimizes or avoids "junk foods" and other products with artificial colors and preservatives, fresh water fish and foods heavy in animal fats. They include red meat and dairy products such as high-fat cheese, butter and cream. But many low fat meats, such as chicken, turkey, and veal, also contain high levels of pesticide contamination, as well as antibiotics, drugs, and female hormones intended to speed growth. Toxins tend to concentrate in liver, sweetbreads and other organ meats.

Studies have shown that vegetarian mothers have much lower levels of pesticides and PCB's in breast milk than the general population. Thus, women of child-bearing age should be especially careful about their diets since many toxic chemicals accumulate in the body's fat reservoirs. They are then released in concentrated form in breast milk to an extremely vulnerable infant whose immature liver is incapable of de-toxifying the chemicals.

But much of our daily exposure to toxins is beyond the control of the average person. For relief, we must look to our political leaders who have the responsibility for passing and implementing the laws that supposedly protect us from dangerous chemicals. Enforcement of the various toxic chemical laws already on the books would go a long way toward reducing the American populace's exposure to these substances. Ultimately, the most important action you can take to protect yourself from chemical contamination is to pressure the bureaucratic and political establishment, and persuade our leaders to do what is in their power to safeguard the public.

Neither industry nor government will act unless forced to. *We* must force them.

Hundreds of millions of pounds of pesticides that are banned or restricted for use in the U.S. are exported each year to underdeveloped nations, where they are often carelessly handled and poison people by the hundreds of thousands. Eventually, these dangerous, often cancer-causing toxins
return to us as contaminants on bananas, tomatoes, coffee, tea, sugar, and
other imported food. This photo shows pesticides being mixed by hand in Sri
Lanka (Ceylon), a common procedure that often results in the chemical being
absorbed through the skin. *Photo courtesy of Michael Scott/OXFAM America.
Medicine cabinet: Photo courtesy of Michael Scott/OXFAM America.*

Part Three

Chemicals That Might Kill Us: Case Studies Of Major Toxics Scandals

A plane spewing pesticides sprays over a pick-up truck on a public road in Kay County, Oklahoma, 1979. Pesticide poisoning is a major problem to residents of many rural and suburban areas. *Photo courtesy of Larry Miller/Caldwell, Kansas.*

Chapter Eight

Tris: Killing Half A Million Children?

Considering the estimated 45 million children now exposed to Tris-treated sleepwear, a 1% expected cancer induction would ... translate to 450,000 cancer deaths.
Dr. Robert H. Harris, Environmental Defense Fund, December 1976[1]

In 1975, it was learned that tens of millions of American children were wearing nightclothes treated with a substance that was a powerful gene-altering and cancer-causing agent. Yet it was another year and a half before the government acted to ban the chemical, known as Tris. By then, some 60 million children had been exposed to it, and it is estimated that hundreds of thousands of children, perhaps millions, may contract cancer as a result of exposure to Tris.[2] There is no incident in modern American history more indicative of the incompetence of the bureaucracy, and its subservience to an unconscionable industry, than the Tris affair. The tragedy culminated with millions of the belatedly-banned garments being shipped abroad to poison children overseas, despite a personal plea by Congressmen to President Carter to prevent this. Not surprisingly, because the garments were never seized and destroyed, they are continuing to turn up for sale in retail stores throughout this country.

How 60 Million Children Were Exposed

In the early 1970's, the U.S. government issued regulations requiring that sleeping garments for children be made largely fire-proof. In response, manufacturers began adding the flame-resistant chemical Tris to children's sleepwear. Ironically, in trying to protect children from one danger — flammable night-wear — the government inadvertently caused some 50 to 60 million children[3] to be exposed to a substance which it is now feared may cause up to 70,000 deaths a year.

Known chemically as tris (2,3-dibromopropyl) phosphate, Tris was never tested for safety before being marketed, even though, like many pesticides, it is an organic phosphate, the same group of substances developed during the second world war for use as chemical warfare agents. These chemicals are absorbed through the skin and attack the central nervous system. One does not have to eat or breath them to be affected; exposure to the skin is often sufficient to cause death or illness.

Thus, children in sleepwear impregnated with Tris were virtually certain to absorb it through the skin; combined with mouthing, chewing, or sucking on their pajamas, infants, over the course of their childhood, could accumulate large amounts of this toxic chemical. Studies have demonstrated that children who had worn garments treated with Tris — whether washed or unwashed — had measurable amounts of a breakdown product of the carcinogen in their blood and urine. [4, 5]

By October 1975, the Consumer Product Safety Commission (CPSC) had received the results of studies showing Tris to be a potent mutagen (capable of altering genes), which usually indicates a potential to cause cancer as well. But CPSC did not make this information public or even notify the clothing manufacturers. In January of the following year, these findings had been confirmed by other researchers; and one scientist, determined to publicize Tris' potency, placed sleepwear treated with Tris in a bowl of goldfish. By the next day, all the fish had died. Tris was found to be lethal to goldfish at levels of one part per million.[6]

The chemical companies manufacturing Tris, including Velsicol, Dow, Stauffer, Tenneco and Great Lakes, also did not act on the information available on Tris hazards. Michigan Chemical Corporation — which was taken over by Velsicol —

had been producing Tris since 1958 for such products as cast acrylic sheets and polyurethane and polystyrene foam, the latter of which is used for the interior of automobiles, insulation in building construction, wire coatings and plastic resins. Other uses for Tris included wigs, draperies, stuffed animals, dolls, doll clothing, toys, canvas camping equipment, aprons used in hospitals, and numerous others. In fact, less than half the U.S. consumption of Tris was for children's nightwear.

Velsicol learned in October or November of 1975 that Tris had been shown to be mutagenic in the Ames test, and the finding was confirmed by another laboratory in early 1976. But the firm did not withdraw Tris from the sleepwear market until 3 February 1977, over a year later. The other manufacturers, who also presumably knew or should have known of the determination of Tris' mutagenicity (as did the CPSC), also failed to notify clothing manufacturers. They waited until February 1977 to withdraw the chemical from the apparel market, after testing by the National Cancer Institute (NCI) showed Tris to be a potent carcinogen.[7]

It should also be pointed out that although CPSC knew that Tris was a mutagen and a probable carcinogen a year and a half before issuing its ban, the commission made no apparent efforts to get industry to eliminate Tris or find an alternative, nor did it even bother to inform the children's sleepwear manufacturers of Tris' toxicity.[8]

When it became clear that the government had no intention of taking action on Tris, the Environmental Defense Fund (EDF) on 24 March 1976, submitted a legal petition to CPSC requesting that use of the chemical be restricted. Compiled by attorney Robert Rauch, and Drs. Joseph Highland and Robert Harris, the brief warned, to no avail, that the health of a generation of children might depend on the government's taking quick action on Tris. By this time, it is estimated that 45 million children were wearing Tris-treated garments. Tris was being used in some 40 to 60 percent of all children's nightwear, and an estimated 240 million pieces of such clothing were on the market.[9]

By the end of 1976, the results of additional research were released showing that Tris was mutagenic, causing sterility and genital abnormalities when it was applied to the skin of male rabbits. Again, CPSC refused to act. Then, in February 1977, the NCI released new scientific studies showing that when Tris was given to laboratory animals, even in low doses, it produced in two separate species an extraordinary number of tumors of the kidney, lungs, liver and stomach. In its petition, EDF cited the fact that "data obtained from the National Cancer Institute indicated that Tris . . . was 100 times more powerful as a cancer-causing agent than the carcinogens in cigarette smoke."[10]

The Ban On Tris: Too Little Too Late

Finally, on 7 April 1977, two and a half years after first learning of the chemical's hazards, the CPSC agreed to act on EDF's petition and to prohibit the manufacture and sale of Tris-treated garments. At the time of the ban, CPSC estimated that 120 million Tris-treated garments might be in consumers' hands, and that another 16 to 18 million were available for sale, in addition to 7 million square yards of Tris-treated fabrics in the hands of garment producers. Between 1972 and 1977, some 50 to 60 million children are estimated to have worn such garments and been exposed to Tris. By mid-1977, over 40 percent of all children's sleepwear being sold was Tris-treated.[11]

But incredibly, the CPSC "ban" was worded in such a way as to include only *unwashed* children's clothes. Dr. Robert Harris called this distinction "unwise" in light of the fact that "after repeated washing, Tris remained in the fabric and can be absorbed through the skin, and after 50 washings can cause a rash on sensitive individuals, indicating that it continues through the life of the garment." Nor did the ban include Tris-treated fabrics or yarn, except for those intended for sale to consumers. CPSC also failed to ban the chemical Tris as a hazardous substance, allowing it to continue to be produced and sold for adult clothing and other uses other than children's nightwear.[12]

Despite the then widely-accepted evidence of Tris' toxicity, the garment industry was determined to dispose of its stocks of children's nightwear, and went to court to overturn the CPSC ban. In May 1977, a Fort Mills, South Carolina textile firm, Spring Mills, represented by attorney Emmett Bondurant, then of the prestigious Atlanta law firm of Kilpatrick and Cody, filed suit against CPSC. On 23 May, the judge in the case, Robert F. Chapman, struck down CPSC's ban.

Some people had serious doubts about the judge's impartiality in ruling for the textile industry on a matter directly affecting its livelihood, since his family has been in that business for three generations. His two brothers ran Inman Textile Mills, founded by his grandfather, and the judge owned stock in the firm. Although it did not manufacture Tris-treated garments, the company was at the time a member of the American Textile Manufacturers Institute (ATMI), which had taken a strong stand in court and before Congress against banning the Tris nightwear. (ATMI's representative at the Congressional hearing was Dan Byrd, vice president and general counsel of Spring Mills.) The judge's views on the testing of potentially harmful chemicals, such as Tris, were colorfully expressed to columnist Jack Anderson, who reported him to exclaim, "I never heard of Tris until ten days ago. Some rat died somewhere, and the whole world goes kerplunk."[13]

Fortunately, after the ban was successfully challenged in court, CPSC continued to keep additional contaminated sleepwear off the market by filing civil actions under the Federal Hazardous Substances Act.

Still, the lawsuit inspired some enemies of the environmental movement as a blow against such "mortal enemies of progress." An editorial in the 6 June, 1977 issue of *Barron's* magazine by Robert M. Bleiberg "hailed" and "saluted" the action:

> Emboldened by their easy victories to date, environ-
> mentalists and other mortal enemies of progress are
> readying hit lists of chemicals used in textile finishing,
> upon which, sooner or later, they also seek to clamp a ban.
> If they succeed ... permanent press and other marvels of
> technology will go the way of Tris; what the well-dressed
> man (and woman) will wind up wearing is sack cloth.

850,000 Children May Get Cancer

The further Tris was studied, the deadlier it was shown to be. Studies commissioned by the CPSC came to the conclusion that from 6,000 to over 50,000 cancer deaths *per year* could be expected for every million children exposed to Tris for one year.

Dr. Bruce Ames, a cancer expert from the University of California at Berkeley and a member of the President's National Cancer Advisory Board, was the source of the higher estimate, and found Tris to be *more carcinogenic than cigarettes!* If this proves correct, Tris may end up killing as many people *in one year* from cancer as cigarettes do — over 100,000.[14]

These estimates, however, may be conservative. According to Dr. Robert Harris (then of EDF; from 1979 to 1981 he served as one of three members of the President's Council on Environmental Quality):

> ...in the Ames test, Tris is a more potent mutagen than benzidine, which has been observed to induce a high instance of bladder cancer in those occupationally exposed. Even at comparable exposures to the lower estimated exposure for Tris, the resultant cancer instance from exposure to benzidine would be expected to be approximately 1%. Considering the estimated 45 million children now exposed to Tris-treated sleepwear, a 1% expected cancer induction...would translate to 450,000 cancer deaths.[15]

In its 1 June 1977 *Federal Register* notice announcing its expanded ban on Tris-treated children's garments, CPSC cites as a basic for its action four different "risk assessments" based on the NCI data for the tens of millions of children who were exposed to Tris-treated nightwear. The highest figure cited, if one assumes that 50 million children were exposed for only one year or more between 1972-77, projects that Tris would cause 850,000 children to develop cancer.[16]

Since many cancer-causing chemicals have been shown in animal tests to be more potent in infants than in adults, and young children are highly susceptible to certain chemicals,[17] the incidence of cancer caused by Tris may be still far greater than anyone can predict.

Tris may also have a profound impact on generations of unborn children. Dr. Bruce Ames found that Tris is capable of causing damage to DNA synthesis in human cells, and that the mutagen is "likely to cause genetic birth defects in the offspring of children exposed to it."[18] Since the scrotum is especially absorbent and presumably "sponged up" an extraordinary amount of Tris, Ames fears that many of the boys who were exposed to it may not only become sterile and have atrophied testicles, but may also incur a high level of damage to their sperm cells.[19]

Refusing To Ban Tris Exports

When the Tris ban went into effect, manufacturers were faced with a major economic problem: what to do with tens of millions of these garments that could not legally be sold in the United States. Realizing the probability that the banned Tris products would be sent abroad, The Center for Law and Social Policy twice petitioned the CPSC to ban Tris exports, with attorney Leonard C. Meeker laying out in detail the legal justification and necessity for such action.

Ironically, the actions of the CPSC and other government agencies made it possible, and even encouraged, the shipment abroad of this nightwear to foreign purchasers. Had not human lives been at stake, the bungling and buck-passing by the bureaucracy, resembling a Marx brothers comedy, would have been humorous.

In June of 1977, the Chairman of the CPSC wrote to Secretary of Commerce Juanita Kreps requesting that her Department take action to control and provide information on such exports, under the Export Administration Act. However, on the advice of the State Department, Commerce refused to take such action after being advised that "controls on Tris-treated garments were not necessary to further significantly the foreign policy of the United States."[20]

And in a classic "Catch 22" situation, the State Department took the position that it could not bilaterally notify any foreign governments of the Tris ban "because we had not received the CPSC's request to do so."[21] (So much for the administration's "human rights" policy!) On 20 October 1977, the Commission itself announced that it had interpreted the inconclusive language of the Federal Hazardous Substances Act to mean that it could not control Tris exports.

On 8 February 1978, the Chairman and a senior member of the House Subcommittee on Commerce, Consumer and Monetary Affairs — Congressmen Benjamin Rosenthal (D-NY) and Henry Waxman (D-Calif) — wrote to President Carter appealing to him to have the administration take action to block Tris exports while there was still time. They informed him of the bureaucratic buck-passing that had taken place between CPSC, Commerce and State. Noting that the CPSC was "paving the way for the export of garments that will cause cancer in the foreign children who wear them," the Congressmen "urgently requested . . . that all means be exhausted to discourage those who would export it to foreign markets at the expense of innocent children . . . We need your help to stop it."

On 26 May, over 2-½ months later, a reply was sent by the White House, signed by the President's Special Assistant for Consumer Affairs, Esther Peterson. She promised no action to halt the Tris exports, but instead gave the classic bureaucratic response to such a situation: she announced that a committee had been formed ("we have convened an *ad hoc* interagency working group") to study the problem![22]

"An Export Trade In Death"

With the White House, the CPSC and other agencies unwilling to take action to prevent the export of Tris products, U.S. companies were eagerly unloading their supplies of the cancer-causing nightwear.

This initial erroneous decision by CPSC was later compounded by a subsequent Commission announcement that helped speed the rush to send the banned nightwear abroad. As the House Government Operations Committee report described the scenario:

> In April 1978, when evidence of Tris exports began mounting, the Commission announced that it would probably reverse its October 10, 1977 export decision, thereby exacerbating the rush to export before the Commission acted. When the Commission finally reversed its interpretation of the Act on May 5, 1978, would-be exporters of Tris had had over a year to sell their inventories abroad.

The Committee also points out that an unknown number of such garments may also have been reimported into the United States.[23]

CPSC's April '78 announcement could hardly have come at a worse time. On 1 May 1978, news reports indicated that hundreds of thousands of Tris-treated garments had already been exported, and that negotiations were underway for the export of millions of dollars worth of additional sleepwear, numbering over 5 million pieces. On that date, an article in *Women's Wear Daily* by Mark Hosenball reported:

> American apparel men are rushing to conclude export
> deals for millions of dollars worth of Tris sleepwear, as
> U.S. government moves to ban overseas sales of the
> controversial articles gain momentum in Washington.
> Hundreds of thousands of dollars worth of children's
> sleepwear treated with Tris...have already been exported
> by American manufacturers.[24]

Thus, CPSC's announced intention to ban such exports had the result of accelerating them.[25] The Commission's actions allowed and encouraged U.S. manufacturers and brokers to engage in a massive way in what columnist James Brody, in *Advertising Age*, called "An Export Trade in Death."[26]

Tris Exports Finally Banned

By mid-1978, a change in membership at CPSC made it possible to overrule the previous export policy.

When CPSC finally did vote in June 1978 to invoke the export ban, Chairman John Byington and Commissioner Barbara Franklin continued to oppose such action and formally dissented, claiming that the Commission did not have such authority.[27] Although they were outvoted, they could take comfort from the fact that their adamant opposition had been effective in holding up action long enough to permit industry to unload much of its inventory abroad.

During the one year period in which Tris garments were banned for sale in the United States but not for export, millions of pairs of pajamas are thought to have been shipped abroad and sold there. According to Mark Hosenball, "experts in the trade" estimate that tens of millions of pairs of these lethal pajamas were shipped abroad. The government's Inter-Agency Working Group on Hazardous Substances Export Policy puts the figure at "approximately 2.4 million pieces valued at $1.2 million."[28] Whatever the actual amount, the number of cancer deaths caused by these exports can be expected to number in the tens or hundreds of thousands.

Have We Learned Any Lessons?

Tris-treated material may also continue to present a risk to the U.S. public, apart from the possibility of re-labeled garments being reimported from abroad. Leftover stocks may have found their way into circulation and could be exposing unsuspecting workers and consumers. Approximately 20 million Tris-treated garments were thought to be in storage in the United States in 1979, and their ultimate use and destination remain unclear.[29] [30] As of the end of 1980, CPSC estimated that about half of the garments and other goods had been disposed of by dumping in secure landfills, burning in hazardous waste incinerators and by being recycled for seat covers and cut up and used for industrial wiping cloths. One CPSC official admitted that sale as cloths presented "some potential hazard," but that such use was permitted because the agency had made no assessment or determination of these risks.[31]

Four years after CPSC's ban, Tris-treated clothes were still turning up in retail stores and being shipped abroad. In February 1981, CPSC announced that some 100,000 Tris-treated children's sleepwear articles had been offered for sale in and around Philadelphia, Pennsylvania and nine southeastern states since August of 1980. Another 75,000 garments also turned up in North Carolina, and additional sales were discovered to have taken place in New York City, Atlanta, Detroit, Chicago, Indianapolis, Fort Lauderdale and several other states.

The Commission filed a complaint against a Philadelphia industrial rag firm, The A & B Wiper Supply Co., charging that it "purchased more than 700,000 Tris-treated children's sleepwear garments to be cut into rags, but then sold an estimated 210,000 of the garments to other distributors without notifying the Commission..." CPSC also charged the Crown-Tex Co., a New York City manufacturer and distributor of women's and children's clothes, with sending 18,000 yards of Tris-treated fabric to an exporter who shipped the material to Ghana to be made into clothes.[32]

In March 1981, CPSC announced that it would seek fines of half a million dollars against each of three firms it accused of sneaking Tris garments into stores because the clothes were too expensive to dispose of or to store. "In all three cases, they knew what they were doing," CPSC stated in its complaint lodged against John R. Lyman Co., Inc., of Chicopee, Massachusetts; S. Schwab Co. Inc. of Cumberland, Maryland and Hollywood Needlecraft Inc., of Los Angeles. CPSC charged Hollywood with distributing sleepwear in packages labeled "no Tris."[33]

CPSC has warned consumers to avoid purchasing clothes without labels or those made from acetate or tri-acetate, "since such garments were manufactured before 1978 when such fabrics almost always were treated with Tris." "Buyers beware" is good advice, since such incidents are bound to continue to occur. As the CPSC pointed out in a February 1981 news release, some 2.3 million Tris-treated garments have been cut into industrial wiping rags, and "about 6.5 million Tris-treated children's sleepwear garments remain warehoused across the country."[34]

Also in early 1981, Dr. Ralph C. Dougherty, of Florida State University (FSU), revealed that a form of the chemical Tris, tris (1,3-dichloro-2-propyl)phosphate, had been found to be present in the semen of 28 percent of the students tested at FSU. This compound, known commercially as Fyrol FR-2, has not been banned even though it is known to be capable of altering genes. It has been used as a flame retardant in dacron fabric, polyurethane foam and in making automobile seat cushions and other consumer products.[35] Ironically, manufacturers switched over to this chemical as a substitute in making children's clothes fire resistant after the other form of Tris was banned.

Meanwhile, the Reagan administration has slashed the budget of CPSC, making it impossible for the agency to carry out its mission. In May 1981, the administration announced that it intended to abolish CPSC entirely (see pp. 376-77).[36]

What You Can Do

The Tris tragedy might never have occured if chemicals were required to be studied and cleared for safety on the basis of quick and inexpensive laboratory tests such as those utilizing bacteria cultures and *in vitro* systems (cell or tissue cultures in test tubes or other containers). Perhaps the best is the Ames Test which detects mutations in a colony of bacteria (as it did early on with Tris) and is 90 percent reliable in predicting whether a chemical is a carcinogen.

With millions of Tris-treated garments potentially still available for sale, and a significant percentage of new children's sleepwear garments treated with possibly dangerous fire-retardant chemicals, consumers must still exercise caution in purchasing children's nightwear. In addition to the information given in the last part of this chapter, it should be noted that many garments are not chemically treated at all. According to the CPSC, the following types of fabrics (including their blends) are inherently fire-resistant and do not generally require chemical additives:

> Modacrylic (brand names SEF, Kanecaron, Verel)
>
> Vinal or Vinyon (brand name leavil)
>
> Matrix (brand name Cordelan)

In addition, some 100 percent polyester fabrics are not treated with flame-retardant chemicals, and may be so indicated on the label. Cotton and nylon fabrics are generally treated with a chemical (other than Tris) to make them flame-retardant.[37]

The Tris affair was not an isolated case, but was one of many such situations that have taken place over which the CPSC has the power to act. With the agency undergoing drastic and crippling budget and personnel cuts, it is probable that many similar situations will occur, and perhaps go undetected, in the years ahead. The best way of preventing such problems is to have effective health and safety agencies committed to protecting the public. As repetitions of the Tris tragedy recur, perhaps public pressure will force such a policy to be implemented.

Chapter Nine

PCB's: Present In Every Living Creature?

PCB's are among the most widespread — and most poisonous — chemicals ever produced. Virtually the entire U.S. population has been exposed to them, and all Americans appear to have accumulated significant quantities in their bodies. A family of petro-chemicals, PCB's are found in almost all freshwater fish, as well as in other food, water and even the air; they are present in the body tissue, sperm, and mothers' milk of the entire populace. Since the tiniest doses — a few parts per million or even billion — have caused birth defects, loss of hair and death in test animals, the presence virtually everywhere of these chemicals has grave implications for the American people, and for much of the world's population.

The Toxicity of PCB's

It is impossible to know just how toxic PCB's may be to humans, since scientists cannot find a level of exposure in test animals that does not produce adverse health effects. Researchers feeding PCB's to primates have been unable to find a safe level, and indications of carcinogenic and mutagenic effects have occurred at doses as low as one-half part per million, and even 25 parts per billion. (It is estimated that PCB levels in the fatty tissue of humans are on the order of one to ten parts per million. Half the American people carry levels of 1 part per million or more.)[1]

Dr. James R. Allen, who has conducted extensive research on these chemicals, concluded that "There's no question that

PCB's are a carcinogenic agent. And without doubt, high levels are transmitted through mothers' milk."[2]

Similar adverse effects have been observed in humans who have been exposed to high concentrations of PCB's. In 1968, thousands of residents of the Japanese island of Kyusha ate rice oil into which PCB's from a factory pipe had leaked. They developed what came to be called *yusho* (rice oil) disease, the symptoms of which included: loss of memory, hair and sex drive; extreme fatigue; a "cheeselike" discharge from the eye; sore throats, coughs, headaches, stomachaches, and nausea; dizziness; bone and joint deformities; jaundice, impotence, and lack of development of teeth in children; temporary loss of vision; increased incidence of respiratory infection; swelling of eyelids; numbness, menstrual difficulties and a darkening of the skin, nails, and around the mouth and eyes.[3]

According to EPA, a study of deaths among the "yusho" victims shows an abnormally high rate of cancer, especially of the liver and stomach, as well as indications that PCB's may cause skin cancer and cancer of the pancreas.[4] In August 1976, the *New England Journal of Medicine* published a study giving further evidence of the carcinogenicity of PCB's and indicating that workers exposed to high levels of PCB's are much more likely to contract cancer than the average person.[5]

Unborn fetuses on Kyushu were also affected. Nine out of ten of the babies live-born to pregnant women who had ingested the rice oil had abnormally dark, grayish skin, and most were much smaller than normal. Similar symptoms appeared in the baby rhesus monkeys whose mothers were fed PCB's in amounts of only 2.5 and 5 parts per million: retarded growth, learning disabilities, skin lesions, hyperactivity and swollen eyelids. Interestingly, the Japanese babies received a *smaller* dose of PCB's than does the average American baby, although the yusho PCB's were contaminated with a dioxin (TCDD)—like chemical known as dibenzofurans.[6]

Well over a decade after consuming the contaminated rice oil, more than 100 of the victims are reported to still suffer from many of the original symptoms, including chronic headaches, fatigue, numbness in their limbs and inability to gain weight.[7]

As if PCB's were not themselves dangerous enough, federal researchers recently discovered that they have spawned another obscure but extremely dangerous chemical that is thought to be up to 1,000 times more toxic than PCB's. Like PCB's, these compounds, polychlorinated dibenzofurans (PCDF's), have also become widespread, and are thought to be capable of causing cancer and birth defects.

It is unclear how widespread these substances have become, since they can only be detected with highly sophisticated equipment that has just recently been developed. But where researchers have looked for them, they have often been present. PCDF's have turned up in smokestack emissions from power plants in St. Louis and Cincinnati, and are presumably present in other discharges from plants, waste disposal sites, and numerous facilities that burn trash containing PCB's. Fish contaminated with the compounds have been found in Michigan lakes and major rivers throughout the country, including trout, salmon, carp, and catfish in the Ohio, Hudson, and Connecticut Rivers, and in Lake Michigan.[8]

Toxicity Known 5 Decades Ago

Manufactured by Monsanto, PCB's have been known for their extremely toxic qualities since production of them began — around 1930. By October of 1933, 23 out of 24 workers in the plant manufacturing the chemical had contracted serious cases of chloracne, with disfiguring skin eruptions on their faces and bodies.

Ten years later, in August 1943, the New York State Department of Labor published a report on outbreaks of cloracne, dermatitis and deaths due to liver damage among workers handling equipment containing PCB's and chemically similar compounds. One of the conclusions reached by the study was that these closely-related chemicals were "in general highly toxic compounds and must be used with extreme care. Industrial hygienists should make every effort to see that such exposures are controlled, in so far as humanly possible."[9]

Thus, both Monsanto and the government were aware of the extremely dangerous nature of these chemicals four or five decades ago, but nothing was done to restrict their use until the late 1970's. Both industry and government forfeited an opportunity to prevent our present environmental crisis with PCB's, at untold cost in lives lost, people's health ruined, and the environment poisoned. Some forty years — and 1.5 billion pounds of PCB's — later, our country is faced with a toxic chemical contamination problem of immense proportions that may, in the end, prove insoluble.

PCB's In Mothers' Milk

PCB's are among the most stable substances known, which makes them useful to industry, but extremely dangerous to have released in the environment: once there, they stay. When PCB's enter the environment and the food chain, they do not degrade but remain for decades, even centuries.

Since the 1930's, some one-and-a-half billion pounds of PCB's have been manufactured, mainly for use as high-grade fluids in electric transformers, where their extreme resistance to heat and fire made them ideal for these and other purposes. As a result of their stability and widespread use, PCB's have become widely dispersed throughout the environment. They are found in virtually every river in America, and are even present in the tissue of Arctic polar bears at levels up to eight parts per million.[10] They are found in the Pacific islands, and in the snows of the Arctic and the Antarctic. Incredibly, these compounds may be present to some degree in every living creature!

As Dr. George Harvey, of the Woods Hole Oceanographic Institution, wrote in 1975:

> PCB's have been found in all organisms analyzed from the North and South Atlantic Oceans, even in animals living under 11,000 feet of water. Based on all available data, it seems safe to conclude that *PCB's are present in varying concentrations in every species of wildlife on earch.* (emphasis added)[11]

PCB's now occur in the milk of nursing mothers around the globe. A 1975-76 EPA study found that the average American infant being breast fed was ingesting ten times the maximum daily intake level recommended by the FDA. The study found detectable levels of PCB's in virtually all of the samples tested — over 99 percent contained the chemical! Only 9 out of 1,038 samples collected and analyzed in 1975 contained no traces of PCB's.[12] In February, 1981, another study was released showing that PCB's had been found in the breast milk of *all* 1,057 Michigan mothers tested, with half of the women having levels near or above the federal limit for cow's milk![13] The median level found was 7 times higher than the FDA permits in baby food![14]

Because of a baby's small size and dependence on breastmilk as the sole or major source of nourishment, the average nursing infant receives 100 times more PCB's on a body-weight basis than an adult.[15] The amount of PCB's found in the average mothers' milk — 1.8 parts per million — is seven times what the FDA allows in cow's milk.[16] It would thus be illegal to commercially market mothers' milk in interstate commerce.

According to Dr. Joseph Highland, Chairman of the Environmental Defense Fund's Toxic Chemicals Program, "in the first year of life, the averaging nursing infant whose mother's milk is contaminated with PCB's will ingest what you or I would ingest in a decade of life."[17]

The June 1980 report published by the Senate Committee on Environment and Public Works stressed the dangers PCB's pose to future generations:

> Second generations are believed at higher risk since fetuses and nursing infants may consume more PCB's per unit of body weight than adults. PCB's ingested by human mothers are found, and to some extent concentrated, in breast milk...

> Dogs and sows exposed to PCB's *in utero* had a significant rate of fetal resorption and a variety of birth defects. Nursing infants are apparently at greater risk than their mothers when the mothers are exposed to PCB's, since the PCB's biomagnify in the milk. Rhesus monkeys fed PCB's averaged PCB levels 20 times higher in maternal milk than in maternal blood. Their nursing infants' blood levels of PCB's average 2 to 3 times that of the maternal blood levels.[18]

Some scientists believe that difficult-to-detect damage may already be showing up in the form of babies with lower IQ's or body weight, or who are hyperactive.[19]

EPA's National Human Monitoring Program found detectable levels of PCB's in the fat tissue of nearly all those sampled, and the agency has estimated that over 90 percent of all Americans have PCB's in their flesh. By now, the figure may be closer to 100 percent. Both the numbers of people and the levels of contamination have continued to increase in recent years.[20] EPA's program of monitoring samples of human fat tissue for

PCB's has documented the extraordinary increase from 1973 to 1976 in the number of Americans with PCB's in their bodies. The percentage of people with no detectable level of the chemicals fell sharply during this period: 1973 — 21.4 percent; 1974 — 9.3 percent; 1975 — 5.6 percent; 1976 — 1.9 percent.[21]

Citizens of some other countries such as Israel, Austria and Germany, have levels several times those found in the United States.[22]

There are indications that the presence of PCB's and other toxins in our environment may be inhibiting the ability of humans to reproduce. A study done in 1929 found that half of the males tested had 90 million sperm or more per milliliter of fluid. But in a 1978 study conducted by Dr. Ralph Dougherty, a professor of chemistry at Florida State University (FSU), the median sperm density had dropped by one-third, to 60 million (see page 257). Dougherty attributed about one-fourth of the decline to PCB's in the samples.[23]

A recent government study found PCB's present in 100 percent of the human sperm samples tested, with a correlation between high PCB levels and low sperm count. The increasing prevalence of PCB's in human tissue has paralleled studies showing that the average sperm count of American males is down 30 percent over the last three decades.[24]

PCB's In Food Fish

A major source of PCB contamination in humans is the eating of fish from lakes and rivers into which these substances have been discharged.[25]

Fish that migrate through, spawn, or live in waters containing PCB's quickly absorb the chemical, and it has been estimated by EPA that fish can accumulate as much as 9 *million* times the amount of PCB's in the water. One species of minnow bioconcentrates PCB's at 230,000 times the level in the water. Birds and mammals that eat fish concentrate PCB's at even higher levels than the fish they consume. A 1974 survey by the U.S. Fish and Wildlife Service found PCB's to be present in 90.2 percent of freshwater fish throughout the United States.[26]

In September 1981, the Michigan Department of Public Health warned the public against eating trout or salmon from the Great Lakes (except as skinless filets) because of PCB contamination. A month later, New York's Department of Environmental Conservation reaffirmed its ban on commercial fishing for striped bass in the Hudson River for the same reason. The Department also released a study showing dangerously high levels of PCB's in ducks and other waterfowl in the state. The State Health Department recommended against consuming too many ducks during the hunting season, and warned that mergansers should not be eaten at all.[27]

General Electric: Pollution Is Our Most Important Product

A classic case of PCB pollution is that of New York's legendary Hudson River, into which 1-½ million pounds of PCB's were dumped, leaked, or spilled for over two decades. Most of the contamination was caused by two General Electric plants, which acted in flagrant violation of state water pollution laws, although the last eight months of the dumping was conducted with a PCB discharge permit from EPA.[28]

As a result of a suit brought against GE by the New York State Department of Environmental Conservation and the Natural Resources Defense Council, the company was forced to admit that 65 of its employees in two plants using PCB's had become ill possibly because of exposure to these chemicals. Symptoms included asthmatic bronchitis, dizziness, nose and eye irritation, nausea, dermatitis and acne. A 1976 study of workers at the two plants found that 45 percent had skin abnormalities, and 40 percent had ear and eye problems. Other studies show a sustantially greater incidence of colon and rectal cancer among residents of Poughkeepsie, New York, which draws its drinking water from the Hudson.[29]

One rock bass taken downstream from a GE discharge contained what was then the highest level of PCB's of any fish ever tested — 340 parts per million, which was 70 times above the allowable level established by the FDA.[30] One eel later taken from the river contained 559.25 parts per million; if consumed, it would give an adult over half the allowable *lifetime* safety limit set by FDA, and a child would ingest double the limit.[31]

GE was eventually forced to pay a fine of $4 million, even though the cleanup costs are estimated at 50 times that amount — in the hundreds of millions of dollars.[32]

In the meantime, the PCB contamination continues to pose a serious health threat to the over 150,000 upstate residents who use the river for drinking water, as well as to the existing commercial fisheries in the Hudson's estuary.[33] Although some 300,000 pounds of PCB's have been dredged out of the river, twice that amount still remains, much of it not readily recoverable.[34] PCB's have thus contaminated this historic river and its fish for the foreseeable future, perhaps forever.

General Electric has been responsible for polluting and damaging other rivers and lakes throughout the country. From 1954 to 1978, a GE facility in Rome, Georgia allowed PCB's to contaminate the famed Coosa River (which flows from Georgia into Alabama) and Weiss Lake, a reservoir in Alabama created by the Coosa. The lake supplies water to communities of several thousand people and supports a $10 million a year recreation industry (boating, swimming, water skiing, fishing) which attracts millions of visitors. State and federal tests conducted in 1976 found PCB levels in fish to be up to twenty-five times higher than FDA tolerance levels.[35]

Banning PCB's: Again, Too Little Too Late

In 1972, Monsanto, the sole U.S. producer of PCB's, agreed to limit its sale for use only in closed-system transformers and capacitors. But the PCB problem continued to mount, and EPA in 1977 issued further restrictions on the chemicals, essentially banning new uses of them as it was required to do under the Toxic Substances Control Act (TSCA). However, this action was largely nullified when the agency granted "hardship exemptions" to those responsible for 99.3 percent of the 758 million pounds of PCB's in use before the "ban."[36] The Environmental Defense Fund later went to court to oppose EPA's blanket exemption of almost all commercial uses from its "ban."[37]

In April 1979, EPA finally got around to banning the manufacture of PCB's, but the regulations for disposal were drafted to apply only to about 60 percent of the estimated 758 million pounds still in use.[38]

(Although the U.S. and western Europe have generally ceased production of PCB's, they are still manufactured in the Soviet bloc, specifically in Czechoslovakia and East Germany.)[39]

Over A Billion Pounds To Be Disposed

PCB's are present in all fluorescent lights and have been used in such products as television sets, motors, drills, machines, air conditioners, carbonless copy paper, paints, ironing board covers, cereal boxes and as sealants in grain storage silos.[40]

EPA estimates that each year some 10 million pounds of PCB's escape into the environment, mainly through dumping, vaporization, spills and leaks.[41] More than 20 tons of PCB's pour into the Great Lakes each year, mostly from the air; the influx is expected to increase over the next few years. A major source of this pollution are municipal incinerators burning electrical capacitors and other items containing PCB's, which are then released into the atmosphere and fall to earth with precipitation.[42]

Besides the 758 million pounds of PCB's currently in use, EPA estimates that 290 million pounds of PCB's are now in dumps and landfills, and 150 million pounds are in the water, soil, sediment, and air.[43] The agency also says that there are over 30 million electric transformers full of PCB's on utility poles scattered throughout the nation.[44]

The ubiquity of these industrial PCB's poses a serious threat to the public, as was demonstrated by two recent near-tragedies. On 5 February 1981, during a fire and explosion in an 18-story office building in Binghamton, New York, an electrical transformer containing 120 gallons of PCB's ruptured, causing the ventilation system to spread the chemicals and fine ashes laden with it throughout the building. Estimates show that it may take several years and millions of dollars to clean up the contamination. Hundreds of people, including policemen and firemen, were exposed to the toxic pollution. But New York Governor Hugh Carey claimed that the dangers had been exaggerated, and offered to "walk into any part of that building and drink a glass of PCB's." Fortunately for the Governor, he never did so.[45]

A few months later, in August, 30,000 people had to be evacuated from 19 buildings in San Francisco's financial district after a busted pipeline sent PCB vapors "roaring into the air like a geyser." State officials warned people who had been hit by the spray to throw away their clothes and shoes.[46]

Most ominous of all, EPA projects that as PCB's escape from dump sites into the ground, the water and the air, the levels of these poisons in the environment will more than triple, increasing from 150 million pounds to 540 million![47]

Dumping PCB's

The illegal disposal of PCB-contaminated wastes continues to be a problem of significant if undetermined proportions, since most of this dumping goes undetected. The few major instances that have come to light have proven to be disastrous for the local residents and authorities.

One night in August 1978, for example, a tank truck driver simply opened the vehicle's discharge pipe and dumped some 31,000 gallons of waste oil containing PCB's along 210 miles of North Carolina roadside. Afterwards, residents of the area noticed an extremely strong and unpleasant odor in the air, and many became ill, complaining of burning eyes, headaches, sore throats and nausea. State officials delivered four thousand letters to people living along the highway, warning them not to market or eat vegetables or livestock raised within 100 yards of the contamination. Some farmers burned their crops, and others stopped using their wells.[48]

There is now suspicion that the incident may have contributed to the apparent increase in birth defects that physicians have notice among residents of the contaminated area. One local doctor, Brenda Armstrong, said in October 1980 that there had been "a significant upsurge in congenital defects in the last twelve months."[49]

Eventually, a man named John Burns and his two sons were arrested and charged with the illegal dumping. (Burns has been implicated in other PCB dumping incidents as well, in Pennsylvania and New York.)[50] They eventually admitted their guilt and testified that the Ward Transformer Company in Raleigh, North Carolina had paid them $75,000 to dispose of the PCB's.[51] Some four years after the incident, most of the PCB's still lie where they were dumped.

In fact, PCB's are so difficult and expensive to dispose of that one enterprising dumper simply filled a tank truck up with oil wastes laden with the chemical and abandoned it in Jamestown, New York beside the road.[52]

But when PCB's are dumped into marshes, streams, forests, lakes, sewers, vacant lots, the basements of abandoned buildings, or other such isolated areas, as is reported to occur with some frequency, the problem often remains undetected. The true extent of our regular exposure to PCB's may never be known, but the evidence available indicates that it is increasing.

Research is underway into methods to de-toxify PCB wastes and convert the chemical into harmless substances, and some of the results to date seem promising.[53] However, there appears to be little hope of ever finding a way to break down and remove PCB's contaminating our rivers, lakes, air, food, wildlife — and our own bodies.

PCB Contamination Of Food

In light of the known hazards of PCB's and their ability to spread throughout and remain in the environment, it is hardly surprising that they are continuing to show up in our food, air and water.

After some 200 gallons of PCB's leaked from an electrical transformer at the Pierce Packing Company of Billings, Montana, about a million pounds of meat meal used in animal feed and produced between June and September 1979 became contaminated. This required the destruction of over 300,000 chickens, a million eggs, 75,000 frozen cakes, and 16,000 pounds of pork in Montana, Idaho and Utah, at a cost of millions of dollars.

Because of delays on the part of the U.S. Department of Agriculture, no public notice of the problem was given until over five weeks after Agriculture had discovered PCB-contaminated chickens at the huge Ritewood Farms Complex in Idaho, one of the largest egg producers in the United States. PCB-laden feed had been shipped there from the Billings, Montana plant, and one sample tested was found to have so much PCB it could not be measured — over 1,000 parts per million! By the time the public announcement was made, contaminated feed, chickens and eggs had been shipped throughout the country. Such major food companies as Swift & Company in Iowa and The Campbell Soup Company in Minnesota had bought unsafe chickens; and over 12,000 cases of contaminated Pepperidge Farms frozen strawberry cakes turned up in Utah. The contaminated food and feed products, which had to be destroyed after being tracked down in at least 19 states and two foreign countries, was worth some $2.7 million.[54]

Such contamination has been showing up in our food supply for over a decade. In 1970, 146,000 chickens in New York State had to be killed and disposed of when high levels of PCB's were discovered by the Campbell Soup Company. The following year, 88,000 chickens and 123,000 pounds of egg products had to be destroyed because PCB's had leaked from a heating system in a North Carolina plant into fish meal eaten by the chickens.[55]

In 1977, a fire in a Ralston Purina warehouse in Ponce, Puerto Rico resulted in the PCB contamination of 800,000 pounds of fish meal. The contamination was not discovered until nine months later, when 2.5 million pounds of feed containing the fish meal was recalled, but by this time, most of the meal had already been consumed by farm animals. According to the FDA, some 400,000 chickens had to be destroyed, and millions of eggs had to be withheld from the market. In addition, an unknown number of contaminated chickens and eggs were consumed by the public.[56]

It can be assumed that such examples represent a tiny fraction of the number of actual incidents, and that most cases of PCB contamination go undetected and/or unreported. Thus, most of the PCB's that contaminate our food ends up being consumed by the public.

Because of the pervasive nature of the contamination, FDA has been extremely reluctant to restrict the levels of PCB's that can be allowed in fish and other foods. In view of the agency's refusal to act, two public interest groups — the Environmental Defense Fund and the Natural Resources Defense Council — petitioned FDA as far back as 1975 "for an immediate and substantial lowering of the tolerances for PCB's in food because of the human health hazards posed by dietary exposure to the chemicals."[57]

It was not until June 1979 — six months after the agency had been taken to court by these groups — that FDA finally published these lowered tolerance levels. But three months later, the reduction in tolerances for fish and shellfish, from 5 to 2 parts per million, were cancelled as the result of objections from an industry trade association, the National Fisheries Institute, which claimed that FDA had underestimated the loss of food this would cause.[58]

Living With PCB's

Despite the government's unwillingness to face up to the PCB problem, it cannot be ignored; it will not go away. These chemicals will be with us for a long, long time, perhaps outlasting our society. It is still unclear what the ultimate long-term effects will be of living in constant contact with these extremely toxic substances. The results may well turn up in the form of dead and deformed children, and a greatly increased rate of cancer and sterility.

Perhaps the greatest irony of the PCB tragedy is that we appear to have learned nothing from it. In hindsight, it is clear that we could have avoided this disaster if PCB's had been banned or restricted when they were first shown to be so toxic in the 1930's. Yet, today hundreds of other toxic chemicals — potential PCB's — remain in widespread use, and over a thousand others, mostly untested, are introduced each year. We seem determined to repeat this painful experience with dozens of other chemicals.

What You Can Do

If we are to prevent the massive leakage of the hundreds of millions of pounds of PCB's still in use or in dumps into the environment, extreme care must be taken in handling these substances. Strict enforcement of TSCA is essential, although this has never been undertaken in the history of the Act and present policy constitutes virtual non-enforcement of it.

Several of the de-toxification methods being developed to chemically convert PCB's into harmless compounds appear promising. And the closed burning of PCB's in high intensity incinerators on ships may prove to be effective in disposing of some of the remaining wastes.

In the meantime, the best way to minimize exposure to the ever-increasing amounts in the environment is by generally avoiding fresh-water fish and those that spend part of their life cycles in fresh water and near coastal areas. Fish with especially high levels of PCB's include those from the Great Lakes such as trout and carp; fish with a high fat content, such as salmon and eels; and bottom feeding fish from estuaries, such as flounder, sole, and catfish. Fish with the lowest levels of PCB's would be those from unpolluted high-altitude lakes and streams, and deep ocean fish such as cod, pollack, haddock, and halibut.[59]

Birds that feed on fish can also be expected to have high concentrations of PCB's. Ducks and geese have recently been found to have dangerously high levels of these compounds and they should be avoided. And since fish contaminated with PCB's, such as menhaden, are ground into protein meal to be fed to chickens and turkeys in order to speed growth, poultry can also contain high PCB levels.

In fact, a diet low in animal fats can considerably reduce PCB levels. Vegetarians have been found to have substantially lower levels of the compound as compared to the general population. Since PCB's often concentrate in mothers' milk, women who plan to breast feed their babies should be especially careful about what foods they consume.

Chapter Ten

Lindane: The Deadly Indoor Pesticide

While most pesticides are manufactured for use on farm crops, forests, swamps and other outside areas, lindane is the most widely utilized chemical intended for use on and around humans: in the home, including the kitchen, and even as an anti-louse shampoo on people, their pets and livestock. Despite these widespread uses that have exposed virtually the entire American population to lindane, it has been known for its potent ability to cause cancer since the early 1970's, and for its other severe toxic effects since the 1950's. Just three uses of lindane — in dog shampoos, shelf paper and floor wax — have in recent years exposed a massive number of Americans each year to this extremely toxic chemical — 126 million, according to a preliminary estimate by EPA.[1]

Over 126 Million Americans Exposed

In early July 1980, EPA announced its proposal to ban most uses of lindane, one of the most toxic and widely-used pesticides, on the grounds that it causes cancer, harms fetuses and brings about central nervous system disorders. Marketed by the Hooker Chemicals & Plastics Corporation of Niagara Falls, New York, over 900,000 pounds of lindane are used each year in some 700 products. According to EPA's 1980 Position Document on lindane, just one use — in floor wax — has subjected an

307

estimated 100 million Americans to exposure.* EPA's announced intention to take action came more than four years
after starting the review process that precedes an actual ban or
restriction, and decades after lindane's extreme toxicity to
plants, animals and humans had been repeatedly demonstrated.

Lindane is heavily used to kill insects on and in a wide
variety of products and applications: hardwood logs and
lumber; seed treatment of corn and small grains such as barley,
winter wheat, lentils and peas; livestock; avocados; ornamental
plants; Christmas trees; pecan orchards; commercial forests;
home exteriors; pineapples and household pests. Many anti-flea
collars for pets contain lindane, which is continuously released
for the 6 to 8 weeks during which the collars claim to remain
effective. Many dog shampoos contain the chemical, and veterinarians bathe dogs in lindane solutions, resulting in exposure of
130,000 veterinarians, and in post-application exposure of an
estimated 15 million people, according to EPA. In some pet
products, the amounts necessary for control of fleas are acutely
toxic to cats, but not apparently so to dogs.[3] (The seemingly
very high rate of cancer among pet dogs cannot be scientifically
verified, since no such studies are known to have been
conducted).

Lindane is used in moth, insect, and general household
sprays; on beds, carpets, and draperies and in smoke fumigation
devices known as "bug" or "cloud bombs." Since the label warns
that people and pets should leave the house during the fumigation, EPA believes that most exposure from these devices
occurs later through vaporization. According to its June 1980
Position Document on lindane, exposure takes place

> ...via respiration of residual vapor in the household
> environment during the 2-week post application period. It
> is important to note that lindane is absorbed onto the
> surface of many materials during vaporizer use. Exposure
> is therefore continuous as lindane is revolatized from these
> surfaces. A study by the U.S. Public Health Service

*In December 1981, an EPA official stated that voluntary cancellation of some such
products had reduced public exposure to lindane, and that the Agency might revise
downward its published estimate of current exposure; but she could not say by how
much.[2]

demonstrated that 20% of the lindane residues initially deposited on painted walls during vaporizer use were present in these surfaces after 179 days.[4]

The agency also estimates that 11 million people are exposed to the chemical through inhaling vapors from lindane-impregnated shelf paper, which is registered for use in closets, drawers, and other non-food storage areas. Approximately 40 percent of the lindane is vaporized at a constant rate over a 105-day period. But the greatest exposure is from floor wax laced with lindane, to which EPA says 100 million Americans have been exposed![5]

Like other chlorinated hydrocarbons, lindane is absorbed through the skin. This makes it effective in killing "crabs" and other body lice, and is used in the prescription drug Kwell, which accounts for about half of the sales of anti-louse products on the market. According to the authoritative "Medical Letter on Drugs," lindane "can be absorbed, and convulsions have followed the treatment of scabies when too much was applied to the whole body, or for too long."[6]

Fortunately, another effective and far less dangerous product — pyrethrum — has been available for decades to kill all types of insects, including lice.[7]

Lindane's Extreme Toxicity

In announcing its proposal of July 1980 to prohibit most uses of lindane, EPA cited at great length evidence of its toxicity, including liver cancer, birth defects, miscarriages, stillbirths, and central nervous systems disorders. It referred to numerous studies in which laboratory animals fed lindane experienced an extraordinarily high rate of liver cancers and mortality.[8]

EPA pointed out:

data have come to the agency's attention which show that
exposure to lindane, both in test animals and in humans,
can cause adverse effects manifest in particular as
symptoms of central nervous system (CNS) stimulation. In
this connection, there is evidence also to suggest that
children are especially sensitive to the acute toxic effects of
lindane ... Studies have come to the agency's attention
which indicate that lindane bioaccumulates in the CNS,
triggering symptoms of acute poisoning.

In one experiment, lindane produced convulsions in all groups of baby rabbits treated with it, but not in the adults.[9] There are also numerous cases on record of children being poisoned — sometimes fatally — by lindane, including the death of a 2-½ month-old child who was treated for scabies with a commercial product. Lindane was found in the infant's blood, urine and brain tissue, and doctors concluded that the pesticide was the probable cause of death. In another case, a 32 month-old Illinois boy was diagnosed as suffering from aplastic anemia, a critical blood disorder similar to leukemia, from exposure to a lindane vaporizer.[10]

However, it is reasonable to assume that lindane poisoning — which is characterized by such common symptoms as headache; nausea; depression; ear, nose and throat irritation; hyperirritability; convulsions and respiratory and heart failure — is rarely recognized or properly diagnosed. A perhaps typical example concerned a little boy who developed aplastic anemia. The child's symptoms vanished after he was hospitalized, but they reappeared when he returned home. Eventually, his ailment was traced to the presence of lindane on the coat of the family's pet dog that was being treated for mange by being dusted with a lindane powder. In playing with his dog, the boy was apparently breathing the chemical into his lungs and absorbing it through his skin, from whence it migrated to the bone marrow and interfered with the production of red blood cells.[11]

Children are also at greater risk because they tend to consume a large proportion of dairy products, which contain relatively higher amounts of lindane.

EPA stresses the dangers lindane poses to the unborn fetus, and cites several studies showing such "adverse reproductive/ fetotoxic responses" as stillbirths, retarded sexual maturation and depressed sexual function. In one set of experiments, there was a progressive increase in the proportion of still-births in successive generations. After its more than two-year study of lindane, EPA concluded that normal human exposure to lindane can harm the fetus without any apparent symptoms appearing in the mother.[12]

A major factor in lindane's toxicity, EPA notes, is its persistence. As evidence of lindane's endurance, EPA cites instances of its being found in areas far, far removed from where it has been used, such as in fox and seal tissues from the Arctic, and birds, penguins, fish and krill from the Antarctic. It has been found in rainwater samples from several countries, and up to 8 percent of what was originally applied to soil was found to be present 15 years later.[13]

In conclusion, EPA emphasizes that its "concern regarding the potential for lindane to produce both short- and long-term effects in humans is heightened by the persistence of lindane in the environment:"

> As residues of lindane can remain in the environment for a long time period, the potential duration of lindane exposure following application of the chemical presents a significant problem. The persistence of lindane poses a particular concern for home uses, such as structures, household uses, and pets, where residents can experience substantial long-term exposure to lindane residues. Moreover, in general, there is concern about the bioaccumulation of lindane in humans and the burden this places on the body's systems in view of the widespread use of the compound and its mobility to sites far removed from treatment areas.[14]

In October 1980, EPA's Scientific Advisory Panel completed its review of lindane and came out with a series of baffling and contradictory recommendations that served more to confuse than clarify the situation. Stressing that "lindane is substantially more toxic to young than adults in both humans and domestic animals and that chronic exposure can sometimes result in disastrous blood dyscracias," the Panel urged that certain uses be "cancelled immediately." "These included home shelf paper and floor wax ("an unwarranted risk to the householder"); in consumer pet products such as flea collars, dog powders and shampoos; and "ornamental applications for unrestricted use by the homeowner." However, the Panel also stated that commercial nursery workers should continue to have access to lindane, in spite of its hazards:

> Ornamental uses restricted to commercial operators should be continued with full warning label cautions about the hazards of cancer, fetotoxicity, and central nervous system effects and a caution that women of child-bearing age and

children must avoid exposure. Full protective clothing must be worn.

Other uses the Panel recommended be continued include application to Christmas trees, avocados, pineapples, pecans, livestock, hardwood logs and lumber, seed treatment, and to control beetles.[14b] The Panel did not explain why these uses should not be considered hazardous.[15]

It should be emphasized that EPA's proposal to restrict lindane is just the *beginning* of the fight, which could amount to a several years' long process. Whatever EPA decides to do about lindane, the chemical has already joined a long list of other cancer-causing pesticides that will remain for the foreseeable future in our environment.

Dangers Known For 30 Years

Almost all of the data compiled by EPA on lindane's hazards had been known for years or decades; environmentalists and scientists had warned of its toxicity and tried to have it banned. They almost succeeded in 1978, when the registration of products containing the closely-related pesticide benzene hexachloride (BHC), also produced by Hooker Chemicals, was supposedly cancelled. (Lindane is the "gamma isomer" of BHC. By the time BHC was cancelled as a powerful cancer-causing and fetus-deforming agent, it had been detected in 99 percent of all U.S. human tissue samples tested.) However, many BHC registrations, instead of being cancelled, were amended to become lindane registrations.[16]

Apparently in anticipation of prohibitions being placed on lindane, Hooker Chemicals stopped producing lindane in the U.S. in May 1976 after EPA began a formal review of the chemical. Since then, Hooker has supplied its customers by importing lindane from abroad, according to EPA.[17]

The remarkable thing about EPA's decision to consider moving against lindane is that it took the federal government so long to act. Medical evidence of lindane's threat to public health goes back to the 1950's, when it was first registered with the federal government. This data was considered so strong in the 1950's that the American Medical Association, through its major publication, waged a lengthy campaign against such products as lindane vaporizers. These emit lindane fumes and are used to kill insects in homes, restaurants and offices. The 9 October 1954 issue of the *Journal of the American Medical Association* carried an article entitled "Abuse of Insecticide Fumigating Devices," warning that lindane can accumulate in the brain and liver, and can cause "profound and long lasting effects on the central nervous system."[18]

In *Silent Spring,* published in 1962, Rachel Carson recited the scientific evidence that had been amassed against lindane, including its association with neurological disease, leukemia and other blood disorders.

There are countless horror stories, going back decades, of lindane's effect on humans. A not untypical one is cited by the Senate Subcommittee on Environment in its 1974 pesticide oversight hearing record.

In the late 1960's, the Phillip Taylor family in Fresno, California had their house fogged with lindane and the attic treated with DDT. Upon returning to the house after a 4 to 8 hour evacuation period for each treatment, the family suffered a variety of ailments, including headaches, stomachaches, vomiting, blurred vision, respiratory difficulties, and skin irritation. The son became gravely ill and temporarily lost his vision. Five years after the spraying, the house was found to have extremely high levels of lindane and DDT, and the Taylors were advised by a toxicologist to vacate their home and get rid of their possessions.

The Taylors spent over $5,000 trying to clean the house of pesticide residues, including having a new heating system and ducts installed; but their ailments returned and their son again lost his vision. One toxicologist told them that the lindane could "never" be removed and that the only way to get rid of the vapors and odor would be to "burn" the house and its furnishings, since the chemical had been absorbed into the wood and continued to vaporize when the house was heated. Eventually, they disposed of their books, furniture and most of their clothes and other possesions, moved into a mobile home, and sold their house at a loss (with a provision in the contract relieving them of liability for any illness resulting from the pesticide contamination). Since they could not find an older home that had not been sprayed, they had to buy a new house. Afterwards, the former symptoms disappeared, but Mrs. Taylor had to have both of her breasts removed because of cancer.[19]

In 1969, the Mrak Commission's report to the Secretary of HEW called for "corrective action" to be taken on lindane and other chlorinated hydrocarbons.[20] Yet despite these scientifically-based warnings over the decades of lindane's toxicity, lindane has continued to be used so widely that it has been almost impossible to avoid some exposure to it.

Even if it is someday restricted or banned, lindane will continue to be pervasive throughout the environment. It has already turned up in wells in California at levels over five times higher than that set by EPA for drinking water.[21] And the infamous Love Canal has been contaminated with over 13 million pounds of lindane dumped there by Hooker Chemicals.[22]

Under the Reagan administration, EPA has virtually halted its regulation of toxic chemicals. Thus, action to restrict lindane is extremely doubtful in the forseeable future.

Given the decades-long lag time between exposure to a carcinogen and the onset of cancer, the price we will pay for our failure to regulate lindane may prove to be a heavy one for millions of Americans.

What You Can Do

It is very dangerous and also unnecessary to use lindane on humans and in the home. There are numerous alternatives to treating lice infestations with prescription drugs containing lindane, such as Kwell, which are especially dangerous to children. There is far less risk in using over-the-counter anti-louse products containing pyrethrin, which are very effective in killing lice and their nits (eggs). (Pyrethrin collars can also be used on pets to keep off fleas.) Shampooing with a coconut oil-based soap is also useful and kills adults and some eggs. Perspiration from heavy exercise repels body lice, and "habitat modification," such as braiding hair, can be helpful. Hand-picking of lice or removal with a fine-toothed comb can also eliminate them, but the procedure must be followed for 12 days since the eggs continue to hatch for up to 11 days.

However, no product can prevent reinfestation if a child is again exposed to lice at school. This can usually be prevented by simple precautionary measures such as not wearing or using other children's clothes, brushes, or combs, and not storing clothes on a surface or next to another garment from which lice can be transmitted. And after treatment for lice, the house should be thoroughly vacuumed and cleaned, and clothes, bedclothes, combs, and other personal items boiled.

The best way to keep one's home free of insects is to make it as difficult as possible for them to enter or survive inside. Dirty dishes, grease, oil, food (including pet food) should not be left out. Unrefrigerated food should be kept tightly covered in metal or glass containers. Leaking pipes and dripping faucets should be repaired. Crevices through which insects can enter the house and live should be filled.

If one's home is so infested with cockroaches or other pests that chemicals must be used, the most effective and least toxic options are pyrethrin sprays, boric acid, or borax powder. The insects take the latter back to their lairs on their bodies, thus affecting the entire colony. Dessicant powders, such as Dri-Die, Perma-Guard, or SG-67 are also effective and can rid a house of flea infestations by "drying-up" the insects.

Chapter Eleven

DBCP:
The Poisoning
of California

Any exposure to DBCP would create a hazard of sterility and cancer.
Dr. Robert Johnson, California State Health
Official, 1977[1]

*Dow Chemical Company chose not to provide its workers with any
health hazard data regarding the sterilant, mutagenic, or carcinogenic
properties of DBCP.*
Peter Weiner, California Department of
Industrial Relations, 1977[2]

F or twenty years — from 1957 until August 1977 — the
Dow and Shell chemical companies manufactured DBCP, an
extremely toxic pesticide whose production was halted only
after it was found to have made workers producing it sterile.
Because the chemical (dibromochloropropane) was so widely
used, millions of consumers throughout the nation were also
exposed to it. As a result, EPA estimates that each year, thousands of Americans will contract and die of cancer from such
exposure. Ironically, Dow and Shell knew of the toxicity of
DBCP in 1957, but chose not to provide this information to the
public or even to its own employees.

The Sterile Workers

The first indications that DBCP was making workers sterile
were not discovered by the companies involved or by government agencies, but were stumbled upon by the workers them-

selves. After several employees at the Occidental Petroleum Company's Lathrop, California plant, casually talking among themselves, complained about their inability to get their wives pregnant, some of the men underwent sperm tests in mid-1977. The findings released in July — that they had no sperm, or so little as to be infertile — were remarkably similar to the results obtained from exposing laboratory animals to DBCP in experiments conducted in 1958. Of 114 exposed workers at the Occidental plant, 35 were found to be infertile.[3]

These findings of sterility among DBCP workers in Lathrop alarmed state and federal, as well as industry, officials and a search was begun to locate other workers throughout the country who may have been harmed by the chemical. Sure enough, workers at other plants were also found to have been made sterile.

An investigation of the Dow Chemical Company plant in Magnolia, Arkansas, found that 62 of the 86 employees tested were sterile or had extremely low sperm counts. A medical report from there indicated that exposures to one-half part per million of DBCP could cause sterility.[4] At a Velsicol Chemical Company plant in Eldorado, Arkansas, which produced DBCP for two months in 1975, about half of the 24 workers who handled DBCP were found two years later to have abnormally low sperm counts.[5] Sterility among workers was also found at a Shell Chemical plant in Alabama making DBCP. In all, a third of the 432 DBCP exposed workers tested in Arkansas, California, and Colorado were determined to be infertile.[6]

On 11 August 1977, Dr. Robert Johnson, of California's Occupational Health Branch, reported:

> The effect of DBCP, in producing sterility in human males, has been established beyond question. DBCP is also a very potent carcinogen, producing not only primary cancer, but rapidly spreading lesions as well.

He also expressed the opinion that "any exposure to DBCP would create a hazard of sterility and cancer."[7]

In mid-August, 1977, California suspended registration and sale of DBCP, and the federal Occupational Safety and Health Administration (OSHA) asked all manufacturers of the chemical to halt production voluntarily. Later, the state and OSHA set the exposure standard at one part per billion, which was one one-thousandth the level that had been recommended 20 years earlier by scientists at Dow and Shell.[8]

OSHA also required factory workers to be given training about the hazards of DBCP; to wear clothing covering the entire body when exposure to it may occur; to have daily showers, changes of clothes, special facilities to dispose of dirty work clothes in closed containers, and even warnings for those who wash the clothes; and to undergo medical surveillance. Although DBCP can be inhaled in the field as well as in the factory, EPA never set any such requirements, even for those who actually applied the chemical.[9]

Citing the exorbitant costs of providing their workers with such stringent protection, Dow and Shell decided to halt manufacture of the chemical.[10]

But such action came a little too late for the many workers who were made sterile and/or exposed to a greatly increased risk of having cancer or children with birth defects. Years after the plant stopped producing DBCP, the affected workers at the Lathrop facility were still experiencing frequent headaches, nausea, bleeding noses, loss of smell, mottled skin, and continued inability to have children.[11]

DBCP In Wells And The Water Table

DBCP has been widely and heavily applied throughout the United States, especially in such farm states as Arizona, Florida, Hawaii, South Carolina and Texas, with an estimated 10-12 million pounds being used each year.

Before being banned in California in 1977, some 800,000 pounds of DBCP were being used annually by growers in the state. State officials took action after determining that the pesticide was capable of causing cancer and sterility at extremely low levels, and warned that its presence is especially dangerous in drinking water, because such large amounts are consumed daily.

In 1979, two years after California banned DBCP, state health officials found dangerous levels of the chemical in *half* of the irrigation and drinking wells tested in the San Joaquin Valley. Extremely high levels were also found in parts of California's vast Central Valley, an area the size of New York

State stretching over some 50,000 square miles. As revealed in private company correspondence, Occidental's Lathrop plant deliberately violated state law by intentionally dumping DBCP and other chemicals it knew to be hazardous at the rate of five tons a year, thereby polluting the local groundwater with poisonous wastes. One internal memorandum told of a pet dog that came into contact with some of this waste water, licked itself and died.[12] (This situation is discussed in detail in the chapter on groundwater contamination, see pages 179-82).

By the fall of 1979, DBCP had been found in some 200 wells and in community water sources supplying a dozen counties and various cities, such as Riverside and Anaheim. In some of these areas, DBCP had not been used for years. Even the famed Disneyland's water supply was found to be contaminated.[13]

The state estimated that some 955,000 Californians were drinking water containing DBCP at concentrations greater than the allowable level set by EPA. At least 16 California wineries were found to be producing wine with dangerous levels of DBCP, and at one point, five million gallons of wine from the Delicato Vineyards were embargoed.[14]

DBCP is also beginning to turn up in ground water supplies throughout the country,[15] indicating that the problem may be national in scale. Thus, millions of Americans may face a threat of sterility or damage to their sperm from the widespread use of this chemical. EPA considers water with as little as one part per billion to be unsafe, which is the equivalent of one drop in 12,000 gallons![16]

The Threat Of Cancer

In its deliberations on whether, and how much, to restrict DBCP, EPA has conducted "risk assessments" on roughly how many people might be expected to contract cancer as a result of exposure to the chemical. EPA's Carcinogen Assessment Group has stated that "there is strong evidence that DBCP is likely to be carcinogenic to man," and cited its ability in animal tests to produce in short periods of time "high levels" of stomach cancer as well as mammary cancers, which spread "to a wide variety of other organs."[17] A 1979 EPA estimate for certain uses of DBCP on fruits and vegetables projected that about 5,040 additional cancer cases would be caused each year.

These estimates do not include non-food residue sources of exposure, such as occupational and from drinking water. But it is clear that in the over quarter-century of DBCP's use on most fruits and vegetables, if EPA's estimates are anywhere near accurate, tens of thousands of cancer deaths many occur just from residues of the chemical on certain crops.[18]

Hazards Known For 20 Years

The health problems caused by DBCP should not have come as a surprise to the chemical industry. As far back as 1957 and 1961, both the Shell and Dow chemical companies conducted and published studies of the effects of DBCP on test animals which indicated that the chemical caused serious reproductive effects, such as damage to sperm cells.[19]

DBCP's potential for damaging humans was first discovered in 1957, two years after it came into agricultural use, when Shell's studies, conducted by Dr. Charles Hine of the University of California, linked the chemical to atrophied testicles, sterility and abnormal production of sperm.[20]

In 1961, Hine and a Dow researcher jointly published their findings in a leading scientific organ, the *Journal of Toxicology and Applied Pharmacology*, reporting that this severe damage to the testicles and sperm production were "the most striking" conclusions of the research, and recommending that occupational exposure to DBCP be severely limited. They warned that "The compound can be absorbed through the skin in toxic amounts ...To prevent ill effects from inhalation, DBCP should be handled in closed systems or with respiratory protection unless adequate ventilation is provided." According to OSHA, the authors also recommended that employers conduct close medical observations of employees until the ill effects of the chemical were known. This recommendation was not followed by Dow or Shell.[21]

Dr. Hine has also come under criticism for what some consider at least the appearance of a potential conflict-of-interest situation in serving as a University of California scientist, professor and medical researcher on the one hand, while at the same time serving as a consultant for, and accepting money from, companies involved in manufacturing hazardous products which were undergoing laboratory testing. A pharmacologist and medical doctor, Hine accepted consulting contracts from, and gave expert testimony in court on worker compensation cases in behalf of, such controversial firms as Dow, Shell, American Smelting and Refining Company and even cigarette companies. By 1977, Shell alone had paid Hine and the university's School of Medicine some $400,000 to conduct research. Yet, in setting regulations for DBCP, the government relied to a large extent on the studies done by Dr. Hine.[22]

DBCP's ability to cause cancer has also long been known; as OSHA has pointed out in one of its positions papers: "In addition to the sterility problem, there are also strong indications that DBCP may cause cancer in humans. This conclusion is based on testing performed as early as 1972 for the National Cancer Institute and others. These tests showed that DBCP caused stomach cancer in both sexes of rats and mice and caused mammary tumors in rats."[23]

None of this information, however, had the effect of creating any hesitancy for government regulatory agencies to allow DBCP to remain in widespread use. After Shell and Dow applied to the U.S. Department of Agriculture for tolerances to be established on the chemical, they were quickly granted. From 1963 to 1973, both the FDA and EPA allowed Dow and Shell to register new DBCP products.

The agencies granted, and sometimes even increased, chemical tolerances resulting from DBCP use on various food crops, including almonds, broccoli, Brussels sprouts, cabbage, cauliflower, eggplants, lima beans, melons, plums, prunes, peanuts, peppers, melons, pineapples, strawberries, soybeans, tomatoes and citrus fruits.[24]

Peter Weiner, Chief Counsel of California's Department of Industrial Relations (DIR), which investigated the situation, has described how critically important information was withheld from the workers:

> Dow Chemical Company chose not to provide its workers
> with any health hazard data regarding the sterilant,
> mutagenic, or carcinogenic properties of DBCP...
> Injurious properties of DBCP were excluded from the
> health hazard sheet furnished to workers, and no attempt
> was made to inform them of their potential risk.[25]

Shell adopted similar policies of non-disclosure of information about DBCP's hazards, according to Weiner.[26]

And officials from Occidental, which used DBCP purchased from Dow in several of its products, told the DIR that Dow did not warn them of any sterility problems associated with the chemical.[27] Yet in an interview appearing in a documentary film shown on public television, "Song of the Canary" (1977-78), a spokesman for Occidental, when asked about the 1961 Dow Study, made the following reply:

> ... it did not show sterility in rats. What it showed was with very high doses of DBCP you could get testicular atrophy, if you will, the shriveling up of the testicles. I've talked to two scientists who are familar with the work, and they both say, "Heck, we just, we just didn't draw the conclusion that there'd be sterility from the fact that the testicles were shriveling up."[28]

After DBCP was banned, Occidental's leftover supplies were reportedly shipped to the Philippines to be used on crops there.[29]

The Reinstatement And "Banning" Of DBCP

Even after DBCP had been conclusively shown to cause cancer and sterility, companies not only continued to manufacture it, but fought, every step of the way, the efforts by the Health Research Group, the Environmental Defense Fund and EPA to have its registration cancelled. In the midst of the fight to ban DBCP, Ronald B. Taylor reported in the *Los Angeles Times:*

> Given the record thus far, one might expect DBCP to be fading away. After all, California authorities banned it entirely two years ago, and the federal government has imposed a host of restrictions on its use elsewhere as evidence of DBCP's dangers has mounted.
>
> ... the pesticide is far from in retreat. Powerful corporate and agricultural interests — backed by the U.S. Department of Agriculture, segments of EPA, and the California Department of Food and Agriculture — are engaged in a major struggle to preserve and extend DBCP's use.[30]

After the public uproar in 1977 over DBCP's having made workers sterile, EPA effectively banned major uses of the chemical. But tremendous pressure to reverse this decision was brought to bear on EPA by the agribusiness and chemical industries and various politicians from farm states.[31]

Because of the higher costs involved in producing DBCP under the stricter safety standards that had been promulgated, both Dow and Shell had halted production. Consequently, production of the chemical was shifted across the border to two plants in Mexicali, Mexico, which began to export large amounts of the chemical into the United States. But 21 out of 23 Mexican workers tested also were found to have similar sterility and low-sperm problems, so the Mexican plants, too, were closed down in late 1978.[32]

From the fall of 1977 until January of 1979, the DBCP produced in Mexico was imported back into the United States by the Amvac Chemical Corporation of Los Angeles and distributed to its customers throughout the country (except for California, where the state ban remained in effect). After the Mexican plants were shut down, Amvac decided to begin producing DBCP, the company reported, because a "vacuum existed in the market." A few weeks before production began in February 1979, Amvac was fined $3,520 for various health and safety violations, to which it pleaded "no contest," including not issuing proper protective clothing to workers, failing to monitor their health, and not reporting a major spill of DBCP on its loading dock and other spills within the plant.[33]

State officials also learned that Amvac was discharging the chemical from the plant into the air outside at 500 times the permitted level in the workplace, and forced the firm to halt this practice.[34] Nevertheless, in February 1979, Amvac began producing DBCP itself at a rate of about 2,500 gallons a day.[35]

After two months of public hearings and 7,300 pages of testimony, an EPA administrative law judge ruled in October 1979 that further use of DBCP constituted an "imminent hazard" to public health. Shortly thereafter, on 29 October, EPA Administrator Douglas Costle formally cancelled the registration for remaining uses of the chemical in all states except Hawaii, stating:

> [DBCP]...has caused sterility among workers who
> produce it and has been shown to be a suspected cancer
> agent and a possible source of human chromosome damage
> ...Residues of DBCP on certain food crops, in some
> drinking water wells, and in the air near treated fields
> threaten serious harm to people exposed in these ways.[36]

Presumably, these same considerations would apply to the people and environment of Hawaii, but for reasons that remain unclear, it can still be used there.

Two 1980 University of Hawaii studies, released in early 1981, indicate the damage being done to the people of Hawaii. One study showed that alarming amounts of DBCP were seeping into local water supplies. A well on the major island of Oahu was found to contain 10 times the recommended "safe" level of the chemical. In the town served by the well, twice as many men had abnormally low sperm counts compared to a control group. Another study showed that 40 percent of the pineapple workers tested had dangerously low sperm counts.[37]

But pineapples are not the only crop that may continue to expose U.S. consumers to DBCP. Despite the restrictions on DBCP in this country, reports persist that Amvac Chemical Company continued to export freely huge quantities of DBCP to foreign, usually less-developed countries. (Amvac has refused to say if it is currently doing so). One of Amvac's best customers is the Castle & Cooke Company, which is among the largest landholders in Latin America. The illiterate peasants who work for Castle & Cooke use this sterility-causing carcinogen on the firm's huge banana and pineapple plantations, which grow these fruits largely for export to the United States under the "Dole" brand label.[38]

After an award-winning exposé on the export of banned pesticides appeared in the November 1979 issue of *Mother Jones* magazine, revealing that Castle & Cooke was using DBCP on its bananas, the firm stopped importing the chemical directly from Amvac. Instead, it began obtaining it more discreetly through local importers, according to David Weir and Mark Schapiro in *Circle of Poison*. Other agribusiness giants such as Del Monte and United Brands also reportedly use DBCP on products intended for U.S. consumption, such as pineapples, bananas and citrus crops.[39]

And the Dow Chemical Company still stands to make money from the chemical: since it owns the patent on DBCP, Dow makes a three percent royalty on all DBCP sold by Amvac.[40]

The Advantages of Being Sterilized

DBCP does have its defenders, and one group of growers went so far as to cite the benefits of having been made sterile. This incredible defense of DBCP is offered in a letter to OSHA from Robert K. Phillips, executive secretary of The National Peach Council, an association of peach growers in 35 states, which argued against restricting DBCP, and stated:

> While involuntary sterility caused by a manufactured chemical may be bad, it is not necessarily so. After all, there are many people now paying to have themselves sterilized to assure they will no longer be able to become parents.

The 12 September 1977 letter to Dr. Eula Bingham, then head of OSHA, listed some of the benefits of having workers sterilized by DBCP:

> If possible sterility is the main problem, couldn't workers who were old enough that they no longer wanted to have children accept such positions voluntarily? Or could workers be advised of the situation, and some might volunteer for such work posts as an alternative to planned surgery for a vasectomy or tubal ligation, or as a means of getting around religious bans on birth control when they want no more children.

The Peach Council concluded that some balance was needed in considering DBCP's effects on workers: "We do believe in safety in the workplace, but there can be good as well as bad sides to a situation."[41]

In any event, in light of the widespread use of DBCP and its presence in wells and groundwater, millions of Americans may someday owe a debt of gratitude to the chemical and agribusiness industries for giving them the chance to become sterile. But those wishing to undergo the modern chemical approach to voluntary infertility may also have to accept the gift of cancer as well.

What You Can Do

Nematodes can be held in check without poisoning the food, soil, and water tables. As CEQ points out in its publication *Integrated Pest Management,* "Crop rotation is one of the oldest and most important measures for controlling plant-parasitic nematodes." These creatures can also be controlled by "trap crops", according to CEQ: "The practice of attracting pests to small plantings of crops which are then destroyed or sprayed with a toxicant has been quite successful against some plant nematodes."[42]

Pathogens (microscopic disease-causing agents) such as bacteria are also used to kill nematodes. CEQ describes some tests of this method as "very promising...No adverse effects in its use are anticipated."[43]

But unless nematodes are causing damage, there is no need to kill them. They have their place in nature's plan, and can even be beneficial to man. Some species (such as roundworms) kill mosquito larvae, and in field tests they were able to infect from 50 to 85 percent of the larvae.[44]

The ultimate proof of the fact that DBCP is not essential to U.S. agriculture is that its general ban has not resulted in either significant losses of or price rises in the crops on which is was being used. If other toxic chemicals still in widespread use were subjected to such restrictions, similar results could be expected.

Chapter Twelve

Endrin: Silent Summer

The chemical endrin provides an instructive example of how the government often appears to take action to protect the public from a dangerous substance, while, in fact, few if any meaningful results are achieved.

Endrin has long been known as one of the most toxic of all insecticides. It is closely related to DDT — both are chlorinated hydrocarbons — but much more poisonous. Rachel Carson wrote of it in *Silent Spring:*

> It makes ... DDT seem by comparison almost harmless. It is 15 times as poisonous as DDT to mammals, 30 times as poisonous to fish, and about 300 times as poisonous to some birds. In the decade of its use, endrin has killed enormous numbers of fish, has fatally poisoned cattle that have wandered into sprayed orchards, has poisoned wells, and has drawn a sharp warning from at least one state health department that its careless use is endangering human lives.[1]

She also relates an incident in which a cockroach-infested house was sprayed with endrin, after which the floors were washed and a year-old baby and pet dog were brought back in about six hours later. An hour later, the dog vomited, went into convulsions and died; that night, the baby also had convulsions and later became a "vegetable," unable to see or hear.

Poisoning Wildlife — And Humans

Since *Silent Spring*, endrin has continued to devastate fish and wildlife in numerous areas. In the early 1960's, massive fish kills in the Mississippi River were traced to discharges from a

Velsicol Chemical Corporation plant in Memphis that was producing endrin. Two of these spills (into the Memphis sewage system) in 1960 and 1963 are estimated to have killed up to 13 million fish. In the wake of these incidents, endrin's use began to decline, but with the 1972 ban on most uses of DDT, endrin regained popularity and its use greatly increased, especially on cotton.[2]

The following year, after a large-scale spraying of endrin, Cherokee County, Alabama experienced what news reports called "an unusually silent September . . . 'We just flat don't hear any birds,' said one resident." The local birds, along with squirrels, rabbits, thousands of fish and several cows were apparently killed off by the insecticide. A survey by the Alabama Game and Fish Division found "no visible indication of any birds or squirrels" in wooded areas, ponds "completely destroyed" and numerous reports of dead birds and mammals. An EPA team found a lake contaminated with lethal levels of endrin. Farmers also became ill, and doctors reported intensified stomach problems and hay fever symptoms in patients.[3]

In 1974 and 1975, endrin, along with toxaphene, was implicated in the poisoning of 30 to 40 percent or more of the population of 400 or so brown pelicans inhabiting Louisiana's Gulf Coast. Ironically, this endangered species is the state bird. The die-offs followed heavy rains and flooding, which washed large quantities of pesticides into the Mississippi River and the ocean.

Although endrin had never been known to have been used or imported into the Persian Gulf city of Doha in the shiekdom of Qatar, the pesticide poisoned hundreds of people there in the summer of 1967. After eating bread from flour that had been contaminated with the chemical, some 700 people experienced convulsions, severe stomach cramps, nausea and mental disorientation, and had to be hospitalized; 24 of them died.[4]

The 1969 Mrak Commission report cited endrin as a pesticide that posed a serious potential threat to humans. Laboratory experiments since then have established the chemical's ability to cause cancer in test animals; small amounts of endrin have caused tumors and birth defects. A quarter ounce, even if absorbed through the eyes or a cut in the skin, is thought to be sufficient to kill a 170 lb. man in under an hour. Lesser amounts can damage the central nervous system and brain.[5]

Manufactured mainly by the Shell and Velsicol Chemical companies, endrin has been widely used on various crops, including cotton, cereal grains such as wheat and on fruit trees.

In 1964 and 1969, EPA cancelled some uses of endrin, such as on tobacco and ornamentals, but has blocked attempts by the National Audubon Society, beginning in 1975, to have the pesticide banned. Most uses of endrin have been allowed to continue as before, but under bizarre rules and regulations that are so complex, inconsistent and incomprehensible as to guarantee that they will be widely misunderstood and disregarded. For instance, on some crops, such as apple orchards, endrin cannot be sprayed "within 50 feet of lakes, ponds, or streams . . . (or) areas occupied by unprotected humans."[6] But on other crops, such as cotton, small grains, alfalfa and clover seed, the EPA rules state:

> Do not apply this product within 1/8 mile of human habitation . . . (or) 1/4 mile or by ground within 1/8 mile of lakes, ponds, or streams. Application may be made at distances closer to ponds owned by the user but such application may result in excessive contamination and fish kills. Do not apply when rainfall is imminent. Apply only when wind velocity is between 2 and 10 m.p.h.[7]

The label on endrin admits that wildlife will be killed even if it is used properly, and warns farmers to keep livestock off of sprayed fields for at least a year. But when aerial sprayers are hired to treat fields, farmers often never see the label. These lengthy EPA regulations — which are generally neither enforced nor enforceable — also note that the use of this extremely dangerous and harmful pesticide may actually be self-defeating:

> Unnecessary use of this product can lead to resistance in pest population and subsequent lack of efficacy.[8]

There is, of course, an obvious question that is raised by this remarkable set of regulations: if a chemical is so dangerous (and potentially ineffective) that it requires such extreme protective measures, why not ban it altogether instead of putting such complex restrictions on its use that they are bound to not be followed?

The 1981 Endrin Tragedy

In the fall of 1981, the dangers of allowing continued use of endrin received widespread attention when it was revealed that perhaps millions of migrating ducks and other waterfowl had been seriously contaminated by heavy spraying of endrin in Montana and other western states.

In the spring of that year, the chemical was sprayed — apparently legally and according to label instructions — on over 120,000 acres of wheat fields in Montana to combat cutworms, as well as on 140,000 acres in Wyoming, South Dakota and Colorado. Later, tests revealed that migrating waterfowl passing through Montana had accumulated dangerously high levels of endrin, with some birds having three to four times the amount permitted in poultry. One EPA toxicologist warned that levels being found in some ducks were so high that a 60 pound child eating a single serving could have convulsions, nausea, headaches and other symptoms of poisoning.[9]

The head of the Montana Department of Health, Dr. John Drinan, said: "The birds should not be consumed. The hazard is real, especially to a woman who might become pregnant...she could have a deformed child.[9b]

Health authorities called, unsuccessfully, for the hunting season to be cancelled for all states of the Central and Pacific flyways, including Arizona, California, Colorado, Idaho, Kansas, Montana, Nebraska, North Dakota, Nevada, New Mexico, Oklahoma, Oregon, South Dakota, Utah, Texas, Washington and Wyoming. It was feared that the spraying would cost the lives of hundreds of thousands of the birds; concern was also expressed for several endangered species using or migrating through the area, including bald eagles, peregrine falcons and even whooping cranes, of which less than one hundred now survive in the wild.

The spraying of the endrin was overseen by the Montana Department of Agriculture, which was accused of covering up the incident for several crucial weeks. The Department's Director, W. Gordon McOmber, replied that there had been "no need to notify the public immediately". But on 4 September, McOmber issued an emergency suspension of all endrin uses, stating: "...an imminent emergency exists which threatens public health, safety, and welfare..."and pointing out that "acceptable sustitutes are available..."[10]

Since almost half the endrin applied can remain in the soil for 12 to 14 years, it will persistently be passed up the food chain — from wheat and soil, to worms, slugs, snails, and grasshoppers, to mice and birds and, inevitably, to humans. Thus, this 1981 spraying episode will continue to present a threat to humans and to the environment for years to come. Ironically, EPA has admitted in an official "position document" that "market and consumer impacts are expected to be negligible if endrin is cancelled for use on wheat."[11]

Interestingly, this was not the first such incident in recent years involving endrin. In 1976, the chemical was sprayed over 5½ million acres in Kansas and Oklahoma, causing widespread mortality among pets, livestock, and wildlife, including bald eagles. In Kansas, two million fish and 20 cattle are known to have been killed, and 8,000 pounds of milk had to be discarded. Damages were estimated at up to $5 million. In Oklahoma, livestock, wildlife and domestic animals were killed, and additional known damages amounted to between $500,000 and $1 million.[12]

Yet a few weeks after the 1981 incident came to light, Montana Agriculture Department Director McOmber announced that it would permit the use of endrin, "if necessary," again the following year.[13]

Commenting on the incidents, Chris Madson, Editor of *Kansas Wildlife*, described the choice we have to make on endrin: "We can keep it in the lab until we know more about its effects, or we can leave it on the market and do the research ourselves. After all, we have half a continent for a laboratory, and our own children for white rats."[14]

In response to the outcry over endrin spraying, Montana Governor Ted Schwinder set up a Pesticide Advisory Council. However, *Montana* magazine (March-April, 1982) describes it as "seriously slanted against wildlife and recreational interests." One of its appointees is Harold Schultz, a Billings chemicals agent who distributed much of the endrin used in 1981. Another member is Ronald Albright, an aerial pesticide sprayer who stated, "We haven't really established there is an endrin problem."

What You Can Do

Since only a very small amount—less than 10 percent—of U.S. wheat acreage is treated with pesticides, the use of endrin on such grain crops is obviously unnecessary. This is especially true since numerous substitutes are available that are not nearly as toxic and persistent as endrin. Wheat is a generally pest-resistant crop that is rarely attacked by cutworms. When the occasional outbreak does occur, considering the enormous damage caused by endrin, it would be cheaper to let the worms run their course and reimburse the farmers for losses.

In order to avoid consumption of foods contaminated with endrin, waterfowl from the western and central flyways should not be eaten. Some of the endrin from the 1981 spraying incident in Montana and elsewhere in the west will probably be stored in the fat tissue of birds and may persist for years. Since deer and livestock often forage in areas that have been sprayed, a diet that avoids meat and food high in animal fats can minimize exposure to this and other dangerous chlorinated hydrocarbon pesticides.

Chapter Thirteen

Toxaphene: 100 Million Pounds A Year

Toxaphene is the nation's most heavily used insecticide (over 100 million pounds a year), accounting for one-fifth of all pesticides used in the United States. It produces cancer in laboratory animals, genetic abnormalities in other tests and is suspected of being similarly toxic to humans.[1]

Because of its ability to cause cancerous tumors in animals and other serious health effects, as well as its severe toxicity to wildlife, toxaphene is undergoing review by EPA to determine if it should be banned or further restricted.[2] In the meantime, it continues to be used in massive amounts, threatening the environment and the health of millions of Americans.

Uses and Dangers

Since being introduced in 1947 by Hercules, Inc., over a billion pounds of toxaphene have been produced and used, mainly on such crops as cotton, rice, cranberries, grains and vegetables, primarily soybeans, peanuts, wheat, corn, tomatoes, beans, oats, barley, radishes, potatoes, and peppers. It is also used on cattle, hogs, sheep, and other livestock to kill parasites. A chlorinated hydrocarbon, toxaphene has been used by farmers at an annual rate estimated at between 40 and 100 million pounds.

This widespread use of toxaphene has contaminated the nation's food, air, water and wildlife. It has been found in fish, milk, vegetables, birds, rivers and lakes throughout the country, and even in rainwater in Bermuda. According to EPA, toxaphene has been "found in the atmosphere over the North Atlantic many miles from any direct source of contamination," possibly as a result of cotton spraying. In 1975 and 1976, almost half of the drinking water samples tested by EPA contained detectable residues of toxaphene, in contrast to no samples that tested positive prior to 1975.[3]

Once toxaphene is used, its contamination persists for a lengthy time period. According to EPA, a greenhouse test showed that:

> 66% of the total applied remained after 34 months. In the same study, a significant reduction in the yield of turnips ...resulted when they were planted in soil that had been contaminated by toxaphene, even though the turnips were never sprayed...[4]

The pesticide can remain in soil for up to 14 years; residue levels have shown a steady rise in leaf and stem vegetables and fish, demonstrating that "it is apparent...toxaphene contaminates vegetables and fish, both basic food commodities."[5] A National Cancer Institute (NCI) study reports that the chemical "appears readily transported from its site of application either by water or by air...it is persistent in soil and water and accumulates at increased concentration in aquatic life." Thus, as toxaphene moves up the food chain in increasingly concentrated levels, it poisons off fish, birds that feed on fish, and other animals. Some of the highest levels are found in fish, waterfowl and other food consumed by humans.

When EPA announced its review of toxaphene in May 1977, it cited a 1976 NCI study revealing that "a significant increase in the incidence of cancerous growths did in fact develop" in laboratory animals fed toxaphene.[6] Two years later, NCI reported the results of six years of research: toxaphene had definitely caused liver cancers in test animals, and probably thyroid cancers as well.

Dr. Melvin Reuber, a specialist in toxicology and pathology who serves as a research scientist for NCI and as a consultant to EPA, said after reviewing and analyzing toxaphene study results going back almost 30 years that there is no question "toxaphene is such a carcinogen it boggles the mind."[7]

Another expert pathologist, Dr. Adrian Gross of EPA, formerly associate director of the Food and Drug Administration's scientific investigations office, said that the test data makes it "abundantly clear that toxaphene is an extremely potent carcinogen in rats as well as mice...I have never encountered an agent purposefully introduced into the environment...which had a carcinogenic propensity as clearly marked and as pervasive."[8]

There are numerous reported incidents of toxaphene poisoning occuring among farm workers and farmers, and it has been held responsible for at least six deaths.[9] The true numbers of such poisoning cases — only a fraction of which are ever reported — can be assumed to be much larger.

As EPA points out in its 1977 "Position Document" on toxaphene, the chemical may be causing subtler but no less harmful effects in humans:

> ...studies demonstrate that exposure to toxaphene
> produces changes in growth and bone composition in
> fathead minnows, brook trout, and channel catfish, as well
> as in black ducks and rats...[I]t may be presumed that
> current exposure is sufficient to produce these, or similar,
> chronic or delayed effects in humans.[10]

Toxicity To Animals

Toxaphene's most devastating effects have occurred in wildlife and livestock. According to EPA, "the pesticide has been implicated in large kills of waterfowl (that) have occurred in California, Arizona, South Dakota, and Texas ... These kills included white pelicans, cattle egrets, blackcapped night herons, greater blue herons, and various duck species. In a recent episode in Louisiana, dead and dying Brown Pelicans, a rare and endangered species, were found. When chemical analysis was completed, significant residues of toxaphene were found."[11] Both toxaphene and endrin were implicated in the 1974-75 pesticide poisonings that decimated the Gulf Coast population of the endangered brown pelican, killing at the very least 30 to 40 percent of the population of 350 to 450 birds in and around Louisiana.[12] Heavy rains and flooding had caused large runoffs of pesticides from farm fields into rivers and the ocean.

Toxaphene is also extremely toxic to fish and has caused hundreds of major fish kills. Small amounts of toxaphene in the water have caused weakened backbones to develop in fish, and even the minutest levels — parts per trillion — result in lower reproduction.[13]

Each year, about a million cattle are treated with several million gallons of toxaphene solutions to control parasites that cause mange or scabies. Under this U.S. Department of Agriculture (USDA) program, cattle as well as sheep are dipped in, or sprayed with, the toxaphene solution. This is required by state and federal law if any herd has been exposed to mange-infested cattle, even though there have been numerous instances of the chemical being implicated in killing or sickening livestock. In the West and Midwest, several thousand calves, cows and steers died either while being treated with toxaphene by USDA veterinarians, or immediately afterward. Thousand of others aborted their fetuses or gave birth to calves that died soon after birth.

In one recent incident, the poisoning of a farmer's cattle created a national controversy over the use of toxaphene. In December 1978, veterinarians from the California Department of Food and Agriculture and USDA treated the 850-head herd of Chico, California farmer George Neary, who urged them in vain not to use toxaphene. Some of the pregnant cows soon showed signs of poisoning, and a few weeks later, between 70 and 95 cows died, and 500 others either aborted their fetuses or gave birth to calves that died shortly thereafter. A dog that ate some flesh from a dead cow walked a short distance and dropped dead.[14]

The government veterinarians not only routinely used illegally high concentrations of toxaphene, they also improperly dumped the waste solutions on the ground, thereby polluting nearby streams, fields and Neary's corral. Furious that the state and federal veterinarians had poisoned his cattle and his land, Neary became so upset that he lost 30 pounds, vomited blood every day and eventually had to sell off his remaining cattle and get out of the livestock business. Ironically, it turned out that his cattle never had mange.[15]

This cattle treating program also presents the threat of exposing consumers of beef to residues of toxaphene, which can be absorbed into the skin of the cows when they are dipped or sprayed.

Because it had shown up regularly in milk samples in Arizona for over three years, toxaphene was largely banned by the state pesticide control board in early 1979 at the request of the dairy industry.

The only action EPA or any other federal agency has taken to restrict the use of toxaphene is a 1969 regulation cancelling such products for use on lettuce and cabbage except at certain rates of application where there is a warning statement, "Do not apply after heads start to form."[16] In 1977, EPA began the years-long process of reviewing the chemical's suitability for continued registration, but no final action is expected in the near future. In the meantime, millions of Americans continue to be exposed to this cancer- and mutation-causing chemical.

As EPA notes in its official "Position Document" on the pesticide, since toxaphene bioaccumulates to several thousand times the levels found in the environment, "Continued use of toxaphene . . . will undoubtedly lead to greater contamination of humans and foodstuffs by [this] environmentally recalcitrant pesticide . . ."[17] The term "recalcitrant" could also be applied to EPA's refusal to ban this destructive compound.

What You Can Do

If the available Integrated Pest Management (IPM) techniques (discussed in detail earlier in this book) were more widely utilized on crops, use of toxaphene and other dangerous pesticides could be drastically reduced at great savings to the farmer and no loss of crop yields.

Since toxaphene is so widely used and is found in food, air, and water, exposure to it can be minimized but not avoided. The recommendations for reducing the intake of other chlorinated hydrocarbon pesticides, (e.g., avoiding red meat and food high in animal fat) will minimize the exposure to toxaphene.

Chapter Fourteen

DDT: The Silencer Of Spring

Over increasingly large areas of the United States, spring now comes unheralded by the return of the birds, and the early mornings are strangely silent where once they were filled with the beauty of bird song. This sudden silencing of the song of birds, this obliteration of the color and beauty and interest they lend to our world have come about swiftly, insidiously, and unnoticed by those whose communities are as yet unaffected.[1]

Silent Spring, 1962

Twenty years ago, with the publication of Rachel Carson's classic work *Silent Spring*, the public first became acquainted with DDT, a pesticide then in widespread use that was killing off birds, fish and other wildlife at an alarming rate, in some cases to the point of near extinction.

Although DDT's ability to cause cancer in animals was also well-known at this time, and was noted in Carson's book, it took over a decade of efforts by environmentalists to persuade the government to ban most uses of the chemical. During this period, DDT was so massively and widely used that virtually the entire population of America became contaminated with detectable residues of DDT.

While there can be no question of its extreme toxicity to humans and wildlife, it is similar in many ways to many other members of its chemical family — the chlorinated hydrocarbons — that are still in widespread use today. In fact, after DDT was largely banned, it was in many cases deliberately replaced by much more acutely toxic chemicals, a development which both the government and the chemical industry made clear would happen at the time the action on DDT was taken.

And contrary to popular opinion, DDT has not been totally outlawed; since the partial ban, hundreds of millions of pounds have been manufactured in the U.S. for use domestically and throughout the world.

A Threat To The Earth's Life Cycles

When we moved here six years ago, there was a wealth of bird life; I put up a feeder and had a steady stream of cardinals, chickadees, downies, and nuthatches all winter, and the cardinals and chickadees brought their young ones in the summer. After several years of DDT spray, the town is almost devoid of robins and starlings; chickadees have not been on my shelf for two years; and this year the cardinals are gone, too.

It is hard to explain to the children that the birds have been killed off... "Will they ever come back?" they ask, and I do not have the answer.[?]

This passage from a letter written in 1958 by an Illinois house-wife mirrored the despair felt by many people throughout the country as they watched the bird life of their communities disappear because of DDT spraying.[3]

DDT has been held responsible for the drastic reduction in population of many bird species, especially those high on the food chain such as brown pelicans, gulls, falcons and eagles. Not only is DDT fatal to birds at certain concentrations, it also interferes with calcium production and causes birds to lay thin-shelled eggs that break before hatching. This phenomenon initially began to occur around 1947, when DDT first came into widespread use, and continued for 25 more years until the chemical was restricted.

The impact of DDT on predatory birds is perhaps most vividly demonstrated by the plight of the endangered American peregrine falcon, a spectacular hunting bird that can attain a diving speed of 175 miles an hour. This magnificent creature once nested throughout Canada and the United States, as far south as Georgia. But shortly after the large-scale introduction of DDT, the birds began to experience reproductive problems, and by 1947, eggshell thickness was found to be reduced by 15 to 20 percent.

By the early 1970's, the peregrine had disappeared from most of its former range, and seemed on the verge of extinction outside of Alaska and northwestern Canada. In 1973, the U.S. Department of Interior's Fish and Wildlife Service announced:

> ...as of 1970, there were less than 65 known active nesting sites for the peregrine falcon in western Mexico, the United States, and southern Canada...Today, their eastern limit is along the eastern slopes of the Rocky Mountains...In the southern Rocky Mountains, from 14 active nests in 1973, only three young were fledged...The major reasons for the decline of these falcons has been DDT and (its metabolic component) DDE.[4]

Other predatory and fish-eating birds fared no better under the onslaught of DDT and other chlorinated hydrocarbons. In Texas, the brown pelicans declined from an estimated 80,000 in the 1930's to about 20 in 1967. By mid-1973, the first year of relatively DDT-free farming along the Gulf Coast, the number had increased to 55, including 11 fledglings, the largest group of young birds in many years.[5]

All along the shores of the East Coast, empty nests of the fish hawk, or osprey, became a common sight. Even America's fierce and proud national symbol, the bald eagle, had by 1973 been decimated, in large part by DDT, with the disappearance of this now endangered species from most of its former range aptly symbolizing the devastating impact of chlorinated hydrocarbons on our country's environment.

DDT is also lethal to fish and shellfish at extremely low concentrations, in part because of the chemical is stored and bioaccumulates in organisms fed on by them. Shellfish that filter water, such as mollusks, are especially vulnerable to DDT saturation. Oysters, which filter as much as 40 quarts of water every hour, can, in a month, accumulate DDT at levels 70,000 times the amount in the water.[6] When such an organism is consumed by another creature, such as a bird or fish, the load of DDT is passed on up the food chain. This is why those predatory species at the top of the chain — such as birds of prey, or humans — are likely to end up ingesting and accumulating extraordinary amounts of DDT and other persistent chemical pollutants.

Since over 100 million pounds of DDT have ended up in just the oceans of North America, the implications of poisoning this vital food-producing system are ominous. Moreover, there is evidence that DDT levels in the sea in some areas have affected the tiny phytoplankton — a major source of the world's oxygen supply — significantly enough to reduce their photosynthesis. Such a development, while still not clearly understood, obviously has the most critical importance in its possible implications for the survival of life on earth.

These microscopic plants called phytoplankton are the basis for the food chain in the sea. They are of paramount importance to the health and stability of the world's oceans, ecosystems that are fundamental to the survival of the earth's natural systems and its life forms. One of the unique and indispensable functions of phytoplankton is to "trap" and convert sunlight, through photosynthesis, into an edible type of energy, or food. Phytoplankton are fed upon by zooplankton, which are eaten by young fish and krill, which in turn are consumed by large fish, whales, squid, sea birds and so on up the food chain to its last link, human beings. Thus, any disturbance of the fundamental component in this cycle, the phytoplankton, can disrupt and destroy the entire balance of life in the sea.[7,8]

DDT: In All Human Tissues And Breast Milk

DDT's ability to cause cancer has been known since the late 1940's, when the U.S. government's own testing showed it to be carcinogenic. It produces tumors in laboratory animals and is presumed to be capable of doing so in humans as well.[9] This fact has grave implications for the American people, since virtually every resident of the U.S. carries DDT in his or her body.

In 1963, when use was at its peak, some 188 million pounds of DDT were produced in the U.S., mostly for export abroad. Domestic use was greatest in 1959, when almost 80 million pounds were applied. The Library of Congress estimated in mid-1980 that over 4.4 billion pounds of DDT have been used for insect control since 1940, one pound for every human on earth. It is thought that some 110 million pounds of DDT have ended up in the oceans of North America.[10]

DDT is so persistent and pervasive that it has been found in the most remote regions of the world, thousands of miles from where it has ever been sprayed. Significant levels have been detected in the bodies of penguins and seals in Antarctica, seals in the Arctic Ocean and frogs living above 11,000 feet altitudes in the Sierra Nevada mountains. Dr. Charles Wurster, one of the pioneers in the fight to ban DDT, has called our experience with the chemical "a biological experiment of truly colossal proportions, using the entire world's biota as an experimental organism."[11]

In *The Darkening Land* (1972), William Longgood describes how the spread of DDT and similar pesticides to the furthest corners of the earth is accomplished mainly through wind currents:

> DDT circles the globe, going from one hemisphere to another, often riding on jet streams that wing along at 250 miles an hour. It is estimated that in the atmosphere at any given time, there is more than one billion pounds of DDT and its metabolites[12] ... it is reasonable to assume that of the 126,000 tons of chlorinated hydrocarbon pesticides sold annually, more than half enters the atmosphere[13] ... Studies of British rainwater indicated that in the United Kingdom, one inch of rain would deposit a ton of pesticides ... It is incredible how far and fast it travels. The man spraying his roses in Kansas City today can contaminate a polar bear in the Arctic tomorrow.

Longgood warns that "the real price of our folly" is that "once DDT is in the atmosphere, there is no escaping it. It becomes part of our environment ... It becomes part of the texture of meat, vegetables, milk, water."[14]

As the President's Council on Environmental Quality wrote in its 1975 annual report,

> Since the beginning of the National Human Monitoring Program in 1967, DDT residues have been found in 99+ percent of all human tissue samples: the figure in 1973 was 100 percent.[15]

DDT was detected in human breast milk as long ago as 1951. In EPA's survey of the milk of nursing mothers around the country, virtually all samples contained detectable levels of DDT.[16] As pointed out by Dr. Charles Wurster, mothers' milk would not be permitted to cross state lines in any other container.

By the early 1970's, it was estimated that the American people, collectively, had some 20 tons of DDT in their bodies.[17]

The Fight To Ban DDT

By the late 1950's, the enormous destruction of birds and other wildlife caused by DDT began to bring about a public outcry, especially when humans themselves were directly sprayed with the chemical. In 1957, when the U.S. Department of Agriculture was conducting a spraying compaign against gypsy moths in New York state, three million acres were treated, including such heavily populated communities as Long Island and Westchester County. Airplanes spewing out a solution of kerosene laced with DDT sprayed dairy farms, gardens, lakes, children and even commuters waiting for their trains — in some cases, on several occasions.

The legal fight to ban DDT nationally began when the Environmental Defense Fund (EDF), on behalf of the National Audubon Society, the Sierra Club, the Izaak Walton League and the Western Michigan Environmental Council, petitioned the Agriculture Department (USDA) in October 1969 (before the Environmental Protection Agency had been set up) to prohibit DDT. When it became clear that the massive spraying of DDT would continue, the environmentalists filed suit to halt its general use. After jurisdiction was transferred to EPA in December 1970, USDA entered the case on the side of the pesticide manufacturers, as did Congressman Jamie Whitten (D—Miss.), Chairman of the House Appropriations subcommittee that handled funding for EPA.

Congressman Whitten was known as an ardent proponent of pesticides and a friend of the chemical and agribusiness industries. He even "authored" a 1966 book, *That We May Live*, extolling the virtues of insecticides and attacking *Silent Spring*. (The book was based on a report written by staffers of his House Appropriations Subcommittee on Agriculture.)[18] According to *The Washington Post*, three pesticide companies subsidized the "sales" and distribution of the Whitten book.

The fight to ban DDT thus pitted the environmentalists against a massive array of powerful opponents, including farm interests, the chemical industry, USDA and various industry lobbyists and members of Congress. Although the scientific evidence was clearly on the side of those wishing to ban DDT, one quite prominent and distinguished scientist spoke out repeatedly in defense of the chemical. Dr. Norman E. Borlaug, who was awarded the 1970 Nobel Prize for his work on the "Green Revolution" in helping to develop fast-growing food plants, insisted that DDT was not only safe but necessary to prevent massive world starvation. He referred to those opposed to DDT as "hysterical environmentalists."[19]

Fortunately, EPA made its decision not on the basis of emotional rhetoric and propaganda, but of hard data and scientific evidence. The final decision to cancel most uses of DDT took almost three years and came after seven months of hearings and some 9,000 pages of testimony from 125 witnesses.[20] Despite a recommendation from the hearing examiner, so fallacious that it was summarily overruled, that DDT be largely retained, EPA found that the chemical was so dangerous and destructive that its continued widespread use could not be tolerated.

In announcing the ban on *most* uses of DDT, which took effect on 31 December 1972, EPA Administrator William Ruckelshaus made it clear that the chemical posed an unacceptable risk not just to wildlife and the environment, but to humans as well:

> ...Human beings store DDT...DDT is a potential human carcinogen. Experiments demonstrate that DDT causes tumors in laboratory animals...DDT presents a carcinogenic risk.[21]

Shortly after the EPA decision to largely ban DDT, the prophets of doom began predicting disaster for the country. Ronald Reagan echoed these arguments in one of his October 1978 radio messages, blaming "these insect plagues" on *Silent Spring*, EPA, and the fact that "the most effective pesticide, DDT, was outlawed...as fast as substitutes came on line, they were outlawed." He expressed outrage and astonishment that a chemical that "was harmless to human beings and that properly used . . . posed no threat to animal, bird or marine life," had been "banned by EPA on the theoretical grounds that it might, under some circumstances, someday harm someone or something."[22] On another occasion, Reagan complained that "The world is experiencing a resurgence of deadly diseases spread by insects because pesticides like DDT have been prematurely outlawed."[23]

Half A Million Acres Sprayed

When EPA did act to restrict DDT in June 1972, environmentalists were able to restrain their enthusiasm over what was widely viewed as a major victory for them. The conservationists were realistic enough to see that under the EPA ruling, the fight over DDT would continue to be waged on which uses to allow under the various exceptions included in the general ban. Thus, EPA ended up being sued — unsuccessfully — not just by the manufacturers but also by several environmental groups, including EDF, the National Audubon Society and the Sierra Club. The latter groups recognized that the uses that would still be potentially allowed represented a serious threat to human health and wildlife.

Sure enough, in 1974 the U.S. Forest Service received permission from EPA to conduct a massive DDT spraying program to control tussock moths on almost half a million acres of forest in Idaho, Oregon and Washington.[24] It is estimated that this program killed hundreds of thousands of songbirds in those areas; some 18,000 cattle and several hundred sheep were rendered unmarketable; and hunters in the area were cautioned against eating local game.[25]

It is now clear, and was at the time, that the government's decision to allow the use of DDT was a political and emotional one made under pressure from the timber industry and its allies, and was not based on facts. The program was totally unnecessary and may have actually hampered the dispersal of the moths.[26]

The precedent created by the spraying of the western forests was an unfortunate one, especially in light of the probability that the moth infestation, had it been allowed to run its natural course, would have demonstrated the futility of the spraying program: the disappearance of the moths would probably have occurred at about the same time without the help of DDT.

Indeed, natural dispersal is just what ended an outbreak of gypsy moths in the northeastern United States in the early 1970's, leaving little if any lasting damage. For example, one area of Westchester County in upstate New York, Ward Pound Ridge, suffered "nearly complete defoliation" in 1970, and the air was described by the park naturalist Nicholas Shoumatoff as being so thick with the silken threads of the moth caterpillars "you had trouble breathing." Although the area was not sprayed the following year, by 1972 Shoumatoff said he "could barely find one caterpillar." Thick foliage had returned to much of the park, and there was virtually no visable difference between the sprayed and unsprayed areas. According to Shoumatoff, the moths had caused no lasting damage, but had accomplished some "helpful thinning out" of the forests by eliminating trees that would not have survived in the long run.[27]

Spraying Passenger Airlines

Some other uses of DDT have quietly continued without attracting public attention. But in July, 1977, one of these programs received extensive publicity when newspaper and television reports revealed that the U.S. Department of Agriculture was spraying DDT on the insides of many passenger airlines flying from the East Coast to the West. The purpose of this program, which was begun in the early 1960's, was to prevent the spread of Japanese beetles to the western United States by use of a mixture of DDT and carbaryl (Sevin). (Carbaryl causes deformed fetuses, reproductive difficulties, and possibly cancer in laboratory animals, and has caused numerous cases of illness in humans and animals.[28,29] Carbaryl was one of the teratogenic pesticides that the 1969 HEW advisory panel on pesticides said should be "immediately restricted to prevent risk of human exposure.")

In July 1976, the Food and Drug Administration investigated the situation and determined that "swabs of food preparation surfaces and other food contact surfaces revealed DDT contamination." An FDA memorandum of 3 August

1976, on the subject revealed the DDT contamination inside the airlines of napkins, facial tissue, coffee cups, brewers, cans and pots, meal trays, paper cups, water dispensers, and seats.[30]

In November 1977, the Aviation Consumer Action Project successfully brought suit against the program and had it halted in April 1978. But since this program was never publicized, there may be no way to know for certain if and when it or others like it will be resumed.

DDT Remains With Us

Although it has been nearly a decade since DDT was largely banned, the chemical will remain with us for the foreseeable future with or without new spraying programs. Fish from the Great Lakes and elsewhere continue to contain higher than acceptable levels of this carcinogen. In August 1981, EPA found concentrations of DDT in the drinking water of several cities in southern Texas' lower Rio Grande Valley to be 80 times higher than federal standards. This area is a major producer of fruits and vegetables for the rest of the nation, and experts called the situation "a significant threat to public health."[31]

In Arizona, four years after a moratorium on its use had been declared in 1969, soil showed no perceptible change in DDT levels. It is estimated that DDT's half-life in soil in an arid climate exceeds 20 years.[32]

In America, blood levels of DDT in the general population average about 16.7 parts per billion.[33] Many residents of Guatemala and Nicaragua, where DDT is still heavily used on cotton and other crops, are reported to have over 30 times the DDT level of Americans.[34]

In Triana, Alabama, the Olin Corporation used to manufacture up to 25 million pounds of DDT a year, much of which was discharged into nearby streams. Although production of DDT there was halted in 1970, fish, ducks and other wildlife continue to show the effects of this decade-old pollution. In 1977, a fish taken from a local creek had DDT in its flesh at a level of 500 parts per million, which is about 100 times the tolerance level set by the FDA. The residents of the town, who used to regularly consume large amounts of locally-caught fish, have 10 times the amount of DDT in their blood than the average American. One man, Felix Wynn, an 85-year-old retired farmer, as of 1980 had the distinction of having the highest DDT level — 3,300 parts per billion — that had ever been found in a human, according to the federal Center for Disease Control in Atlanta, Georgia.[35]

Fortunately, since the general ban on most uses of DDT a decade ago, levels of the pesticide have largely begun to decline, in varying degrees, throughout the country. This has allowed eagles, ospreys, and other birds of prey to make somewhat of a comeback in some areas. Nevertheless, because of the persistence of this chemical, it remains spread throughout the environment and food chain, the long term results of which are yet to be determined.

Exports Of DDT Continue

In the years following the U.S. ban on most uses of DDT, the Agency for International Development (AID) subsidized, at taxpayers' expense, the export of about 900 metric tons of DDT, mainly for use on crops, from 1972-74. AID contracted with the Montrose Chemical Corporation in Torrance, California to continue producing DDT for the U.S.' foreign aid program. (Discharges from this plant are thought to have been at least partially responsible for the decline to the point of near extinction in the early 1970's of the California brown pelican.)

Some 44 million pounds of DDT are still produced in the U.S. and exported for use abroad, which is continuing to cause pollution, devastation of wildlife and domestic animals, and other such problems worldwide (see page 227). As the *Global 2000 Report to the President*, written by the State Department and the Council on Environmental Quality, points out, "On a global scale...if world application were phased out, a downturn in bioaccumulation and physiological effects might not occur for decades because of the DDT residue remaining in atmospheric, soil and oceanic reservoirs."[36]

Much of the DDT the U.S. exports is shipped abroad for malaria control programs. In fact, the U.S. ban on the pesticide made it even more attractive to foreign users, since supplies became more readily available at lower prices. But one result of using DDT for these purposes has been a resurgence in many countries of malaria, since numerous populations and species of the disease carrying mosquitos have become resistant to DDT[37] (see pages 82, 226).

Another result has been to subject American consumers to exposure to DDT from imported food. A very tiny percentage of such imports — less than one percent — is ever tested for pesticide contamination, but spot checks have given some indication of the extent of the problem. In 1976, the Agriculture Department refused to allow the marketing of some 500,000 pounds of beef imported from El Salvador and contaminated with DDT, some of which had levels of 19 times in excess of U.S. standards. The spraying of cotton with DDT has resulted in very widespread contamination of food supplies in El Salvador, Guatemala, Nicaragua, and Honduras.[38]

A 22 June 1979 report by Congress' General Accounting Office found that huge amounts of food imported into the U.S. commonly contain dangerously high levels of DDT and other cancer-causing pesticides which, although generally not allowed to be used in the U.S., have been manufactured here for export abroad. DDT is often found on such commodities as bananas and coffee beans.[39]

There are countless horror stories of DDT's use in foreign lands, but few are ever reported or publicized in this country. One such instance involved attempts to wipe out malaria in the jungle village of Malaysin in Borneo, an effort that ended up so upsetting the natural ecological balance that the village was thrown into chaos. When the World Health Organization (WHO) conducted a DDT spraying program, most of the malaria-carrying mosquitoes were temporarily killed off, but the local cockroaches — creatures renowned for adaptability, persistence and the ability quickly to develop immunity to chemical agents — survived, even though many had high levels of DDT that had accumulated in their bodies.

One of the village's major predators on insects was the gecko, an almost transparent lizard found throughout the Asian tropics. After feeding on the DDT-contaminated cockroaches, the geckos began to die off, or to be so weakened that they were easy prey for the village cats. The cats, too, were sickened and killed by eating the geckos that had concentrated such high levels of DDT that they could no longer evade their predators. Ultimately, with the cats and geckos decimated, the village was inundated with animals that these predators formerly kept in check: disease-carrying rats, and caterpillars that devoured the thatched roofs of the villagers' homes.[40]

DDT's More Toxic Substitutes

Perhaps the main reason for the ban on major uses of DDT being somewhat of a hollow victory from a human health standpoint is that DDT has been replaced in many cases by chemicals that are more dangerous than their predecessor. When 26 pesticide manufacturers and processors filed suit in October 1972 to block the impending DDT "ban", one of their major arguments was that this would mean that methyl parathion would be substituted, which, they contended, is even more harmful than DDT.[41] Moreover, when EPA Administrator Ruckelshaus announced in June that the "ban" would not take effect until the end of the year, he said:

> ...the chief substitute for most uses of DDT, methyl parathion, is a highly toxic chemical and, if misused, is dangerous to applicators...The introduction into use of organophosphates has, in the past, caused deaths among users who are untrained in their application...several years ago, four deaths occurred at the time methyl parathion began to be used on tobacco crops...A survey conducted after the organophosphates began to replace chlorinated hydrocarbons in Texas suggest a significantly increased incidence of poisonings.[42]

What all this demonstrates is that government and industry are well aware of the extremely hazardous and harmful nature of many of the pesticides that are, with our government's approval, in widespread use.

Thus, in America's modern chemical society, the concept of safety from poisonous pesticides has been lost. The best argument industry can come up with to oppose banning a toxic, cancer-causing chemical is that its substitute will be even worse!

What You Can Do

Despite dire predictions that American agriculture would collapse and be taken over by the insects, we have managed to survive the last decade without the massive use of DDT. In fact, the use of various Integrated Pest Management (IPM) techniques in place of DDT has proven quite effective at controlling mosquitoes. In California, for example, pesticide use in mosquito control districts was reduced to just one-tenth the former amount, while cutting labor and material costs as well as largely eliminating pesticide contamination. While 615,000 pounds of insecticides were used in 1962, only 63,000 pounds had to be applied in 1976. [43]

Similarly, in five California cities, 16 percent of city-owned shade trees were treated with insecticides before the introduction of IPM programs; afterwards, less than one-tenth of one percent (0.08 percent) had to be treated chemically, thus greatly reducing a major source of urban and aquatic contamination. [44]

The lessons here are clear, and can be applied to the many other toxic pesticides in widespread use that are unnecessarily polluting our environment.

Chapter Fifteen

Aldrin And Dieldrin: Part Of Us All

Like DDT, dieldrin is found in almost every human tissue sample collected. The incidence was about 99 percent in 1973.

President's Council on Environmental Quality, 1975[1]

Two of the most potent and pervasive carcinogenic pesticides are aldrin and dieldrin, which have caused cancer in animals at the lowest doses tested.[2] By the time they were restricted in 1974, dieldrin was being found to be present in 96 percent of all meat, fish and poultry; in 90 percent of all air samples; in the flesh of 99.5 percent of all Americans[3] and later in the milk of 80 percent of mothers tested.[4] Virtually the entire American population, and to a large extent, future ones as well, have been contaminated with one of the most potent cancer-causing chemicals known to man.

Indiscriminate Use

First introduced in 1948, by the early 1950's, aldrin and dieldrin were being heavily applied to a variety of crops, such as corn, peanuts, potatoes, cotton and sugar beets. Like other chlorinated hydrocarbons such as endrin and DDT, they decompose very slowly, taking up to 15 years to break down. In many instances, dieldrin and aldrin were routinely applied to the soil before planting a crop regardless of whether a problem with insects was anticipated. One study revealed that over half of the corn acreage in Iowa and Illinois to which chlorinated hydrocarbons were applied was treated without justification. In Illinois alone, 3½ million acres, amounting to 60 percent of the corn acreage treated with soil insecticide, did not need it.[5]

Once the soil was treated with dieldrin or aldrin, the pesticides remained there for over a decade, contaminating crops that are normally rotated with corn, such as oats, barley, soybeans and alfalfa. When livestock, given feed from such crops, foraged in the fields that had been harvested, or simply lay on the ground, they too picked up those pesticide residues (which can be absorbed through the skin), leading to contamination of poultry, meat, milk, cheese, cream and other dairy products.

Since home gardens and lawns received incredibly high concentrations of dieldrin from household insecticides, many of which were marketed for indoor use, some of the most common sources of exposure were cities and suburbs. Walking barefoot on treated grass, or just breathing city air or drinking the water, provided almost inevitable exposure to these carcinogens.[6]

In light of what is now known about dieldrin's toxicity —and was known for many years before being publicized — its massive and indiscrimate use in the 1950's even in urban areas is appalling. On one occasion — probably, not an uncommon one — dieldrin was sprayed throughout Tennessee by the state to eradicate Japanese beetles. In a suburban park near Knoxville, dieldrin granules were strewn over picnic tables, and state agents had to warn families to clean off the tables before spreading out their food.[7] Most human exposure to dieldrin has been much subtler, but no less dangerous.

Toxicity To Humans And Wildlife

A reliable rule of thumb is that chemicals that kill wildlife, especially birds and mammals, will also harm humans. By this yardstick, dieldrin must be considered one of the pesticides most dangerous to human health that has ever been used on such a massive scale.

In 1962, Rachel Carson wrote that dieldrin is 5 times more poisonous than DDT when swallowed and 40 times more so when absorbed through the skin. When tested on birds such as pheasants and quail, it was shown to be 40 to 50 times more toxic than DDT.

In 1955, dieldrin was sprayed on 2,000 acres of a Florida salt march to kill sand flies; it killed over a million fish of at least 30 species, and for a month no reproduction was observed among those that survived. Enormous numbers of crabs were also wiped out. These chemicals have also been linked to reproductive failure in birds, including bald eagles and other birds of prey that are high on the food chain.

In the late 1950's and early '60's, millions of acres in the southern United States were treated with dieldrin and heptachlor to eradicate fire ants. The ants continued to multiply and spread, but the chemicals did succeed in wiping out songbirds, wild turkeys, quail, raccoons, oppossums, armadillos, livestock, poultry and pets in many areas[8] (see pp. 98-99, 364-65).

In 1958, a year after the federal fire ant eradication program had gotten underway, reports from the deep South told of the ominous disappearance of almost all the birds from many areas in which they had been previously abundant. In the *Audubon Field Notes*, one correspondent wrote that his picture window, "which often used to frame a scene splashed with the red of 40 to 50 cardinals and crowded with other species, seldom permitted a view of as many as a bird or two at a time." A woman in Alabama reported:

> Our place has been a veritable bird sanctuary for over half a century. Last July, we all remarked, "There are more birds than ever." Then suddenly, in the second week of August, they all disappeared . . . There was not a sound of the song of a bird. It was eerie, terrifying. What was man doing to our perfect and beautiful world?[9]

Warnings distributed by USDA to local residents of areas to be sprayed made it clear that the Department was well aware of the highly toxic and dangerous nature of the chemicals being so widely and indiscriminately applied:

> Cover gardens and wash vegetables before eating them; cover small fish ponds; take fish out of pools and wash pools before replacing fish; don't put laundry out; keep milk cows off treated pastures for 30 days, and beef cattle 15 days; cover bee hives or move them away; keep children off ground for a few days; don't let pets or poultry drink from puddles.[10]

In 1968, the Environmental Defense Fund filed suit to block temporarily the spraying of three tons of dieldrin over Berrien County, Michigan to eliminate Japanese beetles. Although EDF's attorney, Charles Wurster, was able to show that between 10 and 80 birds and mammals, including sheep and cats, would be killed for each beetle eradicated, the U.S. Department of Agriculture and the state of Michigan ended up spraying 3,000 acres to control an estimated one beetle per acre.[11]

The toxicity of dieldrin's sister chemical aldrin is demonstrated by the fact that a pill-sized portion of it is sufficient to kill 400 quail. It especially affects the liver and kidney, and sickens or kills the offspring of laboratory animals.

In humans, dieldrin is known for its rapid effect on the central nervous system, causing convulsions as well as severe liver damage. In World Health Organization anti-malarial spray programs, after some mosquito populations became immune to DDT and dieldrin was substituted, numerous workers were poisoned. Some went into convulsions and died, and others had convulsions up to four months after exposure.[12]

But dieldrin's most fearsome quality is its startling ability to cause cancer. By 1962, the carcinogenicity of aldrin and dieldrin had been firmly established in testing by the Food and Drug Administration (FDA), and these results were later confirmed in additional tests conducted by their manufacturer, Shell Chemical Company.[31] Dieldrin has been shown to induce cancer in laboratory animals at the lowest levels for which it was tested, 100 parts per billion.[14] In animal tests, dieldrin causes cancers of the lungs, liver, lymphoid tissues and the thyroid, uterus, and mammary glands, as well as multiple birth defects. Separate studies conducted in Florida and Hawaii in 1968, and in New Zealand in 1967, have found that patients with cancer had high concentrations of dieldrin in their bodies.[15]

The Victims: 99 Percent Of Americans

Despite the known hazards of dieldrin and aldrin, it was not until a dozen years after the appearance of *Silent Spring*, including five years of litigation by the Environmental Defense Fund (EDF), that the EPA finally banned most uses of these chemicals in 1974 as "an unreasonable risk of cancer in man." By this time, studies showed that dieldrin, which is unequivocally carcinogenic in laboratory animals, was present in various foods common in the American diet and in the flesh of practically all Americans. It has been estimated that less than 10 percent of the nation's meat, dairy products and garden fruits was free of such contamination.[16]

In 1970, when about 10½ million pounds were used, 96.5 percent of people tested had detectable levels of dieldrin in their fat tissue; by the following year, the percentage had risen to 99.5.[17]

From 1970-72, 85 percent of the air samples monitored by EPA contained dieldrin, and it was the most frequently found pesticide in water samples. It was even being found regularly in household dust and in woolen products and rugs which had been mothproofed with the chemical.[18]

In a 1976 EPA study of breast milk contamination, 75 percent of the nursing mothers sampled had significant levels of dieldrin in their milk. The levels were so high that the average infant's daily intake would be nine times the FDA maximum levels: in one instance, dieldrin intake was 700 times greater than that permitted![19] In 1977, detectable amounts of dieldrin were found in human breast milk in over 80 percent of the women sampled: one mother from the southeastern United States, where the chemical was most heavily used, had a level of 12,300 parts per billion, the highest recorded.[20]

The Court Fight Over Dieldrin And Aldrin

EDF's time-consuming and costly fight to have the federal government ban dieldrin and aldrin began in December 1970, when, following a recommendation issued by the 1969 Mrak Commission that the chemicals be greatly restricted, the Fund requested that the newly-created EPA take this action.

In 1971, an EPA review of its own files showed that information had for years been on hand demonstrating that aldrin and dieldrin, as well as heptachlor, could cause cancer. After EPA consultant and pathologist Dr. Melvin Reuber studied the tissue slides from test animals and other data that had been submitted and on which EPA had relied in the registration process, he discovered that the results proved not the safety of the chemicals but rather their carcinogenicity.[21] (In 1974, when Dr.

Reuber did a review of the data on chlordane, he obtained similar results.)

Yet despite the discovery that the justification for the original registration of dieldrin and aldrin proved their danger instead of their safety, EPA Administrator William Ruckelshaus refused to remove the chemicals from the market. Later, the agency did begin the hearing process that can lead to banning a chemical.

The cancellation hearings began in August 1973. Arrayed against the EPA, EDF and the National Audubon Society were the pesticides' manufacturers, the Shell Chemical Company (a division of Shell Oil), a team of seven attorneys from the prestigious and politically powerful Washington, D.C. law firm of Arnold and Porter and three lawyers from the U.S. Department of Agriculture.[22] With virtually unlimited financial and legal resources at their disposal, the pesticide producers and users seemed intent on — and likely to — prolong the hearings for an extended period of time.

The "Banning" Of Aldrin And Dieldrin

The entire population of the United States is continually exposed to aldrin and dieldrin.

Administrative Law Judge
Herbert L. Perlman

But after a year of hearings, a surprise occurred. On 2 August 1974, Ruckelshaus' successor, Russell E. Train, suddenly suspended further manufacture of aldrin and dieldrin on the grounds that they posed "an unacceptably high cancer risk" to humans. Saying that he could not subject the public to the risk of any further delay, Train cited the results of the Food and Drug Administration's 1973 market basket survey, which found "measurable amounts" of dieldrin in 96 percent of all meat, fish and poultry; 88 percent of all garden fruits and 83 percent of all dairy products.

Train also explained why the agency had decided to act on the basis that the pesticides posed "an imminent hazard to public health":

Evidence based upon human subjects is virtually impossible to obtain. The general human population is continually exposed to a variety of chemicals. A significant "control group" is thus impossible to establish.

Moreover, to await the 20 to 30 years of exposure necessary to determine the ultimate effect is only to wait until the damage to an entire generation of humans is complete. We reject the "body count" approach to protection against cancer.[23, 24]

Unfortunately, by this time, an entire generation of Americans — in fact, virtually the entire population — had been contaminated.

Meanwhile, the hearings on permanently cancelling the pesticides continued, and Shell maintained its fight to keep them in use despite its own animal studies showing that dieldrin in concentrations as low as 0.1 parts per million caused cancer in several body organs, including the liver.[25] Shell's sales of aldrin and dieldrin in 1974 were an estimated $10 to $15 million. At this time, sales of the parent company, Shell Oil, were almost $8 billion, with net income after taxes of over $620 million. Thus, in order to make sales amounting to less than one-fifth of one percent of the company's total, Shell was willing, indeed determined, to continue to sell a product that was exposing virtually the entire American population to the risk of cancer.

Shortly after EPA's suspension action, the presiding judge and the cancellation hearings reached a similar conclusion. Almost four years after EDF's petition and 28,000 hearing record pages later, the presiding administrative law judge, Herbert L. Perlman, concluded that aldrin and dieldrin represented "an unreasonable risk (to) the American public". The evidence presented showed that humans were being exposed to these pesticides from the time of conception until death, since they circulated in the bloodstream, passed through the placenta and entered the fetus, and were present in mothers' milk.

Expert testimony given during the cancellation hearings revealed that in 1973, 97 percent of the cows' milk samples tested in Illinois contained dieldrin, and that it had been found in breast milk at levels as high as 31 times that recommended for commercial milk by the World Health Organization.[26]

"The entire population of the United States is continually exposed to aldrin and dieldrin," Judge Perlman stated.

In handing down his decision, Judge Perlman also observed that the evidence showed the unavoidability of contamination:

> Surveys conducted by the Food and Drug Administration show that dieldrin is found in as much as 96 percent of all meat, fish and poultry "composite samples" tested, and 85% of all dairy product "composite samples" tested. In addition, EPA surveys indicate that dieldrin is in approximately 90 percent of all air samples taken nationally, and residues of dieldrin have been found in virtually all of the humans included in the EPA human monitoring surveys.

Concluding that dieldrin and aldrin should be largely banned, Judge Perlman based his findings on the characteristics of such chemical carcinogens:

> ...the scientific inability to determine a safe threshold level for man, the fact that the chemicals are carcinogenic at the lowest doses tested, that residues of dieldrin in laboratory species which developed cancer from dieldrin approximate those residues in the American population, the irreversibility of the carcinogenic effect once set in motion by the chemical carcinogen and the long latency period during which the disease has actually set in and is developing but is not yet manifest.[27]

As a result of these findings, EPA finally acted to ban permanently most uses of dieldrin and aldrin in October, 1974. However, existing stocks were allowed to be exhausted; in 1975, some 2½ million acres of corn were treated with these substances that had been determined to be a grave threat to human health.[28]

Continued Uses — And Hazards — Of Aldrin and Dieldrin

These continued uses were allowed despite the overwhelming evidence of dieldrin's and aldrin's ability to cause cancer, and EPA's stated position during the hearings that if these pesticides were used at current levels for another year and a half, as many as 230,000 Americans could ultimately develop cancer as a result.[29] As Peter Crane, editor of the *Environmental Law Reporter*, wrote concerning the dieldrin hearings, "Dieldrin's omnipresence in the diet has made the entire population of the United States into the experimental group for the study of the carcinogenic effects of the chemical."[30]

And although most uses of aldrin and dieldrin have now been cancelled, their use is still permitted by ground insertion for termite control; dipping of non-food roots and tops and in moth-proofing, all of which represent continuing sources of exposure.[31]

After the U.S. ban took effect, the Shell Chemical Company — which holds the patent on dieldrin — did not stop making the product. Instead, some production was switched to a plant in the Netherlands and the company continued to sell these poisons to other countries in which they were not prohibited. In some Third World nations, pesticides banned in the U.S. are sold over the counter and end up in the U.S. on many imported foods.

In the summer of 1975, in the Brazilian village of Bahia, thirteen children living on the same street were apparently accidentally poisoned to death by dieldrin or aldrin. Before dying, they became extremely ill, began to perspire profusely, vomited, foamed at the mouth and went into convulsions. Extraordinarily high levels of dieldrin and aldrin were found in the liver of one of the girls, and it is thought that the victims ingested poisoned meat laced with these chemicals that was put out to kill stray dogs.[32]

Although most production of aldrin/dieldrin was moved out of the U.S. in 1975 after the ban, the following year's exports from this country of these chemicals amounted to the total 1976 U.S. production of 342,025 pounds, with 88,000 pounds planned for manufacture and export in 1977.[33]

It is no secret that aldrin and dieldrin are still being sold and used in underdeveloped nations; indeed, they are widely advertised and promoted. As of 1979, aldrin was being publicized on a "huge, ugly" billboard on the shores of beautiful Lake Titicaca in Peru. The chemical is manufactured there by Shell and is used extensively in that country.[34]

Despite the partial ban on dieldrin and aldrin in the U.S., the pervasive presence and persistent nature of these chemicals make them a continuing hazard to human health in this country. During the cancellation hearings on these pesticides, an incident occurred that demonstrated the dangers of having them so readily available. In March 1974, the Agriculture Department discovered that dieldrin in chicken feed had contaminated some 8.5 million chickens in Mississippi at levels above what the FDA allows as acceptable, so the fowl had to be taken off the market and destroyed. This was done despite behind-the-scenes pressure brought to bear on the government by Representative Jamie Whitten (D-Miss), Chairman of a powerful House of Representatives appropriations subcommittee with jurisdiction over EPA's budget.[35]

Although grocery items are rarely checked for dieldrin, it is still turning up as a contaminant in our food. On 26 June 1980, the U.S. Department of Agriculture warned that turkey products from Banquet Foods Corporation, of St. Louis, Missouri could contain unacceptable levels of dieldrin. Assistant Secretary of Agriculture Carol Tucker Foreman asked the company to recall the two million or so packages of frozen turkey dinners, pies, and other turkey products that may have been contaminated, since the agency did not have the authority to do so. Banquet initially refused to comply with this request.[36] A few days later, the firm agreed to the recall after the situation had received widespread attention in the press.[37]

Thus, dieldrin continues to be a part of our daily lives. It is still produced for use in the U.S. and even more extensively abroad; it remains in human tissue and mothers' milk; at times it turns up in excessive amounts in our food. It is a mistake our generation of Americans will live with for the rest of our lives, that could kill perhaps millions of us and that is being repeated in the use of hundreds of other chemicals to which we are daily exposed.

What You Can Do

The banning of major uses of aldrin and dieldrin clearly demonstrates that we do not have to rely on such extremely toxic chemicals to grow our food. Despite widespread predictions from industry that the restricting of these pesticides would result in huge losses of and price increases in corn and other crops, we are growing more of these foods than ever before. Numerous alternatives to these chemicals have

been successfully used, although in some cases the substitutes have also been very toxic chemicals.

With the general ban on most of their uses, the persistence of aldrin and dieldrin in the food chain can be expected to gradually subside. To avoid contact with uses that are still allowed, you should refrain from wearing garments mothproofed with aldrin or dieldrin.

If dieldrin is used to treat subsurface soil for termites, it is best to drill diagonal holes from outside of the house. Any holes made in the basement floor should be tightly sealed. Above-ground wood should not be sprayed, since the termites daily descend into the ground.

Chapter Sixteen

Chlordane And Heptachlor: With Us From Conception Until Death

...humans are exposed to heptachlor epoxide from the moment of conception on throughout life. This is sufficient basis for grave concern for the possibility that humans...may react to such exposure by producing malignant tumors.

Russell E. Train
EPA Administrator
November, 1974[1]

...heptachlor is carcinogenic at the lowest levels tested (1/2 part per million) in laboratory experiments...significant levels are found in 95% of samples of adipose (fat) tissue taken from this country's populace...

Environmental Defense Fund
October, 1974[2]

As a result of their ruinous effect on wildlife and domestic animals, chlordane and heptachlor were among the first widely used pesticides, along with DDT, to attract national attention. Because of the dangers the chemicals posed to humans and the environment, environmentalists have long tried to have them banned and did succeed in having the

Environmental Protection Agency (EPA) prohibit most uses of them.

In announcing this suspension and later cancellation for major uses of chlordane and heptachlor, EPA Administrator Russell E. Train stated that the action was being taken on the grounds that the chemicals posed "an unreasonable risk to man and the environment", including the "'imminent hazard' of cancer in man."[3] Yet, even after determining that the chemicals were powerful cancer-causing agents, EPA refused to ban them, instead ordering a gradual phase-out of most of their uses. As a result, chlordane and heptachlor continued to be heavily used and to build up in increasingly high amounts in the bodies of virtually all Americans. Their remaining uses, though only a fraction of what existed formerly, continue to present a serious threat to the American public.

Destruction Of Wildlife

In the late 1950's, heptachlor was heavily sprayed on millions of acres throughout the southeastern United States to eradicate fire ants. Although the ants continued to multiply and spread, native wildlife was devastated by the spraying program. The destruction was so massive and caused such a public outcry that the chemical became the subject of a hard-hitting 1959 article in *Reader's Digest* describing the elimination of much of the region's wildlife, one of the first mass-circulation pieces detailing heptachlor's hazards.

One wildlife biologist attributed the 1957-58 spraying of heptachlor in Hardin County, Texas to the destruction of 95 percent of the birds usually seen along the roads. Poisoned birds were seen to be trembling and going into convulsions before dying. In the spring before the spraying, the fields were described as full of birds singing to establish their territories, building nests and laying eggs. Forty-one nests with eggs were counted in one clover field, but within a few days after the pesticide was applied, only three of the nests were still active.

In other sprayed areas, similar results ensued. Hawks, wild turkeys, cardinals, woodpeckers, mockingbirds, doves, quail, woodcocks, meadow larks, blackbirds, and others were largely wiped out over wide areas, and mammals and other animals were killed off as well, including foxes, raccoons, chipmunks, opossums, armadillos, squirrels, rabbits, snakes, turtles and massive numbers of fish.[4] Livestock and pets were also poisoned, and human residents of the sprayed areas were heavily exposed to heptachlor by direct contact and through contaminated food, water and dust.

Silent Spring: A Communist Plot?

Despite these events, the chemicals' manufacturer, Velsicol Chemical Corporation, realizing the potential profitability of chlordane and heptachlor, continued aggressively to market and sell the pesticides. By the early 1970's, they had become two of the most widely used pesticides for home and agricultural uses.

The firm was determined to increase its sales of the pesticides and from the beginning went to great lengths to defend them. Velsicol even had the temerity to try to halt the publication of *Silent Spring* because of the adverse effect it might have on sales of chlordane and heptachlor. After Rachel Carson's exposé of the dangers of pesticides had been serialized in *The New Yorker* in June 1962, Velsicol sent a five-page letter to the company which was about to publish the book urging that it not do so. The 2 August letter, from Velsicol's Secretary and General Counsel, Louis A. McLean, complained to Houghton Mifflin, Inc. that Carson had made "inaccurate and disparaging statements" about chlordane and heptachlor. Velsicol went so far as to suggest that her book was part of a Communist plot to starve Americans:

> Unfortunately, in addition to the sincere opinions by natural food faddists, Audubon groups and others, members of the chemical industry in this country and in western Europe must deal with sinister influences, whose attacks on the chemical industry have a dual purpose: (1) to create the false impression that all business is grasping and immoral, and (2) to reduce the use of agricultural chemicals in this country and in the countries of western Europe, so that our supply of food will be reduced to east-curtain parity. Many innocent groups are financed and led into attacks on the chemical industry by these sinister parties.[5]

The book, Velsicol seemed to be suggesting, was part of a subversive plot to destroy America. But as more and more evidence was turned up on chlordane's and heptachlor's dangers to man and the environment, it became apparent that it was these chemicals, and not their critics, that represented the real danger to the country.

Toxicity To Humans And Animals

Chlordane's toxicity has frequently been demonstrated in laboratory experiments, and was discussed by Rachel Carson in *Silent Spring* some 13 years before EPA finally acted to restrict its use. She points out that only 2.5 parts per million can cause disintegration of liver cells, and writes that it "has all the unpleasant attributes of DDT plus a few that are peculiarly its own." She quotes the chief pharmacologist of the FDA, Dr. Arnold Lehman, as describing chlordane as far back as 1950 as "one of the most toxic of insecticides — anyone handling it could be poisoned."[6]

In more recent years, there have been numerous reports of humans exposed to chlordane and heptachlor developing leukemia and other disorders, including young children of pregnant mothers exposed to the chemicals in houses that had been treated for termites.[7]

One individual who has observed and experienced the effects of chlordane first hand is Dr. Guy Arnold, a veterinarian from Kalkaska, Michigan who was partially paralyzed and blinded after being regularly exposed to high levels of chlordane in Vietnam. There, he worked with scout dogs, and often rubbed the pesticide on them to kill ticks and fleas. A study by the U.S. Army later revealed that in 20 out of 22 dog-scout platoons in Vietnam, dogs died of internal hemmorrhaging. Dr. Arnold believes that his frequent and prolonged exposure to chlordane caused his medical problems.

Chlordane was also widely used to spray barracks for cockroaches; when Dr. Arnold tried to restrict the uses of the chemical, his superiors threatened to have him transferred to more hazardous areas, such as the Demilitarized Zone. And since returning home, his troubles have continued. When he was operated on for his paralysis, he spent 15 weeks in the hospital, five of them in intensive care. In December 1981, he lost his driver's license because of his poor eyesight, making it difficult for him to make a living.[8]

Chlordane can enter the body in a variety of ways: through ingestion of contaminated food, inhalation in the form of dust or spray, or absorption through the skin. Once inside the body, chlordane, like other chlorinated hydrocarbons, builds up. Thus, experimental animals given a diet of food containing 2.5 parts per million of chlordane can cause the accumulation of the substance at levels of 75 parts per million.[9]

Because of misinterpretation or fabrication of the testing data on the pesticides, results showing them to be carcinogenic were presented as just the opposite. Cancerous tumors were reported as benign, though later analysis by pathologists "revealed a high level of" cancerous tumors, with one group of animals showing a cancer incidence rate of over 90 percent.[10]

In 1971, EPA requested that pathologist Dr. Melvin Reuber review the data, including the actual test animal tissue slides, which the agency had accepted to demonstrate heptachlor's safety. In 1974, EPA asked Dr. Reuber to do the same for chlordane. In both cases, he found that the data which had been submitted, and on which EPA had relied in granting registration, actually showed that the chemicals caused cancer.[11] When EPA announced that it had found the chemicals to be hazardous, Velsicol had a research laboratory run additional tests on them. In 1975, Velsicol submitted a report to EPA stating that no tumors had been found in the test animals, but withheld another report which found that there were, in fact, many tumors produced, some of which were malignant. As a result of this misrepresentation, several Velsicol officials were indicted in December 1977 by a federal grand jury in Chicago and charged with concealing from EPA studies showing that chlordane and heptachlor can cause cancer. (Because of irregularities in the grand jury's investigation, these charges were later dropped.)[12] The case was finally resolved in January 1981 when Velsicol, facing a possible fine of $1 million, was fined only $1,000 by U.S. District Court Judge George Leighton in Chicago.[13] This penalty amounted to less than a dollar for every 200,000 Americans contaminated with these cancer-causing compounds.

EDF Vs. The Pesticide Lobby

In October 1974, the Environmental Defense Fund (EDF) petitioned the EPA to ban heptachlor and chlordane, on the grounds that they posed "an imminent health hazard to man:"

> ... heptachlor is carcinogenic at the lowest levels tested
> (1/2 part per million) in laboratory experiments involving
> mice and rats. Further, ... the incidence of heptachlor in
> human food is currently very high in the United States,
> especially in meat, poultry, fish, and dairy products, where
> it approaches 70%. Additionally, significant levels of
> heptachlor are found in 95% of samples of adipose (fat)
> tissue taken from this country's populace, as well as in
> human mother's milk.[14]

At this time, the chemical was also turning up in umbilical cord blood and in human fat at about the same levels found in test animals that were contracting cancer from it — about 1/2 part per million.[15]

In November 1974, EPA Administrator Russell E. Train issued a "Notice of Intent" to cancel most uses of heptachlor and chlordane, thus commencing the years-long process of hearings and appeals to determine if the chemicals should be banned.

At this time, chlordane was one of the most widely used household and garden insecticides, and its popularity was booming. In 1971, some 11 million pounds were used, and the following year, usage increased by almost 50 percent.[16] By 1974, over 21 million pounds were being produced;[17] chlordane and heptachlor had become Velsicol's most profitable products. For this reason, Velsicol fought the cancellation proposal every step of the way, and brought enormous power and pressure to bear to keep the chemicals in use.

The company was represented largely by the politically-powerful Washington, D.C. law firm of Williams, Connolly, and Califano. It was the firm of famed criminal attorney Edward Bennett Williams and Joseph Califano, President Carter's first Secretary of Health, Education, and Welfare, who is best remembered for his campaign to alert Americans to the hazards of cigarette smoking. A total of at least eleven lawyers were listed in the official record as "active participants" representing Velsicol.[18] Also arrayed against EDF attorneys

Jacqueline Warren and William Butler were the U.S. Department of Agriculture (with four lawyers), the National Agricultural Chemicals Association, various crop grower and pest control associations, and the states of Hawaii, Iowa, Missouri and Vermont. EPA, which was represented by six attorneys, agreed with EDF on some issues, and opposed it on others. The only real allies EDF had were the facts: in the end, they proved decisive.

Most Uses Suspended

A year after the cancellation proposal, in December 1975, EPA Administration Russell Train announced an immediate temporary ban on *most* applications of the chemicals, suspending their use on lawns, gardens, turf, and for household pest control. Even though he said he was taking the action on the basis of an "imminent hazard of cancer in man," Train allowed other uses of chlordane and heptachlor to continue, such as on corn, strawberries, pineapples, citrus crops, fire ants, termites and other crops and insects.[19]

In announcing his decision, Train laid out in detail the case against the chemicals, and pointed out that the evidence of their ability to cause cancer — which has been shown at levels of just 1/2 part per million — goes back many years. The 1969 HEW panel — the Mrak Commission — called for "corrective action" to be taken to restrict the chemicals, and judged heptachlor's breakdown product (heptachlor epoxide) to be "positive for tumor induction;" 90 percent of the animals fed the substance developed cancer in one experiment. "A series of careful experiments," in Train's words, have turned up similar results, showing "that chlordane and heptachlor cause cancer" in animals. "There is no known minimum dosage of cancer-causing chemicals below which exposure is safe," Train observed.[20]

These chemicals' ability to cause cancer was considered a particularly grave problem, since by 1974 the entire American population was being exposed to them on a regular basis. In the words of Russell Train, "... humans are exposed to heptachlor epoxide from the moment of conception on throughout life."[21]

By 1974, chlordane and heptachlor had been so heavily used that residues were being found in the tissues of pronghorn antelopes and mountain goats in South Dakota, which, EPA noted, "is indicative of its widespread distribution." In surveys of air samples from around the country, both chemicals had been detected.

Not surprisingly, residues of these chemicals were regularly turning up in food — and in humans. According to Train, "residues of chlordane and heptachlor are found in human tissue, human milk, and human fetuses, in air, soil and water, meat, fish, and poultry, and to a lesser extent, in raw agriculture products for human consumption." He also noted that heptachlor epoxide was being found in between 93 and 97 percent of human tissues sampled in studies carried out in the early 1970's.

Train also cited the alarming fact that the organs of 10 still-born infants at two Atlanta hospitals contained high levels of heptachlor epoxide. The greatest concentrations were found in the heart, liver, and adrenal gland, clearly demonstrating that the chemical is transferred from the mother to the infant across the placenta. The average concentrations exceeded one-half part per million, a level at which the substance has been shown to cause cancer in animals. Breast milk samples have also been found to contain heptachlor epoxide at about this level.[22]

Despite the suspension, Velsicol continued to manufacture and sell large quantities of the chemicals. In 1976 and 1977, over 2-½ million pounds of heptachlor and 300,000 pounds of chlordane were still being produced in the United States, mainly for export to foreign countries.[23] Much of this amount was sent abroad at taxpayer expense by the Agency for International Development as part of the U.S. foreign aid program.

Most Uses Phased Out

Finally in March 1978, after a legal fight lasting over three years, and costing environmentalists and EPA thousands of hours and millions of dollars, EPA finally agreed in March 1978 to phase out most uses of heptachlor by July 1983, and of chlordane by 1980.

By the time the agency acted to phase out major uses of chlordane and heptachlor, they had been so widely used in home gardening and agriculture — mainly on corn grown as animal feed — that EPA says "virtually every person in the United States has residues ... in his body tissues."[24]

As a result of past and continued use of chlordane and heptachlor, dangerous levels of these carcinogens still turn up in food. In early 1980, for instance, the Department of Agriculture's food lunch program sent 20 tons of ground pork contaminated with high levels of chlordane and heptachlor to school systems in Louisiana and Arkansas, 7 tons of which is believed to have been consumed before the other 13 tons could be recalled.[25]

Moreover, EPA's settlement exempted the use of these substances in dipping the roots or tops of non-food plants and in sub-surface ground insertion for termite control,[26] thereby allowing the continued potential contamination of homes, ground water tables, wells and other water sources.

In fact, just having such toxic chemicals so readily available to anyone with the price to buy them presents a potential danger to the public. In December 1980, water mains serving 20,000 people in Pittsburgh, Pennsylvania were deliberately polluted with chlordane and heptachlor, resulting in thousands of residents losing their source of drinking water, and causing many to suffer from nausea, diarrhea, chest pains and stiff necks and other muscles.[27]

Thus, chlordane and heptachlor join a long list of other chemicals known to cause cancer which, though largely banned or restricted on paper, are still allowed to be used in ways that guarantee continued public exposure to them. By drastically reducing the amounts and applications in which these pesticides can be used, EDF's hard-won victory could save the lives of countless thousands or even millions of Americans. Still, if chlordane and heptachlor are able to cause human cancers — which seems indisputable — most of the harm may have already been done. Because virtually the entire U.S. population has been exposed to these chemicals, in many cases for decades, most of the damage may have already been done.

What You Can Do

As with aldrin and dieldrin, the ban on most uses of chlordane and heptachlor has not resulted in the destruction of the U.S. agriculture system and the starvation of the populace.

Yet, all of the above restricted chemicals continue to contaminate our diet as residues of coffee, sugar, and other imported food on which they are used abroad. Thorough washing of imported fruits and vegetables can remove much of the pesticide residues. But the best way to eliminate continued exposure of American consumers to domestically-banned chemicals would be for the FDA and EPA to refuse to permit such residues on imported food. Until such action is taken, these chemicals will remain a threat to the health of the American people.

When chlordane and heptachlor are used for underground termite control, the same precautions recommended for dieldrin also apply.

Epilogue

Crippling The Health And Environmental Laws

Over the last few decades, the "Poisoning of America" has been a bipartisan affair, with both major political parties giving lip service, but little else, to the concept of protecting the public from deadly chemicals.

Under both Democratic and Republican administrations, the Environmental Protection Agency (EPA) and other government agencies have failed to administer adequately the laws designed to stem the flow of toxic chemicals contaminating our environment. Often, when such agencies have moved to restrict such chemicals, powerful Senators and congressmen from both parties have intervened on behalf of industry, and have exerted pressure to prevent effective action from being taken.

The Carter administration, which came into office with lavish promises to protect the environment, actually did very little to remedy the toxic chemical problems it inherited. EPA's record during these years was generally one of obstruction, delay and indifference. The only government agency that assumed a meaningful leadership role was the Council on Environmental Quality (CEQ), which, unfortunately, has no power to make or implement policy.

One example of the administration's inaction concerns an episode involving several agencies that can be expected to cost the lives of perhaps tens of thousands of children in foreign countries. In 1978, because the Departments of State and Commerce and the Consumer Products Safety Commission (CPSC) refused to act, the leaders of a congressional committee urged President Carter to halt the imminent export of millions of children's nightwear garments that had been prohibited for sale in the U.S. because they had been treated with a powerful cancer-causing chemical known as Tris. Carter ignored the request from Congress, and as a result, millions of children in foreign lands were unknowningly exposed to this potent carcinogen that had been banned in America (see pp. 287-90).

In a few areas, such as the "Superfund" waste site cleanup bill, important progress was made; but these accomplishments were few and far between. EPA successfully resisted efforts by environmentalists to have several cancer-causing pesticides banned, and only a handful of the hundreds of deadly chemicals in widespread use were even subjected to significant restrictions. And under rules promulgated by EPA, it has become virtually impossible for environmental, labor, and public interest groups to participate effectively in the cancellation review proceedings for such chemicals (see page 124).

But the Carter administration's record of inaction on toxic chemicals was more a failure of will and competence than a conscious effort to undercut the nation's ability to protect itself. Nothing that took place during those four years matches the actions of the first year of the Reagan administration, during which a determined attempt has been undertaken to cripple or abolish many of the nation's most important health and environmental laws and regulations.

Within days of taking office, President Reagan and his cabinet began a concerted effort to cancel, postpone or weaken dozens of regulations that protect the public from toxic chemicals. The administration has moved quickly to dismantle many of the hard-won environmental gains of the last few decades.

On 17 February, 1981, President Reagan cancelled President Carter's executive order, issued a month earlier, tightening the procedures for allowing U.S. firms to ship abroad extremely hazardous chemicals and other products banned for sale in the U.S. A few months later, the administration drafted plans to eliminate almost all rules requiring that foreign governments be notified of such hazardous shipments on the grounds that the rules "placed U.S. exports at a competitive disadvantage." (see pages 237-40)[1]

In February 1981, in a move admittedly aimed at curtailing the powers of EPA and the Occupational Safety and Health Administration (OSHA), President Reagan signed an executive order restricting the ability of government departments and agencies to issue regulations.[2] Shortly thereafter, Vice President George Bush, as head of the administration's Task Force on Regulatory Relief, announced that EPA would propose easing Clean Air Act pollution control regulations in California dealing with oil refineries and automobile plants. This action was widely viewed as setting a precedent for changing enforcement of the Act that "could have a major environmental impact throughout the country."[3]

Also being prepared, as *The Washington Post* reported, was:

a target list of major health, safety, environmental, and social regulations that the President's aides are determined, in the words of budget director David A. Stockman, to "defer, revise, or rescind." They range from testing of food and drugs to safety and pollution control equipment on autos to industry's handling and disposal of hazardous chemical wastes.[4]

In March, Bush targeted 27 regulations for review and change, many of which comprised some of the government's most important environmental and job safety rules. Included in these were regulations governing the disposal of toxic wastes under the Resource Conservation and Recovery Act.[5] In April, President Reagan issued a statement proposing to ease or eliminate 35 air quality and safety regulations dealing with automobile emission standards and other rules dealing with the automobile industry.[6] In August, Bush announced additional proposed changes in 30 regulations, including allowing more lead in gasoline, which would represent a serious potential threat to children in urban areas, who absorb the toxic metal from the air and dust. Other targeted regulations included those under the Toxic Substances Control Act dealing with notification and testing requirements for the introduction of new chemicals and pesticides.[7]

Industry-oriented people with long records of opposition to environmental protection were appointed to head up key departments and agencies dealing with toxic chemicals and pollution.

Reagan's budget czar David Stockman, head of the Office of Management and Budget (OMB), was given virtual veto power over spending by EPA. The first budget Stockman sent to Congress contained major cuts in conservation, solar energy and environmental cleanup activities, with some programs being eliminated entirely. The budgets of such agencies as EPA and OSHA were particularly hit, especially programs dealing with air pollution, groundwater protection, pesticides (cut by $9 million) and cleaning up toxic waste dumps (cut by $50 million). By late March, the Energy Department had drafted legislation to abolish or drastically curtail almost all of its programs to promote energy conservation and to develop solar and other renewable fuel sources, including programs dealing with solar energy research and development; energy conservation for commercial buildings; consumer education on conservation; residential energy efficiency; ocean thermal and wind energy development; small-scale hydroelectric projects and research on electric vehicles and methane fueled transport.[8]

In contrast, the administration proposed to increase spending on nuclear energy to $1.6 billion, not including military programs.

Funding for the Consumer Products Safety Commission (CPSC) was cut by 30 percent, prompting the acting chairman, Stuart M. Statler, to say that such reductions would be "devastating" and would prevent the agency from carrying out its mission to protect the public from dangerous products.[9] He pointed out that because of past actions by CPSC involving chemicals, "several thousand cases of cancer will not occur." And Commissioner David Pittle said that the cuts "seem designed more to hamper the ability to function than to effect a real savings for the nation's economy." Later, Statler said that as a result of these cutbacks in budget and staff, "hundreds of more people will be killed each year; thousands more will be injured." During this period of sharp budget cutbacks, CPSC's new Chairman, Nancy Harvey Steorts, spent some $10,000 to redecorate her office.[10]

In May 1981, the Reagan Administration recommended that CPSC be eliminated, and OMB Director Stockman told a Senate committee, "Our preference would be to abolish the agency entirely." While the matter is being resolved, CPSC in September 1981 began to reduce its staff by 200 employees and closed half of its ten regional offices and its office of communications. These cutbacks will severely reduce CPSC's ability to deal with the over 30 million injuries and 30,000 deaths that occur each year because of dangerous products.[11]

The administration also indicated initially that it intended to abolish the President's Council on Environmental Quality (CEQ), the one agency that has consistently taken the lead in studying and calling attention to the nation's environmental problems. It ended up cutting CEQ's budget by 72 percent and firing virtually all of its employees, even those hired during the Nixon administration, thereby removing the experience and expertise built up over more than a decade.

One of the most disturbing developments was described in an article in *The Washington Post* of 15 January 1981, by Ward Sinclair, who reported, "the chemical industry has been given an important role in the Reagan Administration's selection of a new chief of the EPA." The report revealed that Paul Oreffice, chief executive of Dow Chemical Company and head of the executive committee of the Chemical Manufacturers Association, had met with White House personnel chief E. Pendleton James and had vetoed the leading candidate for the EPA post on the grounds that the person had not been sufficiently opposed to the "Superfund" legislation.[12]

Shortly thereafter, it was announced that Anne M. Gorsuch, a Denver attorney with a pro-industry, anti-regulation background, had been picked for the EPA post. While a state legislator in Colorado, she had opposed a state toxic waste control program, and had been "instrumental in killing the proposal." At the time of her appointment, she downplayed the dangers of acid rain, stating that its causes and cures were "largely speculative" and could not definitely be linked to power plant emissions. She also critized the required use of sulfur-removing scrubbers in utility plants.[13]

Columnist Carl T. Rowan wrote that "To name Gorsuch to this post ... is like sending the grand dragon of the Ku Klux Klan to stand guard over a meeting of the NAACP."[14]

From the beginning, Gorsuch made no attempt to hide her disdain for the agency she would be running. Following her nomination to the post in February, she refused to move into the EPA building for several weeks, but instead took an office in the Interior Department a few doors away from her mentor, James Watt. Gorsuch would not visit EPA headquarters or even meet with its career employees, except for one awkward picture taking session.[14][b]

Once she assumed office, Gorsuch quickly confirmed these fears by beginning to dismantle many of the agency's health and environmental regulations. To help her run the agency, Gorsuch appointed as its top officials a dozen or so lobbyists, lawyers and consultants who had represented some of the nation's biggest polluters in their attempts to avoid regulation by EPA. Included in these appointments were representatives of the chemical, oil, coal, steel, and paper industries, such as the following:[15]

- Dr. John Todhunter, who opposed pesticide regulation on behalf of industry clients, was named assistant administrator for pesticides and toxic substances.

*• Nolan E. Clark, whose law firm was representing the Dow Chemical Company in its efforts to prevent the cancer-causing and fetus-damaging herbicide 2,4,5-T from being banned, was put in charge of policy and resource management.

- John E. Daniel, a lawyer and lobbyist for the Johns-Mansville Corporation, a leading manufacturer of the health-destroying and cancer-causing product asbestos, was named as Gorsuch's chief-of-staff.

- Robert M. Perry, a lawyer for Exxon oil company, was named EPA General Counsel.

- Kathleen Bennett, who, as a lobbyist for the American Paper Institute and the Crown Zellerbach paper company, fought to weaken the Clean Air Act, was named assistant administrator for air pollution control programs.

- Rita M. Lavelle, a public relations executive for Aerojet-General Corporation, the third biggest polluter in California, which had been charged by the state with illegally dumping 20,000 gallons a day of toxic waste, was named assistant administrator in charge of the "Superfund" hazardous waste cleanup program.

*● Frank A. Shepherd, a lawyer who had represented General Motors, a leading opponent of the Clean Air Act, was placed in charge of law enforcement.

● William A. Sullivan, Jr., an attorney who fought against environmental controls for steel mills, was appointed as Shepherd's deputy. Commenting on his appointment, Sullivan said, "If I were an environmental activist, I'd be scared to death."

By April 1981, EPA had halted its efforts to ban 2,4,5-T, and was discussing a settlement with the manufacturer, Dow Chemical.[17] In June, Gorsuch eliminated EPA's Office of Enforcement,[18] and in August, she asked a federal court to allow the agency to withdraw from a five-year-old agreement that set up the nation's major water cleanup program![19]By the end of 1981, the number of cases EPA was referring to the Justice Department for prosecution had dropped by almost 90 percent compared to the previous year.[20]

Several actions were also taken by EPA to weaken its program to clean up hazardous waste and dangerous dump sites. The Resource Conservation and Recovery Act was dealt a crippling blow when EPA decided to let hazardous waste dumps operate without insurance to cover injuries to the public, thus possibly allowing them to evade responsibility for damage claims.[21]

One of Gorsuch's most memorable moments came in August when, as the nation's chief environmental protection official, she urged several hundred industry representatives meeting in Washington to "light up the switchboards" of Congress in support of the administration's efforts to weaken the Clean Air Act.[22] Environmentalists estimated that EPA's proposed changes in the law, in addition to greatly increasing the death rate from lung and heart disease, would increase pollution levels in the national parks and recreation areas five fold. But in February, 1982, Gorsuch told a Congressional committee these levels would only double.[22b]

That same month, EPA formally proposed to relax or repeal the rule limiting the amount of lead oil refineries can add to gasoline. "If the Reagan administration has its way," said Clar-

---

*In September, 1981, Clark and Shepherd resigned from their posts, reportedly because of their inability to get along with Gorsuch.

ence Ditlow, head of the Center for Auto Safety, "an old environmental hazard that caused brain damage, retardation, and learning disabilities in young children will reappear."[22c]

In January, 1982, EPA assistant administrator, John Todhunter, recommended abolishing the agency's Special Pesticides Review Division, which was act on carcinogenic pesticides. EPA also said it would eliminate and destroy 86 of its inventory of 104 of its publications on such topics as pesticide safety for farmworkers and children.[22d]

And EPA gave the go-ahead to the Drug Enforcement Administration to spray the lung-damaging herbicide paraquat, as well as the fetus-destroying weedkiller 2,4-D, to kill off marijuana plants, thus potentially endangering the health of tens of millions of Americans who use the drug.[22e]

In September 1981, EPA's proposed future budget was made public, revealing that the agency planned to cut the staff and operating budget nearly in half over the next two years. The cuts were described by knowledgeable observers as so massive that they could cause a serious pullback on the major environmental programs of the past decade, crippling the agency's ability to be effective.[23] Later, OMB proposed cutting the agency's 1983 budget even deeper — by over one-third, reducing its staff by a fourth and slashing the pesticide control program by 42 percent and the hazardous waste program by 65 percent.[24] Former EPA assistant administrator William Drayton pointed out that such deep cuts, at a time when the agency's workload was doubling, meant that "there will be doubled exposure to toxics and dirty air and dirty water ... with only the shattered shell of an EPA left, our environmental statutes will be largely meaningless."[25]

In February, 1982, the administration's proposed 1983 budget for EPA ended up 30 percent down from 1981, 39 percent if corrected for inflation.

As John B. Oakes, former Senior Editor of *The New York Times*, wrote on 1 November 1981: "Mrs. Gorsuch is in effect dismantling EPA, making it impossible to administer the antipollution and toxic substance control laws it was designed to oversee."[26]

On 2 February, 1982, Russell E. Train, who served as EPA Administrator from 1973 to 1977 under two Republican administrations, charged that the "budget and personnel cuts, unless reversed, will destroy the agency as an effective institution for many years to come. Environmental protection statutes may remain in full force on the books, but the agency charged with their implementation will be a paper tiger."[26b]

Asst. EPA Administrator John W. Hernandez, shown a draft copy of this chapter, defended the administration's policies as being "fully committed to carrying out EPA's responsibilities for protecting public health and the environment from unreasonable risks posed by pesticides and toxic substances."[26c] However, Hernandez did not challenge the accuracy of a single fact or specific statement in the chapter other than to deny that the administration was "dismantling existing environmental laws."

Other top Reagan appointees fit the same anti-environmental mold. Before becoming Secretary of Labor, Raymond J. Donovan headed the Schiavone Construction Company, which drew 135 citations from OSHA (which is in the Department of Labor) for safety violations in the previous six years. Fifty-eight of the violations were classed as "serious," in which there was "a substantial probability that death or serious physical harm could result."[27]

One of the Secretary Donovan's first acts in office was to announce, in early February 1981, that he was reviewing OSHA's policy on minimizing the exposure of workers to cancer-causing substances and was considering cancelling regulations requiring that exposure to carcinogens be set at the lowest feasible levels.[28]

At the same time, shortly after being urged to do so by Robert A. Roland, president of the Chemical Manufacturers Association, Donovan withdrew a rule proposed only a month earlier under the Carter Administration which would have required the labeling of hazardous chemicals in the workplace. This regulation would have helped workers protect themselves from dangerous chemicals, but industry considered this simple and sensible measure, in Mr. Roland's words, "enormously expensive and unnecessarily burdensome."[29]

Soon after becoming head of OSHA, Thorne G. Auchter ordered the destruction of between 50,000 and 100,000 government booklets describing the dangers of cotton dust because he felt the cover photograph was too sympathetic to the victims of crippling "brown lung" disease, which is caused by the particles. Auchter also ordered that educational films and slides on worker safety and health laws not be distributed because he found them to be "anti-business."[30]

In February 1982, Auchter announced plans to revise the cotton dust safety standard, which had been imposed to reduce dust levels and protect some 300-600,000 textile workers against "brown lung."[30b]

Auchter also announced that OSHA would reexamine all of its existing safety and health regulations limiting workers' exposure to toxic substances, and would subject the rules to a cost/benefit analysis. Later, OSHA specified it would reconsider, in terms of cost to industry, its regulation designed to protect almost a million workers from lead poisoning.

After an appeal from the formaldehyde industry, Auchter withdrew a government warning that fumes from the substance cause cancer. (It is used in plywood, particle board, carpets, permanent press fabrics and, as urea formaldehyde, to insulate some 400,000 to 600,000 homes.) Auchter followed up the withdrawal in the summer of 1981 by trying to fire Dr. Peter F. Infante, an OSHA scientist who had warned of the dangers of formaldehyde, thereby incurring the displeasure of the industry's lobbyists. After Congressional hearings were held on the matter, the agency backed off from its attempts to fire the cancer specialist.[31]

(On 17 February, 1982, EPA reversed a recommendation from its staff and announced that it would not ban or restrict formaldehyde. But a few days later, on 22 February, the CPSC ordered a ban on the sale of formaldehyde home insulation on the grounds that it could cause cancer).[31b]

Another blow against worker protection fell on 4 March 1981, when Dr. Anthony Robbins, head of the National Institute for Occupational Safety and Health (NIOSH), and a frequent target of business groups, was fired after serving only two years of his six year term. He was dismissed by Department of Health and Human Services Secretary Richard S. Schweiker after being attacked in an article published by the U.S. Chamber of Commerce.[32]

Perhaps the most disappointing Cabinet appointment to conservationists was that of Interior Secretary James G. Watt, a Denver lawyer who had spent the previous three years fighting against such environmental initiatives as trying to reduce pollution in Denver, protecting the Colorado River and public grazing lands, minimizing the chemical poisoning of coyotes and other predators, and preserving wilderness areas and national parks. In earlier years, Watt led the fight for the U.S. Chamber of Commerce's effort to thwart federal water pollution standards.[33]

In the late 1970's, Watt gave a speech attacking conservationists and asking, "What is the real motive of the extreme environmentalists? Is it to weaken America?"[34]

In opposing his nomination, the National Audubon Society's Dr. Russell W. Peterson, a former Republican governor of Delaware and head of CEQ under Presidents Nixon and Ford, said that Watt "has one of the most anti-environmental records of anyone I've ever met. To put him in charge of Interior is a crime."[35]

Reagan's choice for Secretary of Agriculture was John R. Block, an Illinois farmer who has fought against the Department's food safety program. He called the inquiry into the possible cancer-causing ability of nitrites, a preservative added to processed food, "an almost ridiculous charade" that was overly costly to farm interests.[36] After his appointment, Block summarized his views on such matters as the presence of cancer-causing pesticides in food by stating, "I think oftentimes the past administration was sidetracked, they were concerned about peripheral issues, and they thought they were helping the consumers by zeroing in on some small thing, like food safety issues."[37] Now, USDA no longer inspects and rates processing plants on their compliance with health and safety regulations.[38] Another important change made at Agriculture was to eliminate its Office of Environmental Quality, which monitored the Department's actions to ensure that it did not violate environmental laws.

None of these actions and appointments should have taken anyone by surprise; indeed, they accurately reflect the policies and philosophy of President Reagan, who has spent much of his public life attacking environmental regulations and defending corporate polluters. He has blamed air pollution on trees and other vegetation (see page 202), and he has criticized the restricting of DDT and other cancer-causing pesticides (see pp. 344-345), saying that EPA's power should be cut back because "we are in the hands of environmental extremists." Reagan has characterized environmentalists as "a tiny minority opposed to economic growth which often finds friendly ears in regulatory agencies for its obstructionist campaigns," and has said "I'd invite the coal and steel industry in to rewrite the Clean Air regulations."[39] (That is, of course, just what he has done in pushing for crippling amendments to the Act.)

While Governor of California, Reagan consistently opposed attempts to clean up toxic pollution. According to the nonpartisan League of Conservation Voters:

> Most environmentalists agree that Reagan was responsible for undermining what could have been the most far reaching air pollution program in the nation... The Reagan administration's air pollution control program was so weak that EPA (under Nixon) rejected it on five counts, with the main objection being lack of enforcement.

The League goes on to note that, concerning toxic substances, "Reagan's record on the Occupational Safety and Health Act is dismal."[40]

This is not to say that all Republicans are against protecting the environment. There are, in fact, various Republican Senators and Congressmen with fine environmental records. Two key Republican Senators who have recently played crucial roles in resisting administration pressures to cripple the Clean Air Act are Robert Stafford (Vt.), Chairman of the Environment and Public Works Committee, and John Chafee (R.I.), Chairman of the Subcommittee on Environmental Pollution.

Indeed, after the 1980 elections, a 14-member Republican task force sent a report to President-elect Reagan urging him to consider "protecting and enhancing the environment" as one of the "major concerns" of his presidency. The group stressed the need for awareness of "the delicate balance of our ecosystems, ...the irreversible harm we can do to our natural surroundings..., the waste of our non-renewable resources and the destruction of these systems on which renewable resources are based."[41]

But given the policies of the Reagan Administration, which is committed to dismantling key environmental laws and regulations, the prospects for action to address the nation's toxic chemical crisis have suffered a major setback. These anti-regulatory actions are ostensibly designed to help the economy, increase industrial production, and fight inflation, high taxes and excessive government spending. But the Reagan administration's policies will, over the long run, have just the opposite effect. Government spending, along with taxes and inflation, will accelerate in future years because of the weakening or elimination of environmental regulations that is now taking place. Health care costs (now estimated at over $250 billion a year)[42] will increase as more and more Americans are struck by cancer, birth defects and other health disasters. The costs of cleaning up toxic chemical pollution will be hundreds or thousands of times greater than it would have been to have prevented such contamination.

As consumer advocate Ralph Nader informed OMB Director Stockman in a 211 page report in August 1981, five government regulatory agencies—including EPA, OSHA, and CPSC—were worth $5.7 billion to the U.S. economy in 1978, providing $37.1 billion in benefits while costing only $31.4 billion through their regulations. Such savings consist mainly of money *not* spent on prevented deaths, injuries and illnesses.

Most tragic of all, as numerous government studies have made clear, millions of Americans will be unnecessarily condemned to an early death, will have their health destroyed or their children deformed or killed, because of the misguided policies of the present administration.

The government is taking this action in full knowledge of the inevitable consequences. The real horror stories may be just now beginning.

What You Can Do

Just as our current environmental crisis has a political basis, the solution is also political in nature. Relief will only come when the American people demand that their President, Senators, Congressmen, and other elected officials at all levels see to it that the public is protected from toxic chemicals.

When enough people make their feelings known, the required changes will be made. There can be no question of how the public feels about such issues. Poll after poll shows overwhelming public support for increased environmental protection at virtually any cost.) For example, in testimony given on 15 October 1981 before the House Subcommittee on Health and the Environment, pollster Louis Harris described a public opinion survey he had just conducted on the Clean Air Act and environmental regulations. Harris found that "by 80 to 17 percent, a sizable majority of the public nationwide does not want to see any relaxation in existing regulation of air pollution." Harris emphasized that "this message on the deep desire on the part of the American people to battle pollution is one of the most overwhelming and clearest we have ever recorded in our 25 years of surveying public opinion." Harris went on to note that "the public will oppose vehemently any measure that might have the effect of reversing some of the environmental gains that have been made in the last 10 years. The American people . . . will not tolerate any reductions in environmental cleanup efforts . . ."

With the Clean Air Act in danger of being seriously weakened by crippling amendments, the problem seems to be that not enough people who support the Act have communicated this to their representatives in Congress and demanded to know how these politicians stand on the issue. The same could be said of environmental issues across the board.

Numerous local and national citizens' environmental and public interest groups are fighting, often on shoestring budgets, against the multi-billion dollar chemical lobby. By joining these organizations, which can now devote some 20 percent of their budgets to lobbying on legislation, one can often keep up with the crucial issues of the moment, and become involved in influencing their outcome.

The solution is for more of the overwhelming majority of people who support conservation to get active and organized. The public must demand that the politicians either support environmental protection or be voted out of office. Only then will progress begin to be made in saving our country and our people from what is one of the greatest threats we have ever faced. Otherwise, we will continue to turn "America the Beautiful" into "America the Poisoned."

Footnotes

Chapter One Herbicides

1 "Effects of 2,4,5-T on Man and the Environment,"Hearings before the Subcommittee on Energy, Natural Resources and the Environment, U.S. Senate, April 1970, p. 1

2 Ibid., pp. 68-69.

3 David Burnham, "Scientist Urges Congress to Bar Any Use of Pesticide 2,4,5-T" The New York Times, 10 August 1974.

4 Myra MacPherson, "The Bitter Legacy and the New Chic of the Vietnam Veteran," The Washington Post, 1 June 1979, p. C6.

5 For an account of such attacks by communist forces, see Sterling Seagraves, Yellow Rain: A Journey Through the Terror of Chemical Warfare, M. Evans and Company, New York, 1981.

6 "Emergency Action to Stop Spraying of the Herbicides 2,4,5-T and Silvex," Press Conference Statement by Barbara Blum, U.S. Environmental Protection Agency, Washington, D.C., 1 March 1979.

7 Associated Press, "9 of 10 Pregnant Women Miscarry; Herbicide Blamed," Billings Gazette, 8 February 1980.

8 Letter from 11 residents of Ashford, Washington to Environmental Protection Agency Administrator Douglas Costle, 22 April 1980.

9 "A Plague on Our Children," NOVA, WGBH Educational Foundation, Boston, 1979.

10 Richard Severo, "Two Studies for National Institute Link Herbicide to Cancer in Animals," The New York Times, 27 June 1980.

11 Ron Nordland and Josh Friedman, "Poison at Our Doorstep," Philadelphia Inquirer, reprint of 23-28 September 1979 series, p. 6; Phil Keisling, "The Spraying of Oregon," Willamette Week, Willamette, Oregon, five part series, December 1979.

12 Effects of 2,4,5-T . . .", supra, note 1; "Involuntary Exposure to Agent Orange and Other Toxic Spraying,"Hearings Before the Subcommittee on Oversight and Investigations, U.S. House of Representatives, June 1979.

13 Associated Press, "Herbicide Showdown Today," Lewiston (California) Morning Tribune, 5 June 1979, p. 1A.

14 "Politics of Poison," News Releases, KRON-TV, San Francisco, California, 13 April 1979; 1 June 1979; 21 May 1979. Margot Hornblower, "Herbicide, Birth Defects Linked for House Panel," The Washington Post, 22 June 1979, p. A8.

15 "Politics of Poison," written, produced, and directed by John David Rabinovitch, KRON-TV, Chronicle Broadcasting, San Francisco, 1979. This forms a chapter of a book by Ralph Nader, Ronald Brownstein, and John Richard, Who's Poisoning America, Sierra Club Books, San Francisco, 1981, pp. 240-66.

16 Ibid.

17 Hornblower, supra, note 14.

18 "Politics of Poison," supra, note 15.

19 Ibid.

20 Ibid.

21 Bill Mandel, "The Orange Death," San Francisco Examiner, 25 April 1979.

22 "Health Effects of Toxic Pollution: A Report from the Surgeon General," and "A Brief Review of Selected Environmental Contamination Incidents with a Potential for Health Effects," Reports prepared by the Surgeon General and the Library of Congress for the Committee on Environment and Public Works, U.S. Senate, August 1980, p. 235.

23 Keisling,supra, note 11.

24 Ibid.

25 Interview with Carol and Steve Van Strum, 7 December 1981.

26 Keisling,supra, note 11; interview with Eve DeRock, 7 December 1981.

27 Interview with Eve DeRock, ibid.

28 Keisling, supra, note 11; interview with Phil Keisling, November 1981.

29 Ibid.; Ronald B. Taylor, "Herbicides: Hot Dispute Stirs Oregon," Los Angeles Times, 30 December 1979.

30 Keisling, supra, note 11.

31 Eric Jansson, "Background Paper to Petition for an Imminent Hazard for 2,4,5-T. . .", and attachments, Friends of the Earth, Washington, D.C., 26 May 1978.

32 Ibid.

33 Sara Polenick, Defenders of Wildlife, Medford, Oregon, pers. comm., 30 September 1979.

34 "A Brief Review . . .; supra, note 22.

35 Letter from 11 residents of Ashford, Washington, supra, note 8.

36 Mother Earth News, November/December 1981, p. 17

37 "9 of 10 Pregnant Women Miscarry . . .", supra, note 7.

38 Associated Press, "Herbicide Use Brings Lawsuit by 6 Women," Billings Gazette, 16 February 1980, p. 8-A.

39 Bryant T. McMahon, "Poison Hits Parish," Hammond (Louisiana) Vindicator, 1 May 1980.

40 Thomas Whiteside, The Pendulum and the Toxic Cloud, Yale University Press, New Haven, 1979, p. 3

41 Eric Jansson, supra, note 31 at 10.

42 Ibid.

43 Chris Kalka, "Wisconsin FOE Group Rallies Against 2,4-D," Not Man Apart, Friends of the Earth, San Francisco, September 1980, p. 12.

44 Dr. Samuel Epstein, "Testimony on Agent Orange," given before the House Subcommittee on Medical Facilities and Benefits, 22 July 1980, p. 23

45 "Better Regulation of Pesticide Exports and Pesticide Residues in Imported Food Is Essential," U.S. General Accounting Office, Washington, D.C., 22 June 1979, p. 53.

46 Epstein, supra, note 44 at 10.

47 Jansson, supra, note 31.

48 Jay Heinrichs, "'T' on Trial, Part 2," American Forests, American Forestry Association, Washington, D.C., March 1979, p. 50.

49 Ronald B. Taylor, "Most Cases of Poisoning Go Unreported," Los Angeles Times, 28 June 1979.

50 Heinrichs, supra, note 48 at 19.

51 "Phenoxy Herbicides in Forest Management: Efficacy and Environmental Effects," Hearing before the Subcommittee on Forests, U.S. House of Representatives, 3 February 1980, p. 119.

52 Keisling, *supra*, note 11.

53 Taylor, *supra*, note 29.

54 Medford (Oregon) Mail Tribune, September 1978; from Polenick, *supra*, note 33.

55 Heinrichs, *supra*, note 48 at 19.

56 "Politics of Poison," *supra*, note 15.

57 Lee Purcell, "Oregon Residents Oppose BLM Spraying of 2,4-D" *Not Man Apart*, Friends of the Earth, San Francisco, June 1980, p. 5.

58 "Herbicide Spraying Halted," *Washington Post*, 24 August 1980, p. A12.

59 Eric Jansson, letter to Douglas Costle, Environmental Protection Agency, and to Langhorne Bond, Federal Aviation Administration, 30 May 1979.

60 Alan Riding, "Free Use of Pesticides in Guatemala Takes a Deadly Toll," *The New York Times*, 9 November 1977.

61 "Phenoxy Herbicides...," *supra*, note 51 at 1, 9, 66.

62 Taylor, *supra*, note 29.

63 Keisling, *supra*, note 11.

64 "Effects of 2,4,5-T...," *supra*, note 1 at 80.

65 Keisling, *supra*, note 11.

66 "Opinion," *BioScience*, American Institute of Biological Sciences, Arlington, Virginia, June 1979, p. 339.

67 "Phenoxy Herbicides...," *supra*, note 51 at 3.

68 Keisling, *supra*, note 11.

69 *Ibid.*; "Phenoxy Herbicides...," *supra*, note 51 at 4-7.

70 *Ibid.*

71 Keisling, *supra*, note 11.

72 *Ibid.*

73 "Politics of Poison," *supra*, note 15.

74 Keisling, *supra*, note 11.

75 *Ibid.*

76 *Ibid.*

77 *Ibid.*

78 *San Francisco Chronicle* (undated), reprinted in *NCAP* (National Coalition Against Pesticides) *News*, Eugene, Oregon, Fall 1979, p. 29.

79 Anne Crittenden, "Farm Research Aims Disputed," *The New York Times*, 3 December 1980, pp. D1, D8.

80 Myra MacPherson, "Two Faces of War," *The Washington Post*, 26 May 1981, pp. B1, B2; MacPherson, "The Bitter Legacy...," *supra*, note 4; Nordland and Friedman, *supra*, note 11; Keisling, *supra*, note 11.

81 "Involuntary Exposure...," *supra*, note 12 at 67.

82 Joanne Omang, "U.S. Lawyers Argue Both Sides of a Deadly Issue," *The Washington Post*, 3 May 1980, p. B8.

83 Paul Lapsley, Office of Pesticides and Toxics, Environmental Protection Agency; from Joanne Omang, *supra*, note 82.

84 "A Plague on Our Children," *supra*, note 9.

85 *Ibid.*

86 *Environmental Quality: 1979 — The Tenth Annual Report of the Council on Environmental Quality*, Washington, D.C., December 1979.

87 "A Plague on Our Children," *supra*, note 9.

88 Severo, *supra*, note 10.

89 "Involuntary Exposure...," *supra*, note 12 at 4.

90 Epstein, *supra*, note 44 at 13.

91 Jay Heinrichs, "'T' on Trial," Part I, *American Forests*, February 1979, p. 54.

92 *Ibid.*, p. 53; Dr. Granville Knight, Testimony in Citizens Against Toxic Sprays vs. Earl Butz, Civil No. 76-438, U.S. District Court for the District of Oregon.

93 Matthew Meselson et. al., "The Evaluation of Possible Health Hazards from TCDD in the Environment," Presentation for Symposium on the Use of Herbicides in Forestry, Arlington, Virginia, February 21 and 22, 1978; from Jansson, *supra*, note 31.

94 Keisling, *supra*, note 11.

95 Ralph C. Dougherty and Krystyna Piotrowska, "Screening by negative chemical ionization mass spectrometry for environmental contamination with toxic residues: Application to human urine," Proceedings National Academy of Sciences, Washington, D.C., June 1976, pp. 1777-81; from Jansson, *supra*, note 31.

96 Rachel Carson, *Silent Spring*, Crest Books, Greenwich, Connecticut, 1962, p. 75.

97 "Involuntary Exposure...," *supra*, note 81 at 69.

98 Epstein, *supra*, note 44 at 16, 21; "Effects of 2,4,5-T...," *supra*, note 1 at 410.

99 Taylor, *supra*, note 29.

100 Keisling, *supra*, note 11.

101 Samuel Epstein, *The Politics of Cancer*, Anchor Books, Garden City, New York, 1979, p. 308.

102 "The Effects of Butyl 2,4-D Treatment of Pastures on the Reproductive Functions of Sheep," *Zhivotnovodstvo* (Animal Breeding), Vol. 34, No. 1, 1972, pp. 73-74; Eric Jansson, Friends of the Earth, Washington, D.C., Letter to Edwin Johnson, U.S. Environmental Protection Agency, Washington, D.C., 27 July 1981.

103 "The Environmental Protection Agency and the Regulation of Pesticides," Staff Report to the Subcommittee on Administrative Practice and Procedure, U.S. Senate, December 1976, p. 15.

104 Howie Kurtz, "Research Lab's Safety Tests Are Questioned," *The Washington Star*, 5 July 1981, p. A3.

105 "Suspended and Cancelled Pesticides," U.S. Environmental Protection Agency, October 1979, p. 5.

106 Joanne Omang, "Park Service to Stop 2,4-D Weedkiller Use...," *The Washington Post*, 15 November 1980, p. A2.

107 Associated Press, "U.S. Applies Herbicide to Oklahoma Reservoir," *The New York Times*, 20 August 1981.

108 Joanne Omang, *supra*, note 82; "EPA Requests Data on Safety of Herbicide 2,4-D," *Not Man Apart*, Friends of the Earth, San Francisco, June 1980, p. 5.

109 Edward Greenspon, "Banned Herbicide Still a Big Seller," (Toronto) Globe and Mail, 28 February 1981; "Canada to Phase Out Some 2,4-D Products," *EPA Weekly Report*, 28 January 1981.

110 "Additional Criteria Announced for 2,4,5-T Use in National Forests," U.S. Department of Agriculture News Release, Washington, D.C., 11 August 1978.

111 "A Brief Review...," *supra*, note 22 at 235.

112 "Emergency Action to Stop Spraying of the Herbicides 2,4,5-T and Silvex," Press Conference Statement by Barbara Plum, Environmental Protection Agency, Washington, D.C., 1 March 1979.

113 "EPA Takes Emergency Action to Halt Herbicide Spraying," Environmental News, U.S. Environmental Protection Agency, Washington, D.C., 1 March 1979.

114 "Involuntary Exposure...," *supra*, note 12 at 20, 221; "Emergency Action...," *supra*, note 112.

115 "EPA Takes Emergency Action...," *supra*, note 113.

116 *Federal Register*, 13 December 1979, p. 72331.

117 *Ibid.*, pp. 72318-19, 72331.

118 *Ibid.*, pp. 72316-72340.

119 *Ibid.*, p. 72325.

120 Keisling, *supra*, note 11; Thomas Whiteside, "A Reporter At large: Contaminated," *The New Yorker*, 4 September 1978, pp. 76-77.

121 Bill Butler, *pers. comm.*, February 1980.

122 "A Plague on Our Children," *supra*, note 9.

123 Margot Hornblower, *supra*, note 17.

124 "A Plague...," *supra*, note 9.

125 Joanne Omang, "EPA and Dow Negotiating Settlement on Herbicide," *The Washington Post*, 11 April 1981, p. A4.

126 MacPherson, *supra*, note 4.

127 Steven D. Jellinek, Testimony before the Committee on Veterans' Affairs, U.S. Senate, 10 September 1980, p. 6.
128 "Search for an Orange Thread,"*Newsweek*, 16 June 1980, p. 56.
129 *"Environmental Quality — 1979," supra*, note 86 at 213.
130 "Pesticide Oversight," Hearings before Subcommittee on the Environment, U.S. Senate, August 1974, p. 22.
131 "Search for an Orange Thread," *supra*, note 128.
132 *Environmental Quality: 1979...," supra*, note 86 at 213.
133 Associated Press, "Agent Orange Detected In an Entire Test Group," *The New York Times*, 15 July 1980, p. A16.
134 "Involuntary Exposure...," *supra*, note 12 at 4.
135 Marlene Cimons, "Veterans Gaining Ground in Agent Orange Struggle," *The Los Angeles Times*, 27 December 1979.
136 Murray Polner, "The Poisoned Battlefield," *Book World, The Washington Post*, 17 August, 1980, p. 9.
137 Margot Hornblower,"A Sinister Drama of Agent Orange Opens In Congress," *The Washington Post*, 27 June 1979.
138 Epstein, testimony, *supra*, note 44 at 8.
139 *Environmental Quality: 1979...," supra*, note 86 at 166.
140 Whiteside, *The Pendulum...*, *supra* note 40 at 22.
141 John W. Finney, "Vietnam Defoliants Study Sees Effect of 100 Years," *The New York Times*, 22 February 1974, pp. A1 and 4; John W. Finney, "U.S. Panel to Study Steps to Heal Herbicide Damage in Vietnam," *The New York Times*, 29 April 1974, p. A2.
142 "Politics of Poison," *supra*, note 15.
143 "Effects of 2,4,5-T...," *supra*, note 1 at 7, 8.
144 "Involuntary Exposure...," *supra*, note 12 at 4, 67-71; Whiteside, *The Pendulum...*, *supra* note 40 at 25-27, 2, 6.
145 "Involuntary Exposure...," *Ibid.*
146 *supra*, note 144.
147 "EPA Takes Emergency Action...," *supra*, note 113.
148 "Effects of 2,4,5-T...," *supra*, note 1 at 1.
149 *Ibid.*, pp. 168, 176-178.
150 *Ibid.*
151 *Ibid.*, p. 468.
152 "Report of the Secretary's Commission on Pesticides," (The Mrak Report), U.S. Department of Health, Education, and Welfare, December 1969, pp. 661, 674.
153 "Effects of 2,4,5-T...," *supra*, note 1 at 9.
154 *Ibid.*, pp. 379-80.
155 *Ibid.*, p. 36.
156 *Ibid.*, pp. 68-69.
157 *Ibid.*, pp. 190-202.
158 *Ibid.*, pp. 405-418.
159 "Pesticide Oversight...," *supra*, note 130 at 100-107.
160 *Ibid.*, pp. 19-25; Morton Mintz, "U.S. Scientist Disputes EPA on Toxic Herbicide," *The Washington Post*, 28 August 1974, p. A2; Burnham, *supra*, note 3.
161 Whiteside, "A Reporter at Large," *supra*, note 120 at 34-35, 42.
162 *Ibid.*, pp. 53, 73; Keisling, *supra*, note 23.
163 Whiteside, "A Reporter at Large," *supra*, note 120 at 36, 29, 46, 55-56, 75; "Involuntary Exposure...," *supra*, note 12 at 4.
164 Taylor, *supra*, note 29.
165 *The Washington Post*, 20 December 1980.
166 Environmental Defense Fund and Robert H. Boyle, *Malignant Neglect*, Alfred A. Knopf, New York, 1980, p. 121.
167 "Six Case Studies of Compensation for Toxic Substances Pollution," by the Library of Congress for the Committee on Environment and Public Works, U.S. Senate, June 1980, pp. 279-95; "A Brief Review...," *supra*, note 22 at 237-39.
168 *Ibid.*
169 Carol Van Strum, "Herbicides: A Faustian Bargain," *The Co-Evolution Quarterly*, Spring 1979, p. 22.
170 Eric Jansson, "The Impact of Hazardous Substances Upon Infertility Among Men in the U.S. and Birth Defects," Friends of the Earth, Washington, D.C., 17 November 1980.
171 Eugene C. Weinbach, "Biochemical Basis for the Toxicity of Pentachlorophenol," *Proceedings of the National Academy of Sciences*, Volume 43, 1957, pp. 393-97.
172 Whiteside, *The Pendulum...," supra*, note 40 at 138.
173 *Environmental Quality 1977: The 8th Annual Report of the Council on Environmental Quality*, Washington, D.C., December 1977, p. 3.
174 "Dispatches," *Outside* (undated), 1980, p. 14.
175 Sheldon W. Samuels, A.F.L.-C.I.O., Washington, D.C., *pers. comm.*, 1 October 1980.
176 Whiteside, *The Pendulum...*, *supra*, note 40 at 141.
177 Associated Press, "Container of Deadly Chemical Empty When Lifted to Surface," *The Washington Post*, 2 August 1980, p. A2; Associated Press, "Large Fishing Area Closed Because of Chemical Spill," *The New York Times*, 5 August 1980; Associated Press, "Fishing Area Shut By PCP May Open for Shrimp Catch," *The New York Times*, 12 August 1980.
178 "Action Proposed on Wood Preservative Chemicals," *EPA Weekly Report*, 11 February 1981.
179 "A Brief Review...," *supra*, note 22 at 260.
180 *Ibid.*, p. 261.
181 Anita Johnson, "Unnecessary Exposure to Toxic Chemicals," 19 January 1978, Environmental Defense Fund, Washington, D.C., p. 10.
182 Whiteside, *The Pendulum...*, *supra* note 40 at 134.
183 Associated Press, "Officials Say Pesticide Put in Water Supply Purposely," *The New York Times*, 17 December 1981; Associated Press, "Water Supply Contaminated in Area of Pittsburgh," *The New York Times*, 10 December 1981.
184 "Hazardous Waste Disposal," Report by the Subcommittee on Oversight and Investigations, U.S. House of Representatives, September 1979, p. 9.
185 Whiteside, *The Pendulum, supra* note 40 at 141.
186 *Environmental Quality: 1979..., supra* 86 at 213.
187 "Hazardous Waste Disposal," *supra*, note 184 at 9; Ron Nordland and Josh Friedman, *supra*, note 11, p. 19.
188 "Six Case Studies...," *supra*, note 167.
189 S.T. Kellogg, D.K. Chatterjee and A.M. Chakrabarty, "Plasmid-Assisted Molecular Breeding: New Technique for Enhanced Biodegradation of Persistent Toxic Chemicals," *Science*, 4 December 1981, pp. 1133-35.
190 "Getting the Bugs Out: A Guide to Sensible Pest Management In and Around the Home," National Audubon Society, New York, 1981, p. 17.
191 Dale R. Bottrell, *Integrated Pest Management*, Council on Environmental Quality, Washington, D.C., 1980, p. 36.

Chapter Two   Pesticides

1 Dale R. Bottrell, *Integrated Pest Management*, Council on Environmental Quality, Washington, D.C. 1980, p. *vi*.
2 *Ibid.*
3 *Ibid.*
4 *Ibid.*, pp. *vi*, 3.
5 *Ibid.*, p. *vi*.
6 Robert van den Bosch, *The Pesticide Conspiracy*, Anchor Books, Garden City, New York, 1980, p. 24.
7 William Longgood, *The Darkening Land*, Simon and Schuster, New York, 1972, pp. 137-38.
8 Rachel Carson, *Silent Spring*, Crest Books, Greenwich, Connecticut, 1962, p. 62.
9 Bottrell, *supra*, note 1 at 14.
10 "Involuntary Exposure to Agent Orange and Other Toxic Sprayings," Hearings before the Subcommittee on Oversight and Investigations, House of Representatives, Washington, D.C., June 1979, p. 154.
11 Bottrell, *supra*, note 1 at *vi*.
12 van den Bosch, *supra*, note 6 at 29.
13 *The Global 2000 Report to the President, The Technical Report*, Vol. 2, prepared by the Council on Environmental Quality and the Department of State, July 1980, p. 287.
14 Bottrell, *supra*, note 1 at 12-13.

15 *Ibid.*, p. 15; Joyce Wadler, "Wax, Wane," *The Washington Post*, 22 August 1980.

16 "Federal Pesticide Registration Program: Is It Protecting the Public and the Environment Adequately from Pesticides?", U.S. General Accounting Office, Washington, D.C., 4 December 1975; Environmental Defense Fund and Robert H. Boyle, *Malignant Neglect*, Alfred A. Knopf, New York, 1979, pp. 152-53.

17 *Ibid.*

18 "Better Regulation of Pesticide Exports and Pesticide Residues In Imported Food Is Essential," General Accounting Office, Washington, D.C., June 1979, p. 32.

19 Food and Drug Administration, "Current levels for natural or unavoidable defects in food for human use that present no health hazard," Dept. of Health, Education, and Welfare, 5th revision, Rockville, Maryland, 1974.

20 Bottrell, *supra*, note 1 at 99.

21 "Involuntary Exposure . . . ," *supra*, note 10 at 175.

22 *Ibid.*, pp. 159-160, 172.

23 Ronald B. Taylor, "Most Cases of Poisoning Go Unreported," *Los Angeles Times*, 28 June 1979.

24 "Involuntary Exposure . . . ," *supra*, note 10 at 153, 160.

25 *Ibid.*, p. 175; Bottrell, *supra*, note 1 at 13; Van den Bosch, *supra*, note 6 at 26-27.

26 Taylor, *supra*, note 23.

27 "Pesticide Oversight," Hearings before the Subcommittee on the Environment, United States Senate, August 1974, pp. 7, 30.

28 Margot Hornblower, *The Washington Post*, 30 June 1979.

29 Ronald Sullivan, "Halt to Aerial Spraying Asked After Death of Boy," *The New York Times*, 6 July 1971.

30 *Ibid.*

31 "Involuntary Exposure . . . ," *supra*, note 10 at 155, 165-66.

32 *Ibid.*, pp. 155, 163-65, 173.

33 *Ibid.*, p. 151.

34 *Ibid.*

35 "Safety Factors for Children Employed as Hand Harvesters of Strawberries and Potatoes," prepared for the Department of Labor by Clement Associates, Washington, D.C., 18 May 1979, p. 1.

36 Diane B. Cohn, (Public Citizen Litigation Group), "Berries, Chemicals and Kids," *The Washington Post*, 6 June 1980, p. A16.

37 "Safety Factors . . . ," *supra*, note 35 at 19, 44-56.

38 *Ibid.*

39 Frank Graham, Jr., *Since Silent Spring*, A Fawcett Crest Book, Greenwich, Connecticut, 1970, p. 250.

40 "RPAR Issued on Captan," Environmental Protection Agency Weekly Report, Office of Pesticides and Toxic Substances, U.S. Environmental Protection Agency, 6 August 1980. 41 "EPA to Investigate Captan, Suspect Cancer Agent," Environmental News, Press Office, Environmental Protection Agency, Washington, D.C., August, 1980.

42 "Captan: Position Document 1," Office of Pesticide Programs, U.S. Environmental Protection Agency, Washington, D.C., 22 July 1980, pp. 3, 5, 10, 36.

43 Notice of Rebuttal Presumption Against Registration (RPAR) and Continued Registration of Pesticide Products Containing Captan," Environmental Protection Agency, Washington, D.C., August 1980, p. 36.

44 Bottrell, *supra*, note 1 at 71-72.

45 "Involuntary Exposure . . . ," *supra*, note 10 at 74-105.

46 *Ibid.*

47 *Ibid.*

48 *Ibid.*, p. 117.

49 *Ibid.*, pp. 104-105.

50 Interview with Suzanne Prosnier, 8 December 1981.

51 William Butler, Maureen Hinkle, Environmental Defense Fund, Washington, D.C., letter to Environmental Protection Agency, 23 January 1978, pp. 4-5; Bottrell, *supra*, note 1 at 81.

52 Ward Sinclair, "Politics Wins Battle on Pesticide to Curb

Fire Ants," *The Washington Post*, 15 February 1979, p. A33.

53 William Butler, News Release, Environmental Defense Fund, Washington, D.C., 31 March 1978.

54 Butler and Hinkle, *supra*, note 51, at 5-6.

55 Robert S. Strother, "Backfire in the War on Insects," *Reader's Digest*, June 1959.

56 Butler and Hinkle, *supra*, note 51 at 3.

57 Bill Richards, "EPA Ponders Approving a New Weapon Against Fire Ants," *The Washington Post*, 14 February 1978.

58 "Lindane: Position Document no. 2/3," Office of Pesticides and Toxic Substances, U.S. Environmental Protection Agency, June 1980, pages I-1; II-63-64, 75; V-18.

59 Lawrence Longo, M.D., "Environmental Pollution and Pregnancy," *American Journal of Obstetrics and Gynecology*, 15 May 1980.

60 Jane Ogle, "Sweet Scents to Repel Pests," *The New York Times Magazine*, 1 June 1980, p. 7.

61 "Carbaryl Decision Document," December 1980; "Notice of Determination Not to Initiate a Rebuttable Presumption Against Registration (RPAR) of Pesticide Products Containing Carbaryl;" Office of Pesticides and Toxic Substances," U.S. Environmental Protection Agency, Washington, D.C.

62 Wayne Biddle, "Restocking the Chemical Arsenal," *The New York Times Magazine*, May 1981.

63 "Hazardous Waste Disposal" Part 2, Hearings Before the Subcommittee on Oversight and Investigations, House of Representatives, 1979, p. 1388.

64 Michael Brown, *Laying Waste*, Pantheon Books, New York, 1979, pp. 253-54.

65 Carson, *supra*, note 8 at 35-36.

66 *Ibid.*, pp. 38, 160, 177.

67 *Ibid.*

68 Jay Mathews, "California High Court Refuses to Block Medfly Spraying," *The Washington Post*, 14 July 1981.

69 "Localities Try to Stop Spraying," *The Washington Star*, July 1981; Jay Mathews, "California Forced to Expand Medfly Quarantine," *The Washington Post*, 25 August 1981.

69-b Jay Mathews, "California Weighs Test of Medfly Spraying Effects," *The Washington Post*, 23 January 1982, p. A4.

70 Myron Struck, "Medfly Fumigant Risky, EPA Scientist Says," *The Washington Post*, 8 October 1981, p. A29.

71 Margot Hornblower, "U.S. Firms Export Products Banned Here as Health Risks," *The Washington Post*, 25 February 1980, pp. A1 and A16.

72 Carson, *supra*, note 8 at 35-37.

73 Graham, *supra*, note 39 at 131.

74 "Safety Factors . . . ," *supra*, note 35 at 25, 128-29, 132.

75 *Federal Register*, 7 July 1972, p. 13374.

76 "Chemical First Strike," Editorial, *The Washington Post*, 17 May 1980, p. A18.

77 "True or False," League of Conservation Voters, Washington, D.C., 1980.

78 "Suspended and Cancelled Pesticides," U.S. Environmental Protection Agency, Washington, D.C., October 1979.

79 *Ibid.*

80 "Pesticides Used in the National Capital Parks, 1977," National Capital Parks Region, National Park Service, U.S. Department of the Interior, Washington, D.C.

81 Michael Frome, "Crusade for Wildlife," *Defenders of Wildlife*, Washington, D.C., April 1980, p. 102.

82 "Pesticides Used in the National Capital Parks, 1977," *supra*, note 80.

83 "Draft Environmental Impact Statement: Federal Aid in Fish and Wildlife Restoration Program," U.S. Fish and Wildlife Service, Washington, D.C., 1978, pp. I-3, 4, 36-37, 52-53.

84 *Ibid.*, pp. III-31, V-1.

85 *Federal Register*, 13 December 1979, p. 72317; Environmental Defense Fund v. Environmental Protection Agency, 548 F. 2nd at 1018 (D.C. Cir. 1976).

86 "Deficiencies in Administration of Federal Insecticide, Fun-

gicide, and Rodenticide Act," Committee on Government Operations, U.S. House of Representatives, 13 November 1969, p. 3-4.

87 Ibid.

88 Ibid., pp. 13-16, 60-62.

89 Ibid.

90 Ibid., pp. 41-42.

91 Ibid., p. 53.

92 Ibid., p. 59.

93 Ibid., pp. 60-62.

94 Ibid., pp. 52-53.

95 Longgood, supra, note 7 at 143.

96 Malignant Neglect, supra, note 16 at 130.

97 "The Environmental Protection Agency and the Regulation of Pesticides," Staff Report to the Senate Subcommittee on Administrative Practice and Procedure, Washington, D.C., December 1976, pp. 4-5, 34.

98 Ibid.

99 Ibid.

100 Ibid.

101 Ibid.

102 Howie Kurtz, "Research Lab's Safety Tests Are Questioned," The Washington Star, July 1981, p. A3; Associated Press, "U.S. Charging 4 Falsified Reports on Drugs in Lab," The New York Times, 23 June 1981.

103 Ibid.

104 Carol Van Strum, "Herbicides: A Faustian Bargain," The Co-Evolution Quarterly, Spring 1979.

105 Malignant Neglect, supra, note 16 at 226.

106 Joanne Omang, "EPA Chief's Farewell," The Washington Post, 14 December 1980, p. A2.

107 Bottrell, supra, note 1 at iv, v, vii, viii, 1, 39.

108 Ibid.

109 "Modern chemicals solve, create problems," Front Lines, U.S. Agency for International Development, Washington, D.C., 12 June 1980.

110 van den Bosch, supra, note 6 at 7.

111 Ogle, supra, note 60.

112 The New York Times, 30 September 1980.

113 Janice Fillip, "American Farmers and USDA Start to Take Organic Seriously," Not Man Apart, Friends of the Earth, San Francisco, September 1980.

114 Ibid.

115 Ibid.

116 Strother, supra, note 55.

117 Ibid.

118 Carson, supra, note 8 at 197, 202.

119 Ibid.

120 Malignant Neglect, supra, note 16 at 22.

121 Louis A. McLean, "Pesticides and the Enviornment," BioScience, American Institute of Biological Sciences, Arlington, Virginia, September 1967.

122 van den Bosch, supra, note 6 at 7.

122b. David Attenborough, Life on Earth, Little, Brown, Boston, 1979, p. 87.

123 Cleveland Amory, pers. comm., December 1980.

124 Helga and William Olkowski, "How to Control Garden Pests Without Killing Almost Everything Else," Rachel Carson Trust for the Living Environment, Washington, D.C., 1977, pp. 2-5.

125 Ibid., p. 5; "Getting the Bugs Out: A Guide to Sensible Pest Management In and Around the Home," National Audubon Society, New York, 1981, p. 17.

Chapter Three Toxic Wastes

1 "Hazardous Waste Disposal," Report by the Subcommittee on Oversight and Investigations, U.S. House of Representatives, September 1979, p. 1.

2 Ibid., pp. 4, 32; Joanne Omang, "DOE Backs Down on Toxics Showdown," The Washington Post, 30 April 1980, p. B1.

3 "Hazardous Waste Disposal," supra, note 1 at 1.

4 "Health Effects of Toxic Pollution: A Report from the Surgeon General," and "A Brief Review of Selected Environmental Contamination Incidents....," Committee on Environment and Public Works, U.S. Senate, August 1980, p. 159; "Coping with Toxic Waste," Newsweek, 19 May 1980, p. 34; Ralph Blumenthal, "Fight to Curb 'Love Canals'," The New York Times, 30 June 1980, p. B1 and B11.

5 Georgette Jasen and Jonathan Kwitny, "Prosecutions of Dumpers of Toxic Waste Underscore the Shortage of Suitable Sites," The Wall Street Journal, 2 September 1980, p. 48.

6 "Last Days for Superfund," Editorial, The Washington Post, 2 October 1980.

7 John McGuire, "How Buried Wastes Return to Haunt Us," The Chicago Tribune, 22 June 1980, p. 14.

8 Ron Nordland and Josh Friedman, "Poison at Our Doorsteps," The Philadelphia Inquirer, 23-28 September 1979. (Reprint of series).

9 Ibid.; "The Poisoning of America," Time, 22 September 1980, pp. 58, 63.

10 Blumenthal, supra, note 4; "Corporate Crime," Subcommittee on Crime, U.S. House of Representatives, May 1980, pp. 14-15.

11 Environmental Quality 1979: The 10th Annual Report of the Council on Environmental Quality, Washington, D.C., December 1979, p. 176: "Corporate Crime," Ibid., p. 16.

12 The New York Times, 6 August 1978: The New York Times, 29 May 1980: Energy and the New Poverty, National Council of Churches, New York, March 1979, p. 8.

13 "Fleeing the Love Canal," Newsweek, 2 June 1980, p. 56; Harry Reasoner, "60 Minutes," CBS Television, New York, 25 May 1980.

14 Kathy Trost, "Love Canal: Despite Warning Signs, Some Residents Stay On," The Washington Post, 23 June, 1980, pp. A1 and A4.

15 "Hazardous Waste Disposal," supra, note note 1 at 15-16.

16 Ibid.

17 "Hazardous Waste Disposal, Part I," Hearings before the Subcommittee on Oversight and Investigations, U.S. House of Representatives, 1979, p. 68.

18 Environmental Quality—1979, supra note 11 at 177; Joanne Omang, "U.S. Sues Chemical Dumpers," Washington Post, 21 December 1979.

19 Robin Herman, "Carey Criticizes Aspects of Study at Love Canal," The New York Times, 19 May, 1980, p. B3.

20 "Corporate Crime," supra, note 10 at 14-15; Michael Brown, Laying Waste, Pantheon Books, New York, 1980, pp. 9-10.

21 "Corporate Crime," supra, note 10 at 18. **22** "Hazardous Waste Disposal, Part I," supra, note 17 at 665.

23 Transcript, "MacNeil/Lehrer Report," WNET Television, New York, 2 July 1980.

24 Morton Mintz, "Product Danger Cover-Up Worthy of Jail, Hill Told," The Washington Post, 15 February 1980; Ralph Nader, Ronald Brownstein and John Richard, Who's Poisoning Amerca, Sierra Club Books, San Francisco, 1981, pp. 298-99. "Hammer Tells Nun to Go Back to Buffalo," Buffalo Courier-Express, 22 May 1980.

25 Kathy Trost, supra note 15; Richard J. Meislin, "Panel Discounts Two Studies on Love Canal Problem," The New York Times, 11 October 1980, p. 25.

26 Nordland and Friedman, supra note 8.

27 Jack Anderson, "Chemical Dumping a Widespread Peril," The Washington Post, 14 June 1980, p. E43.

28 Blumenthal, supra, note 4.

29 Nordland and Friedman, supra, note 8.

30 "Hazardous Waste Disposal," supra, note 1 at 3-4, 9-10.

31 Joanne Omang, "General Chemical Pollution Often Equals Levels at Dumps," The Washington Post, 19 August 1980, p. A2.

32 Janet Staihar, "Federal Agencies Flayed on Hill for Neglecting Chemical Canal," *The Washington Post*, 23 May 1980, p. A15.

33 Victor Cohn, "EPA: 1.2 Million May be Exposed to Toxic Waste," *The Washington Post*, 6 June 1980, p. A9; Philip Shabecoff, "EPA Suspects High Potential for Health Risk at 108 Waste Dumps," *The New York Times*, 6 June 1980.

34 Brown, *supra*, note 20 at 296.

35 Victor Cohn, "Waste Sites May Invade Water Supply, Subcommittee Told," *The Washington Post*, 7 June 1980, p. A2.

36 "The Poisoning of America," *supra*, note 9 at 58-60.

37 *Ibid.*

38 Associated Press, "Nader Group Says Toxins Dumped in Water Supply," *The Washington Star*, 10 July, 1980, p. A8; Steve Kroft, CBS Evening News, 9 July 1980.

39 Erik Larson, "TCE, Solvent Suspected as Cancer Cause, Frequently Shows Up in Drinking Water," *The Wall Street Journal*, 12 August 1980, p. 46.

39b Associated Press, "Old Waste Dumps and New Illnesses May Be Tied in Massachusetts Town," *The New York Times*, 28 December 1981, p. A16.

40 "Hazardous Waste Disposal," *supra*, note 1 at 4-6, 10-11, 17; Robert H. Harris, Member, Council on Environmental Quality, statement before the Subcommittee on Environment, Energy, and Natural Resources, U.S. House of Representatives, 24 July 1980, p. 7.

41 Nordland and Friedman, *supra*, note 8.

42 Jasen and Kwitny, *supra*, note 5.

43 Nordland and Friedman, *supra*, note 8.

44 *Ibid.*

45 Selwyn Raab, "Ex-Owner Says Mob Took Over Chemical Firm," *The New York Times*, 24 November 1980, pp. B1, B6.

46 Nordland and Friedman, *supra*, note 8.

47 Paul Guardino, New Jersey Department of Environmental Protection, "The MacNeil/Lehrer Report," *supra*, note 23.

48 Jasen and Kwitny, *supra*, note 5.

49 "How the Mob Really Works," *Newsweek*, 5 January 1981, p. 39.

50 "The Poisoning of America," *supra*, note 9 at 2.

51 Michael Brown, "New Jersey Cleans up Its Pollution Act," *The New York Times Magazine*, 23 November 1980, pp. 143, 146.

52 Jasen and Kwitny, *supra*, note 5.

53 "Mob Control and Arson Alleged in Toxic Waste Fire," *The New York Times*, 30 November 1980; "Ex-Owner of Toxic Dump Links Fire Last April to Mob," *The New York Times*, 26 November 1980; Ralph Blumenthal, "Hearing Is Told of Crime Tie to Disposal of Toxic Wastes," *The New York Times*, 17 December 1980, pp. B1, B4.

54 Nordland and Friedman, *supra*, note 8.

55 *supra*, note 53.

56 Blumenthal, *supra*, note 53.

57 "The Poisoning of America," *supra*, note 9 at 63.

58 "Drums of Chemicals Explode in New Jersey Industrial Park Fire," *The Washington Post*, 8 July 1980, p. A9.

59 Gladwin Hill, "Disposal of Toxic Waste in Wells Expected to Grow Despite Critics," *The New York Times*, 8 April 1979; William Longgood, *The Darkening Land*, Simon and Schuster, New York, 1972, p. 157.

60 *Ibid.*

61 Jasen and Kwitny, *supra*, note 5.

62 Nordland and Friedman, *supra*, note 8.

63 Brown, *Laying Waste*, *supra* note 20 at 230, 232-33; "Dealing with the Poisoners," Editorial, *The Washington Post*, 20 August 1979.

64 "Hazardous Waste Disposal," *supra*, note 1 at 23.

65 *Ibid.*, p. 14.

65b Peter Lance, "20/20," ABC News, 17 December 1981.

66 "Chemical Dumping Agreement," *The Washington Post*, 11 November 1980, p. A16.

67 Brown, *Laying Waste*, *supra* note 20 at 247-48, 252.

67b "Turning Off the Tap," Editorial, *The Los Angeles Times*, 21 October 1981, p. 10; "Breakthrough On Toxic Waste," Editor-

ial, *The Sacramento Bee*, 21 October 1981.

68 "Hazardous Waste Disposal," *supra*, note 1 at 50.

69 *Ibid.*, p. 46.

70 *Ibid.*, p. 2.

71 *Ibid.*, p. 35.

72 Blumenthal, *supra*, note 4.

73 *Ibid.*

74 "Hazardous Waste Disposal," *supra*, note 1 at 43-44.

75 Blumenthal, *supra*, note 4.

76 "The Poisoning of America," *supra*, note 9 at 60.

77 Michael Knight, "Toxic Wastes Hurriedly Dumped Before New Law Goes Into Effect," *The New York Times*, 16 November 1980, pp. 1, 28.

78 *Ibid.*

79 Nordland and Friedman, *supra*, note 8.

80 *Ibid.*

81 "Hazardous Waste Facilities with Interim Status May Be Endangering Public Health and the Environment," U.S. General Accounting Office, 28 September 1981; Congressman James J. Florio, News Release, 30 September 1981.

81b Philip Shabecoff, "Health Fears Grow as Debate Continues on Toxic Wastes," *The New York Times*, 2 January 1982, pp. 1, 7.

82 *Ibid.*

83 "Hazardous Waste Disposal," *supra*, note 1 at 28-29.

84 Philip Shabecoff, "Waste Cleanup Bill Approved by House," *The New York Times*, 24 September 1980, p. A16.

85 Philip Shabecoff, "Congress Set to Debate 'Superfund' to Clean Up Hazardous Wastes," *The New York Times*, 18 September 1980, p. A24; "A 'Superfund' to Clean Old Dumps," *Not Man Apart*, Friends of the Earth, San Francisco, 1980, p. 11.

86 Kathy Koch, "Chemical and Oil Companies Take Aim at Senate Toxic Substances Cleanup Measure," *Congressional Quarterly*, Washington, D.C., 6 September 1980, pp. 2643-44.

87 News Release, Congress Watch, Washington, D.C., August 1980.

88 Peter Harnik, "Campaign Contributions from 'Filthy Five'," Environmental Action News Release, Washington, D.C., 28 October 1980.

89 *Ibid.*

90 Philip Shabecoff, "DuPont Chairman Backs Cleanup Fund," *The New York Times*, 20 November 1980.

91 Philip Shabecoff, "Senate Votes 78-9 A $1.6 Billion Fund on Chemical Wastes," *The New York Times*, 25 November 1980, p. A1; Joanne Omang, "Senate Approves Fund to Clean Up Hazardous Wastes," *The Washington Post*, 25 November 1980, pp. A1 and A5; George J. Mitchell, "Not a Super Fund," Op-Ed Page, *The New York Times*, 8 December 1980.

92 Ralph Blumenthal, "EPA Restricts Regional Orders on Toxic Wastes," *The New York Times*, 21 June 1981, pp. 1, 17.

93 Cass Peterson, "Executive Notes," *The Washington Post*, 22 February, 1982.

Chapter Four   Contaminating the Water

1 "Interim Report on Groundwater Contamination: EPA Oversight," Committee on Government Operations, U.S. House of Representatives, 30 September 1980, p. 4.

2 "Contamination of Ground Water by Toxic Organic Chemicals," Council on Environmental Quality, Washington, D.C., January 1981.

3 *Environmental Quality, 1979: The 10th Annual Report of the Council on Environmental Quality*, Washington, D.C., December 1979, p. 89.

4 "Resource Losses from Surface Water, Groundwater, and Atmospheric Contamination: A Catalog," by the Library of Congress for the Committee on Environment and Public Works, U.S. Senate, March 1980, pp. ix-x.

5 *Ibid.*, pp. ix, xiii; United Press International, "Great Lakes

Revival Still Some Way Off," *The Washington Star*, 26 December 1979.

6 Eckardt C. Beck, Environmental Protection Agency, Statement Before the Subcommittee on Environment, Energy, and Natural Resources, U.S. House of Representatives, 25 July 1980; Robert H. Harris, Council on Environmental Quality, Statement Before the Subcommittee on Environment, Energy, and Natural Resources, U.S. House of Representatives, 24 July 1980.

7 "Groundwater Protection," U.S. Environmental Protection Agency, Washington, D.C., November 1980.

8 Eckardt C. Beck, "User Beware," Op-Ed page, *The New York Times*, 30 June 1980, p. A19.

9 Beck, 25 July 1980, *supra*, note 6.

10 "Interim Report on Ground Water Contamination...," *supra*, note 1.

11 Harris, *supra*, note 6.

12 *Ibid.;* "Interim Report...," *supra*, note 1 at 2.

13 Beck, "User Beware," *supra*, note 8.

14 "Hazardous Waste Disposal," Report by the Subcommittee on Oversight and Investigations, U.S. House of Representatives, September 1979, p. 13.

15 Harris, *supra*, note 6.

16 Beck, 25 July 1980, *supra*, note 6; "Interim Report...," *supra*, note 1 at 9.

17 Beck, *supra*, note 6.

18 *Ibid.;* Robert Hanley, "Spread of Pollution Feared in Wells Around New York," *The New York Times*, 6 April 1981, pp. B1, B8.

19 Associated Press, "Suit Says Landfill Threatens Water in Atlantic City," *The Philadelphia Inquirer*, 23 September 1981, p. 4B.

19b Hillary Brown, ABC News, 29 December 1981.

20 Donald Janson, "Atlantic City Rushing to Move Wells," *The New York Times*, 4 November 1981, pp. B1, B2.

21 "The Poisoning of America," *Time*, 22 September 1980, p. 60; "Interim Report..." *supra*, note 1.

22 Michael H. Brown, "New Jersey Cleans Up Its Pollution Act," *The New York Times*, 23 November 1980, p. 146.

23 Raymond Bonner, "Toxic Dumping in Niagara River Is Reported," *The New York Times*, 11 October 1981, p. 48.

24 Beck, *supra*, note 6; "Interim Report...," *supra*, note 1.

25 *Ibid.*

26 "Six Case Studies of Compensation for Toxic Substances Pollution," prepared by the Library of Congress for the Committee on Environment and Public Works, U.S. Senate, June 1980, pp. 29, 39, 221-25.

27 *Ibid.*

28 *Ibid.*, pp. 214, 230, 234.

29 Beck, *supra*, note 6.

30 "Interim Report...," *supra*, note 1 at 10-11.

31 Beck, *supra*, note 6.

32 Erik Larson, "TCE, Solvent Suspected as Cancer Cause, Frequently Shows Up in Drinking Water," *The Wall Street Journal*, 12 August 1980, p. 46.

33 Stewart Brand, "Human Harm to Human DNA," *Co-Evolution Quarterly*, Spring 1979, pp. 19-20; John Pekkanen, "Is Decaffeinated Coffee Safe?" *Washingtonian*, January 1981, p. 47.

34 "Resource Losses...", *supra*, note 4 at ix, xiii.

35 Larson, *supra*, note 32.

36 Harris, *supra*, note 6.

37 "Pipe Contamination Puts School on Bottled Water," *The New York Times*, 18 July 1980.

38 Wallace Turner, "Tacoma Pollution on '10 Worst List'," *The New York Times*, 6 November 1981.

39 Larson, *supra*, note 32.

40 Associated Press, "Chemicals Tied to Cancer Found in California Wells," *The New York Times*, 21 June 1980, p. A6; "Pipe Contamination...," *supra*, note 37.

41 "Resource Losses...," *supra*, note 4 at 225.

42 Bill Richards, "Toxic Chemical Found in California Wells," *The Washington Post*, undated 1979; Bill Richards, "Panel of Scientists Urges DBCP Pesticide Be Banned," *The Washington Post*, 30 June 1979.

43 Ronald B. Taylor, "DBCP Found in Water Supplies of 12 Counties," *Los Angeles Times*, 5 September 1979.

44 Congressman George Miller, "Criminal Penalties for Corporate Coverups," *Congressional Record*, 15 November 1979.

45 "Trying to Be Fair About Poison," Editorial, *The New York Times*, 9 August 1979.

46 "Six Case Studies...," *supra*, note 26 at 140-44.

47 *Ibid.*, p. 145.

48 *Ibid.;* Miller, *supra*, note 44.

49 "Six Case Studies...," *supra*, note 26 at 146, 148-51, 159-60.

50 *Ibid.*, pp. 140, 151.

51 Tom Ferrell, "After Ten Years, Earth Day Again: How Well Are We Doing," *The New York Times*, 20 April 1980, p. 18E.

52 Environmental Defense Fund and Robert H. Boyle, *Malignant Neglect*, Alfred A. Knopf, New York, 1979, p. 82.

53 Molly Sinclair, "Water: Safe, Cheap, and Full of Chemicals," *The Washington Post*, 28 September 1980, p. B1.

54 *Malignant Neglect*, *supra*, note 52, pp. 92-93.

55 Associated Press, "Cancer Study Reports Firmer Link of Chlorinated Water to Tumors," *The New York Times*, 17 October 1980.

56 *Ibid.*

57 Science Research Systems, Inc., "Drinking Water and Cancer: Review of Recent Findings and Assessment of Risk," Council on Environmental Quality, Washington, D.C., December 1980.

58 Joanne Omang, "Using Chlorine in Water Raises Risk of Cancer," *The Washington Post*, 18 December 1980, p. 15.

59 *Toxic Chemicals and Public Protection*, A Report to the President by the Toxic Substances Strategy Committee, Council on Environmental Quality, Washington, D.C., May 1980, p. 6.

60 *Malignant Neglect*, *supra*, note 52 at 84, 93-94.

61 *Ibid.*, pp. 97-99.

62 *Ibid.*

63 Robert H. Harris, "Testimony on the Regulation of Cancer-Causing Chemicals in Drinking Water by the Environmental Protection Agency," 29 March 1978, Environmental Defense Fund, Washington, D.C.; Jacqueline M. Warren, "Testimony on Regulation by EPA on Cancer-Causing Chemicals in Drinking Water," 12 July 1978, Environmental Defense Fund, Washington, D.C.; *Malignant Neglect*, *supra*, note 52 at 86.

64 *Ibid.*

65 "Federal Report," *The Washington Post*, 18 September 1981.

66 Joanne Omang, "Industry Told of Easier Clean-Water Rules," *The Washington Post*, 13 January, 1982, p. A8.

67 A Myrick Freeman III, "The Benefits of Air and Water Pollution Control...," Council on Environmental Quality, Washington, D.C., December 1979, pp. ix-xii.

68 Carol Keough, *Water Fit to Drink*, Rodale Press, Emmaus, Pa., 1980, pp. 128-30.

Chapter Five   Air Pollution

1 "Uncertainties Cloud U.S. Debate on Air Pollution Controls," *The Washington Post*, 6 July 1981, p. A6.

2 Robert Mendelsohn and Guy Orcutt, "An Empirical Analysis of Air Pollution Dose-Responsive Curves," *Journal of Environmental Economics and Management*, June 1979, pp. 85-106.

3 *Toxic Chemicals and Public Protection*, A Report to the President by the Toxic Substances Strategy Committee, Council on Environmental Quality, Washington, D.C., May 1980, pp. 5, 121-22.

4 William Longgood, *The Darkening Land*, Simon and Schuster, New York, 1972, p. 81.

5 Mendelsohn and Orcutt, *supra*, note 2.

6 Longgood, *supra*, note 4 at 86-87.

7 "Health Effects of Toxic Pollution: A Report from the Surgeon General, and "A Brief Review of Selected Environmental Contamination Incidents with a Potential for Health Effects," Reports prepared by the Surgeon General and the Library of Congress, for the Committee on Environment and Public Works, U.S. Senate, August 1980, p. 170.

8 *Toxic Chemicals and Public Protection, supra*, note 3 at 4, 121-22.

9 "A Brief Review . . .," *supra*, note 7 at 162-68.

10 Environmental Defense Fund and Robert H. Boyle, *Malignant Neglect*, Alfred A. Knopf, New York, 1979, p. 104; *"Toxic Chemicals and Public Protection . . . ,supra*, note 3 at 121-22.

11 Joanne Omang, "General Chemical Pollution Often Equals Levels at Dumps," *The Washington Post*, 19 August 1980, p. A2.

12 *Malignant Neglect, supra*, note 10 at 210.

13 "Before the U.S. Environmental Protection Agency: Petition of the Natural Resources Defense Council, Inc. for Control of Fine Particles," Natural Resources Defense Council, Washington, D.C., 29 May 1980, pp. 1, 16-17, 20, 33.

14 *Toxic Chemicals and Public Protection, supra*, note 3 at 5.

15 Charles S. Warren, EPA Administrator, Region II, "What the EPA Thinks of New York Transit," Letter to the Editor, *The New York Times*, 19 July 1980.

16 "Mr. Reagan v. Nature," Editorial, *The Washington Post*, October 1980; "The Environment and the Stump," Editorial, *The New York Times*, 22 October 1980, p. A30; Joanne Omang, "Reagan Criticizes Clean Air Act, EPA as Obstacles, Pledges Changes," *The Washington Post*, 9 October 1980, p. A2.

17 *Malignant Neglect, supra*, note 10 at 9, 18, 103.

18 Stuart Auerbach, "Chemical Plant Areas Lead in Cancer Deaths," *The Washington Post*, 20 January 1974, p. E4.

19 Joanne Omang, "Administration May Seek EPA Study of Indoor Air Pollution," *The Washington Post*, 8 September, 1981, pp. A14-15.

20 Howie Kurtz, "OSHA Head Accused in Carcinogen Case," *The Washington Star*, 17 July 1981, p. A3; Joanne Omang, "EPA Staff Finds Formaldehyde No Great Cancer Risk," *The Washington Post*, 19 September 1981.

21 Mendelsohn and Orcutt, *supra*, note 2.

22 *Ibid.*

23 Natural Resources Defense Council petition, *supra*, note 13 at 29-31; "A Brief Review . . .," *supra*, note 7 at 162-68.

24 Associated Press, "Pollution Assessed in Ohio River Valley," *The New York Times*, 2 March 1981, p. A12.

24b Anne LaBastille, "Acid Rain: How Great a Menace?", *National Geographic*, November 1981, p. 675.

25 *Environmental Quality—1979: The Tenth Annual Report of the Council on Environmental Quality*, Washington, D.C., December 1979, p. 71; Joanne Omang, "Acid Rain Doesn't Hurt Most Crops . . . ," *The Washington Post*, (undated), p. D7; "Rain Falls Everywhere," Editorial, *The Washington Post*, 12 July 1980.

26 Dr. Ellis P. Cowline, "Acid Precipitation: A Status Report," U.S. Environmental Protection Agency, Washington, D.C., December 1979, p. 1.

27 Roger Peterson, ABC Evening News, 18 and 19 June 1980; Joanne Omang, "Acid Rain: Push Toward Coal Makes Global Pollution Worse," *The Washington Post*, 30 December 1979, pp. A1 and A8.

28 *The Global 2000 Report to the President—The Technical Report, Volume Two*, by the Council on Environmental Quality and the U.S. State Department, July 1980, p. 337.

29 Congressional Research Service, *Resource Losses from Surface Water, Groundwater, and Atmospheric Contamination: A Catalog*, Committee on Environment and Public Works, U.S. Senate, March 1980, p. xiii; Roger Peterson, *supra*, note 27, 18 and 19 June 1980; Associated Press "High Acidity Is Found in Many New York Streams," *The New York Times*, 3 November 1981; "Conservation Report," National Wildlife Federation, Washington, D.C., 3 October 1980; Philip Hilts, "Acid Rain, Snow Said to kill 50% of High Altitude Adirondack Lakes," *The Washington Post*, 6 January 1981.

30 Gene Mueller, "Acid Rain Eating Away Canadian—U.S. Relationship," *The Washington Star*, 20 July 1980, p. D4.

30b La Bastille, *supra*, note 24b at 657.

31 *Global 2000 Report, supra*, note 28 at 336.

32 Omang, *supra*, note 27; Cowling, *supra*, note 26 at 2; J. Moody, "Acid Rain Fallout," Acid Rain Task Force, Washington, D.C., March 1980; Press Release, National Clean Air Coalition, Washington, D.C., 6 March, 1980, p. 2; "Acid From the Skies," *Time*, 17 March 1980, p. 88; Ralph Pribble, "Acid Rain Threat to BWCA Probed at Forum." *Minnesota Daily*, 4 February 1980; La Bastille, *supra*, note 24b at 662.

32b *Ibid.*

33 Peterson, 19 June 1980, *supra*, note 27.

34 *Environmental Quality—1979 . . . , supra*, note 25 at 71.

35 *Global 2000 Report, supra*, note 28 at 336.

36 Ralph Blumenthal, "Study Finds that Acid Rain Is Intensifying In Northeast," *The New York Times*, 25 January 1981.

37 Lance Gay, "Storm Brews Among Officials On Solutions to 'Acid Rain'," *The Washington Star*, 10 April 1980, p. A2; Press Release, National Clean Air Coalition, *supra*, note 32.

38 "Conservation Report," *supra*, note 29.

39 *Ibid.*

39b La Bastille, *supra*, note 24b at 661.

40 "Report of the Committee on Health and Environmental Effects of Increased Coal Utilization," U.S. Department of Health, Education, and Welfare, Washington, D.C., December 1977, pp. 2233-34.

41 Michael Oppenheimer, Environmental Defense Fund, Letter to *The New York Times*, Editorial Page, 23 September 1981.

42 Peterson, 19 June 1980, *supra*, note 27.

43 Environmental Defense Fund, "Year End Report," January 1978—December 1979, p. 5.

44 Oppenheimer, *supra*, note 41.

44b La Bastille, *supra*, note 24b at 661.

45 "Canadian Official Warns on Acid Rain," *The New York Times*, 31 March 1981.

46 Lance Gay, "Canadian Minister Protests to U.S. About Feared Increase in Acid Rain," *The Washington Star*, 19 April 1980.

47 Associated Press, "Acid Rain Prediction," *The New York Times*, 22 April 1980, p. C4.

47b La Bastille, *supra*, note 24b at 661.

48 Philip Shabecoff, "Canadian, U.S. Witnesses Differ on Acid Rain," *The New York Times*, 7 October 1981; Henry Giniger, "Canada Blames Itself for Half of Its Acid Rain," *The New York Times*, 9 October 1981, p. A9.

50 *Global 2000 Report, supra*, note 28 at 335-37.

51 *Ibid.*, p. 418.

52 Russell W. Peterson, National Audubon Society, Letter to *The New York Times*, Editorial Page, 29 August 1981.

53 Joanne Omang, "Denver Lawyer Reagan's Choice to Head EPA," *The Washington Post*, 21 February 1981, p. A4.

54 *supra*, note 16.

55 Lance Gay, "Buildup of Carbon Dioxide Risky, Hill Panel Told," *The Washington Star*, 3 April 1980, p. A2., Gus Speth, Testimony Before the Senate Energy and Natural Resources Committee, 3 April 1980, pp. 2, 5-8.

56 Gay and Speth, *Ibid.*

57 Robert Reinhold, "Evidence Is Found of Warming Trend," *The New York Times*, 19 October, 1981, p. A21; Knight-Ridder Newspapers," Antarctic Ice Pack Shrinks . . . ," *The Washington Post*, 21 October 1981.

58 *Ibid.*; "The Global Environment and Basic Human Needs," a report to the Council on Environmental Quality by the Worldwatch Institute, Washington, D.C., 1978, pp. 37-38.

59 The World Conservation Strategy: "The World Conservation Strategy in Brief," p. 1; "Forests," pp. 1 and 2; World Wildlife Fund, Washington, D.C., 1980.

60 "The Land-Gambling Billionaire from Michigan," *Detroit Free Press*, 12 April 1981, pp. 1B, 4B.

61 *The Global 2000 Report to the President . . . , Volume One*, by the

Council on Environmental Quality and the U.S. State Department, Washington, D.C., July 1980, p. 37.

62 *Energy and Climate*, National Academy of Sciences, Washington, D.C. 1977.

63 "The Global Environment..." *supra*, note 58 at 38-39; Wallace S. Brocker, "The Coming Superinter-glacial," Lamont Doherty Geological Observatory, July, 1976.

64 "Forests," *supra*, note 59.

65 A. Myrick Freeman III, *Benefits of Air and Water Pollution Control: A Review and Synthesis of Recent Estimates*, Council on Environmental Quality, Washington, D.C., 1980; Charles Osolin, "CEQ News Release," 21 April 1980.

66 "Acid Rain Fallout," *supra*, note 32.

67 Lance Gay, "Costle Sees Threat to Clean Air Act, Industry Planning Attempt to 'Gut' Law, EPA Head Says," *The Washington Star*, 25 June 1980.

68 David Rogers, "Oil Emerges as Leading Hill Patron," *The Washington Post*, 15 September 1981, p. A2.

69 Bill Peterson, "Corporate PAC's Gave $1.9 Million to '80 Favorites," *The Washington Post*, (undated), 1981; "PAC Spending Climbed to $131 Million in '80," *The Washington Post*, 21 February, 1982.

70 Morton Mintz and Merrill Brown, "AT&T Spending on '80 Elections Tops U.S. Firms," *The Washington Post*, 19 April 1981.

71 "True or False," League of Conservation Voters, Washington, D.C., 1980; Joanne Omang, "Reagan Criticizes Clean Air Act, EPA As Obstacles, Pledges Changes," *The Washington Post*, 9 October 1980, p. A2.

72 Denis Hayes, *Energy: The Case for Conservation*, The Worldwatch Institute, Washington, D.C., 1976.

73 *World Prospects for Z.P.G.*, Zero Population Growth, Washington, D.C., 1979.

74 "The Global Environment...," *supra*, note 58 at 2.

75 *Environmental Quality—1979...*," *supra*, note 25; R. Stobaugh and D. Yergin, editors, *Energy Future*, Random House, Inc., New York, 1979.

77 Robert D. Hershey, Jr., "U.S. Study Says Conservation Could Slash Energy Use," *The New York Times*, 15 March 1981, p. 23; "Periscope," *Newsweek*, 13 April, 1981. p. 7.

78 Jackson S. Gouraud, "Our Energy Future," *The New York Times*, 31 August 1980, p. E15.

79 Albert Crenshaw, "Curtains for OPEC?" *The Washington Post*, 29 June 1980, p. C8.

80 T. R. Reid, "An Energy Plank," *The Washington Post*, 7 September 1980, p. A2.

81 Douglas Feaver, "Transport System Needs an Overhaul," *The Washington Post*, 24 August 1980.

82 James Murphy, Natural Resources Defense Council, Letter to the Editor, *The New York Times*, 18 August 1980.

83 Christopher Flavin, "Energy and Architecture: The Solar and Conservation Potential," Worldwatch Institute, Washington, D.C., November 1980.

84 United Press International, "Reagan Ends Temporary Curbs," *The New York Times*, 18 February 1981.

85 *Environmental Quality 1978: The Ninth Annual Report of the Council on Environmental Quality*, Washington, D.C., December 1978.

86 Denis Hayes, *Rays of Hope*, Worldwatch Institute, Washington, D.C., p. 155.

87 *Environmental Quality: 1979, supra*, note 25 at 359.

88 Katherine Seelman, "Energy, Ethics, and Education," *Connexion*, United Ministries Education, New York, October 1979, p. 6.

89 Lee Lescaze, "Reagan Remains Firm on Pushing Oil Decontrol," *The Washington Post*, 26 November 1981, p. A4.

90 Joanne Omang, "Senate Panel Decides to Take Look at Troubled EPA," *The Washington Post*, 1 October 1981, p. A4; Thomas Friedman, "Revolutionary Changes for Solar Field," *The New York Times*, 8 August 1981, p. D1.

91 John M. Berry, "OMB Proposes to Cut Solar Energy Funds," *The Washington Post*, 5 December 1981, p. A3.

Chapter Six   Poisoning The Planet

1 "U.S. Export of Banned Products," Hearings Before a Subcommittee of the Committee on Government Operations, U.S. House of Representatives, July 1978, p. 2; United Press International, "Recall of Dalkon IUD's Suggested," *The New York Times*, 26 September 1980; "The Corporate Crime of the Century," *Mother Jones*, November 1979. (This latter article was later published in expanded form as a book by David Weir and Mark Schapiro, *Circle of Poison*, Institute for Food and Development Policy, San Francisco, 1981.)

2 "Better Regulation of Pesticide Exports and Pesticide Residues in Imported Food Is Essential," U.S. General Accounting Office (GAO), Washington, D.C., 22 June 1979, pp. 6-22.

3 E. W. Kenworthy, "Warnings Urged on Pesticide Use," *The New York Times*, 16 June 1972; Morton Mintz, "World Health Unit to Count Mercury Victims in Iraq," *The Washington Post*, undated.

4 David Weir, "The Boomerang Crime," *Mother Jones*, November 1979.

5 Margot Hornblower, "U.S. Firms Export Products Banned Here as Health Risks," *The Washington Post*, 25 February 1980, pages A1 and A16; R. Jeffrey Smith, "U.S. Beginning to Act on Banned Pesticides," *Science*, 29 June 1979.

6 Frank Graham, Jr., *Since Silent Spring*, A Fawcett Crest Book, Greenwich, Connecticut, 1970, page 131.

7 *Ibid.*, pages 129-130.

8 *Ibid.*, page 131.

9 Erik Eckholm and S. Jacob Scherr, "Double Standards and the Pesticide Trade," *New Scientist*, London, England, 16 February 1978, page, 442.

10 "Report on Export of Products Banned by U.S. Regulatory Agencies," Committee on Government Operations, U.S. House of Representatives, 4 October 1978, page 8.

11 "U.S. Export of Banned Products," *supra*, note 1 at 165.

12 Weir, *supra*, note 4; Weir and Schapiro, *supra*, note 1.

13 "U.S. Export of Banned Products," *supra*, note 11 at 70.

14 *Ibid.*, pages 205-209; "Report on Export of Products...," *supra*, note 10 at 28.

15 Cristine Russell, "Researchers at Harvard Suggest Coffee, Pancreatic Cancer Link," *The Washington Star*, March 1981; "Medicine: Coffee Nerves," *Time*, 23 March 1981; Harold M. Schmeck, Jr., "Study Links Coffee Use to Pancreas Cancer," *The New York Times*, 12 March 1981, p. B15.

16 Hornblower, *supra*, note 5.

17 "Better Regulation of Pesticide Exports...," *supra*, note 2 at 6-7.

18 *Ibid.*, page 8.

19 *Ibid.*

20 *Ibid.*, pages 6-12; "U.S. Export of Banned Products," *supra*, note 1 at 167.

21 "Better Regulation of Pesticide Products...," *Ibid.*, pages iii, 39, 42-43.

22 *Ibid.*, pages 40-41.

23 *Ibid.*, page iii.

24 *Ibid.*, pages ii, 28-29, 34.

25 Robert Richter, "Pesticides and Pills: For Export Only," Public Broadcasting Service, 5 October 1981.

26 "Interagency Working Group on a Hazardous Substances Export Policy," Draft Report, *Federal Register*, 12 August 1980, page 53755.

27 Sheldon W. Samuels and Richard Cleary, "Comment on Proposed Rules for the Assessment of AID-Related Projects," A.F.L.-C.I.O., Washington, D.C., 21 December 1977.

28 Alan Riding, "Free Use of Pesticides in Guatemala Takes a Deadly Toll," *The New York Times*, 9 November 1977.

29 Richter, *supra*, note 25.

30 Instituto Centroamericano de Investigacion Y Techologia

Industrial. (ICAITI), *An Environmental and Economic Study of the Consequences of Pesticide Use in Central American Cotton Production*, Guatemala, January, 1977; Samuels and Cleary, *supra*, note 27.

31 Eckholm and Scherr, *supra*, note 9 at 441-443.

32 *Ibid.*

33 Samuels and Cleary, *supra*, note 27 at 5; Samuels, *pers. comm.*, 1 October 1980.

34 Riding, *supra*, note 28.

35 *Ibid.*

36 *Ibid.*

37 Weir, *supra*, note 4.

38 *Ibid.*

39 Weir and Schapiro, *supra*, note 1 at 5-6.

40 "Better Regulation of Pesticide Exports . . . *supra*, note 2 at 3, 8-9; Smith, *supra*, note 5 at 1392.

41 "Modern Chemicals Solve, Create Some Problems," *Front Lines*, U.S. Agency for International Development, Washington, D.C., 12 June 1980.

42 "Interagency Working Group . . . ,"*supra*, note 26 at 53755.

43 "Modern Chemicals . . . ," *supra*, note 41.

44 S. Jacob Scherr, "Proceedings of the U.S. Strategy Conference on Pesticide Management," U.S. Department of State, Washington, D.C. June 7-8, 1979, pp. 32-34.

45 Hornblower and Smith, *supra*, note 5.

46 Smith, *supra*, note 5.

47 Eckholm and Scherr, *supra*, note 9 at 443; Samuels and Cleary, *supra*, note 27.

48 Samuels and Cleary, *Ibid.*, p. 8.

49 Graham, *supra*, note 6 at 137-38; *Medical World News*, 11 July 1965.

50 Alan Cowell, "Pesticides Are Endangering Wildlife in an African Valley," *The New York Times*, 22 October 1981.

51 Henry Giniger, "Refuge in Spain for Europe's Migrating Birds Becomes Death Stop for Many," *The New York Times*, 2 October 1973, p. 14.

52 Craig Van Note, The Monitor Consortium, Washington, D.C., "Testimony on Elephant Protection Act Before the House Merchant Marine and Fisheries Committee," 25 July 1979.

53 Weir and Schapiro, *supra*, note 1 at 4-5.

54 Eckholm and Scherr, *supra*, note 9 at 443; "U.S. Export of Banned Products," *supra*, note 1 at 90.

55 Weir, *supra*, note 4.

56 Mark Dowie, "A Dumper's Guide to Tricks of the Trade," *Mother Jones*, November 1979.

57 "U.S. Export of Banned Products," *supra*, note 1 at 160.

58 "Worker Safety in Pesticide Production,"Hearings Before the Subcommittee on Agricultural Research and General Legislation, U.S. Senate, 13 and 14 December 1977, pp. 79, 362-63.

59 "Corporate Crime," Subcommittee on Crime, U.S. House of Representatives, May 1980, p. 12; John E. Blodgett and Connie Musgrove, "Summary of Hearings on 'Kepone Contamination'", prepared for the Subcommittee on Agricultural Research and General Legislation by the Library of Congress.

60 *The Global 2000 Report to the President, The Technical Report,* Volume II, The Council on Environmental Quality and the U.S. State Department, July 1980, 305.

61 *Environmental Quality—1976: The Seventh Annual Report of the Council on Environmental Quality*, Washington, D.C., December 1976, p. 30; *Environmental Quality—1977: The Eighth Annual Report of the Council on Environmental Quality*, Washington, D.C., December 1977, p. 15.

62 "Corporate Crime," *supra*, note 59 at 13; "Summary of Hearings . . ." *supra*, note 59 at 355, 365, 373.

63 Connie Musgrove, *Kepone Pollution: A Summary Review*, Library of Congress, Washington, D.C., 27 May 1977.

64 Environmental Quality—1976 . . . , *supra*, note 61 at 30.

65 "Corporate Crime," *supra*, note 59 at 13.

66 "Six Case Studies of Compensation for Toxic Substances Pollution . . . ," A Report by the Library of Congress for the

Committee on Environment and Public Works, U.S. Senate, June 1980, p. 55.

67 Karlyn Barker, "5 Years After," *The Washington Post*, 20 July 1980, pp. C1 and C5.

68 "Corporate Crime," *supra*, note 59 at 14.

69 "Six Case Studies . . .", *supra*, note 66 at 56.

70 "The Poisoning of America," *Time*, 22 September 1980, p. 69.

71 "Worker Safety . . .", *supra*, note 58 at 79.

72 "Better Regulation of Pesticide Exports . . ." *supra*, note 2, pp. 62-63.

73 Hornblower, *supra*, note 5.

74 *The Global Environment and Basic Human Needs*, Council on Environmental Quality, Washington, D.C., 1978, page 23.

75 Peter Milius, Sheldon Samuels, interviews, September and October 1980.

76 *Environmental Quality—1977*, *supra*, note 61 at 3.

77 Peter Milius, "Leptophos Handled at 9 Plants in U.S.," *The Washington Post*, 4 December 1976, p. A1, A8.

78 Peter Milius, "My Spine Is . . . Dissolving," *The Washington Post*, 3 December 1976, pp. A1, A7.

79 *Ibid.*

80 *Ibid.;* "Worker Safety in Pesticide Production," *supra*, note 58 at 173, 183.

81 "The Environmental Protection Agency and the Regulation of Pesticides," Staff Report to the Subcommittee on Administrative Practice and Procedure, U.S. Senate, December 1976, p. 36.

82 Milius, *supra*, note 78; Peter Milius and Dan Morgan, "How a Toxic Pesticide Was Kept Off the Market," "Outlook" section, *The Washington Post*, 26 December 1976.

83 Staff Report, *supra*, note 81 at 35-36.

84 *Ibid.*, p. 36.

85 *Ibid.*, p. 35.

86 Wade Roberts, "Phosvel: A Tale of Missed Cues," *Columbia Journalism Review*, July/August 1977, p. 24.

87 Bill Richards, "Health Officials Cite Hazards at Pesticide Plant," *The Washington Post*, 14 December 1977.

88 Staff Report, *supra*, note 81 at 39-40.

89 *Ibid.*, pp. 36-37.

90 Richards, *supra*, note 87; Milius, *supra*, note 77.

91 *Environmental Quality—1979: The Tenth Annual Report of the Council on Environmental Quality*, Washington, D.C., December 1979, p. 240; Weir, *supra*, note 4.

92 "Interagency Working Group on a Hazardous Substances Export Policy," *supra*, note 26 at 53755.

93 Milius and Morgan, *supra*, note 82.

94 Staff Report, *supra*, note 81 at 42.

95 "Report on Export of Products Banned By U.S. Regulatory Agencies," *supra*, note 10 at 8.

96 "U.S. Export of Banned Products," *supra*, note 1 at 43, 82.

97 Peter Milius and Dan Morgan, "Leptophos Found on Imported Tomatoes," *The Washington Post*, 9 December 1976, pp. A1, A6-7.

98 "Worker Safety . . .", *supra*, note 58 at 172; "And Now, Leptophos," Editorial, *The Washington Post*, 9 December 1976, p. A18; Peter Milius, "Pesticide Sales Halt Laid to Economics, Not Safety," *The Washington Post*, 2 December 1976.

99 "Interagency Working Group," *supra*, note 26 at 53754-53769.

100 *Ibid.*

101 Gaylord Shaw, "Curb on Harmful Exports Delayed," *The Los Angeles Times*, 12 September 1980, pp. 1, 11; Conrad MacKerron, "Does 'Made in USA' Mean 'Let the Foreign Buyer Beware'?" *National Journal*, 18 April 1981, p. 649.

102 MacKerron, *Ibid.*

103 Caroline E. Mayer "Easing of Hazardous Exports Studied," *The Washington Post*, 9 September 1981.

104 *The Global 2000 Report to the President*, *supra*, note 60 at 285-86, 342.

105 Harry Dennis, "The Third World: Our Dump for Dangerous Products," Not Man Apart, Friends of the Earth, San Francisco, December 1979, p. 16.
106 Global 2000, supra, note 60 at 342.
107 Ibid., pp. 286-87.

Chapter Seven   Our Chemical Heritage

1 "Health Effects of Toxic Pollution: A Report from the Surgeon General," and "A Brief Review of Selected Environmental Contamination Incidents with a Potential for Health Effects," Reports Prepared by the Surgeon General, Department of Health and Human Services, and the Library of Congress, for the Committee on Environment and Public Works, U.S. Senate, August 1980, p. 153.
2 News Release, Council on Environmental Quality, Washington, D.C. 29 June 1980.
3 The Global Environment and Basic Human Needs, A Report to the Council on Environmental Quality by the Worldwatch Institute, Council on Environmental Quality, Washington, D.C., 1978, pp. 18-19.
4 Environmental Quality—1975: The Sixth Annual Report of the Council on Environmental Quality, Washington, D.C., December 1975, p. 381; "Aldrin/Dieldrin," Criteria Document, U.S. Environmental Protection Agency, Washington, D.C., 1976; "A Brief Review . . ." supra, note 1 at 173-75.
5 Erik Eckholm and S. Jacob Scherr, "Double Standards and the Pesticide Trade," New Scientist, London, England, 16 February 1978, p. 442.
6 William A. Butler, Jacqueline Warren, "Petition for Suspension and Cancellation of Chlordane/Heptachlor," Environmental Defense Fund, Washington, D.C., October 1974.
7 Environmental Quality—1975, supra, note 4 at 375; Toxic Chemicals and Public Protection, A Report to the President by the Toxic Substances Strategy Committee, Council on Environmental Quality, Washington, D.C., 1980, p. 3; "A Brief Review . . .", supra, note note 1 at 290.
8 The Global Environment and Basic Human Needs, supra, note 3 at 18-19.
9 "Surveillance, Epidemiology, and End Results: Incidence and Mortality Data, 1973-77," National Cancer Institute Monograph 57, U.S. Department of Health and Human Services, National Institutes of Health, Bethesda, Maryland, June 1981, p. 4.
9b Michael Brown, Laying Waste, Pantheon Books, New York, 1980, p. 329.
10 Fatalities for recent U.S. wars are as follows: World War II, 292,000; Korea, 33,000; Vietnam, 47,000. Annual traffic fatalities exceed 50,000.
11 Environmental Defense Fund and Robert H. Boyle, Malignant Neglect, Alfred A. Knopf, New York, 1979, pp. 3-4; Ralph Nader, Ronald Brownstein and John Richard, Who's Poisoning America, Sierra Club Books, San Francisco, 1981, p. 11.
12 Environmental Quality—1975, supra, note 4 at 1.
13 Environmental Quality—1979: The Tenth Annual Report of the Council on Environmental Quality, Washington, D.C., December 1979, p. 188.
14 Toxic Chemicals and Public Protection, supra, note 7.
15 News Release, Council on Environmental Quality, Washington, D.C., 29 June 1980.
16 "Toxic Chemicals and Public Protection," supra, note 7 at 116.
17 Ibid., page 9.
18 News Release, supra, note 15.
19 Joanne Omang, "Cancer Rise Linked to Toxic Chemicals in the Workplace," The Washington Post, 29 June 1980, page A15; Charles Seabrook, "Cancer Rate is Climbing," The Atlanta Journal, 31 August 1980, page 4-A.
20 Joseph Highland, "Testimony of the EDF Before the House Subcommittee on Health and the Environment," 1 March 1978, Environmental Defense Fund, Washington, D.C., pages 5-6.

21 "Health Effects . . . ," supra, note 1 at iii, 1, 8.
22 "A Brief Review . . . ," Ibid., pp. 153, 157.
23 "Surveillance, Epidemiology, and End Results," supra, note 9.
24 Adrian Peracchio, "Agent Orange May Be Just One Garnish In A Toxic Cocktail," The Washington Post, 20 June 1980, page A-4.
25 Stephanie Harris, "Organochlorine Contamination of Breast Milk," Environmental Defense Fund, Washington, D.C., 7 November 1979, page 10.
26 Ibid., p. 2.
27 Malignant Neglect, supra, note 11 at 206-207.
28 A Discussion of the pro's and con's of breast and bottle feeding and how a woman can reduce her levels of toxic chemicals can be found in the book by Stephanie G. Harris and Joseph H. Highland, Birthright Denied, Environmental Defense Fund, Washington, D.C., 1977.
29 Lawrence D. Longo, M.D., "Environmental Pollution and Pregnancy . . . ," American Journal of Obstetrics and Gynecology, Volume 37, No. 2, 15 May 1980.
30 Joel Balbien, Stephanie Harris and Talbot Page, "Diet As A Factor Affecting Organochlorine Contamination of Breast Milk, Environmental Defense Fund, Washington, D.C., undated. 31 Harris, supra, note 25 at 3, 5, 10. 32 Lindane: Position Document No. 2/3," Office of Pesticides and Toxic Substances, Environmental Protection Agency, Washington, D.C., June 1980, page ii, 30, 76.
33 Malignant Neglect, supra, note 11.
34 "Facts, 1981," March of Dimes Birth Defects Foundation, White Plains, New York, 1980, pp. 5-6.
35 Gabriel Stickle, "Defective Development and Reproductive Wastage in the United States," American Journal of Obstetrics and Gynecology, St. Louis, Missouri, 1 February 1968.
36 Chemical Hazards to Human Reproduction, prepared by Clement Associates, Inc. for the Council on Environmental Quality, Washington, D.C., January, 1981, p. III-5.
37 "Facts, 1981," supra, note 34.
38 Toxic Chemicals and Public Protection, supra, note 7 at 9.
39 Longo, supra, note 29.
40 Associated Press, "Survey Discounts Chemicals' Role in Birth Defects," The Washington Post, 7 June 1979.
41 "Politics of Poison," written and produced by John David Rabinovitch, KRON-TV, San Francisco, Chronicle Broadcasting Company, 1979.
42 Stewart Brand, "Human Harm to Human DNA," Co-Evolution Quarterly, Spring, 1979, p. 11.
43 Ibid.
44 Gail Bronson, "Issue of Fetal Damage Stirs Women Workers at Chemical Plants," The Wall Street Journal, (undated), February 1979.
45 Richard Severo, "Should Firms Screen the Workplace or the Worker?", The New York Times, 28 September 1980; "Cyanamid Upheld on Lead Issue," The New York Times, 14 September 1980.
46 Severo, Ibid.
47 Chemical Hazards to Human Reproduction, supra, note 36 at II-3, 12.
48 Philip J. Hilts, Chemicals at Parents Job May Cause Child's Tumor," The Washington Post, 3 July 1981.
49 Severo, supra, note 45; Jane E. Brody, "Lead Persists As Threat To Young," The New York Times, 13 May 1980, pp. C1, C3.
50 "Unplugging the Gene Pool," Outside, September, 1980; Eric Jansson, "The Impact of Hazardous Substances Upon Infertility Among Men in the U.S. and Birth Defects," Friends of the Earth, Washington, D.C., 17 November 1980.
51 Bill Richards, "Drop in Sperm Count Is Attributed in Part to Toxic Environment," The Washington Post, 12 September 1979; Jane E. Brody, "Sperm Found Especially Vulnerable to Environment," The New York Times, 10 March 1981, pp. C1, C3; "A Plague on Our Children," NOVA, WGBH Educational Foundation, Boston, 1979.

52 Jansson, *supra*, note 50; Diane Swanbrow, "Immaculate Conception," *New West*, 25 August 1980.

53 Jansson, *Ibid.*

53b Bambi Batts Young, *Environment & Behavior*, Center for Science in the Public Interest, Washington, D.C., August 1981.

53c Joanne Omang, "Chemicals' Wider Effects On Humans Described," *The Washington Post*, 9 January 1981, p. A4.

53d T. Oliver, Ed., *Dangerous Trades*, E. P. Dutton & Co., New York, 1902, pp. 470-74.

54 *Toxic Chemicals and Public Protection*, *supra*, note 7 at 1.

55 "Environmental Protection Agency Is Slow to Carry Out Its Responsibility to Control Harmful Chemicals," U.S. General Accounting Office, Washington, D.C., 28 October 1980, page 1.

56 *Environmental Quality—1978: The Ninth Annual Report of the Council on Environmental Quality*, Washington, D.C., December 1978, page 178; Thomas H. Maugh, "Chemicals: How Many Are There," *Science*, 199 (4325): 162, 1978.

57 *Environmental Quality—1978* . . . , *Ibid.*, page 178.

58 *The Global Environment* . . . , *supra*, note 3 at 20.

59 *Toxic Chemicals and Public Protection*, *supra*, note 7 at xiv.

60 "Environmental Protection Agency Is Slow . . .," *supra*, note 55 at 1.

61 "Chemical First Strike," Editorial, *The Washington Post*, 17 May 1980, page A18.

62 *Toxic Chemicals and Public Protection*, *supra*, note 7 at 2.

63 "Chemical First Strike," *supra*, note 61.

64 *Toxic Chemicals and Public Protection*, *supra*, note 7 at 131, 133.

65 *Ibid.*

66 *Malignant Neglect*, *supra*, note 11 at 7; *Environmental Quality—1979*, *supra*, note 13 at 198.

67 *Ibid.*, page 196-98.

68 Liebe F. Cavalieri, "Carcinogens and the Value of Life," Letter to the *New York Times*, 20 July 1980.

69 *Environmental Quality—1979*, *supra*, note 13 at 198-199. *Malignant Neglect*, *supra*, note 11 at 212.

70 Jane E. Brody, "Farmers Exposed to a Pollutant Face Medical Study in Michigan," *The New York Times*, 12 August 1976, p. C-20.

71 Ellen Grzech and Kathy Warbelow, *Detroit Free Press*: "State Knew But Did Not Warn Farmers of PBB-Tainted Feed," 14 March 1977; "How State Leaders Ducked PBB Issue," 15 March 1977; "Distribution Hid Facts on PBB Peril," 13 March 1977.

72 "Corporate Crime," Subcommittee on Crime, U.S. House of Representatives, May 1980, pp. 25-28.

73 "PBB Found to Cause Cancer in Lab Animals," *Detroit Free Press*, 17 June 1981, p. 15A.

74 Grzech and Warbelow, *supra*, note 71.

74b Associated Press, "Michigan Study Indicates 97% Have Traces of PBB," *The Washington Post*, 31 December 1981.

75 "Corporate Crime," *supra*, note 72 at 26.

76 Brody, *supra*, note 70.

77 *Environmental Quality—1978*, *supra*, note 56 at 180.

78 *Environmental Quality—1979*, *supra*, note 13 at 196.

79 Rachel Carson, *Silent Spring*, Crest Books, Greenwich, Conn., 1962, p. 196; *Malignant Neglect*, *supra*, note 11 at 6-7.

80 Ben A. Franklin, *The New York Times*: "Toxic Paint Chemicals Raise Alarm as to Threat of Workers," 12 April 1981, pp. A1, A46; "Paint Use Is Linked to Artists' Cancer," 17 May 1981.

81 *Ibid.*

82 *Malignant Neglect*, *supra*, note 11 at 8-9, 47.

83 *Environmental Quality—1979*, *supra*, note 13 at 195.

84 United Press International, "18 Cancer Deaths Are Reported At A Plant in Texas," *The New York Times*, 28 October 1980; "Brain Cancer Discovered in Petrochemical Workers," *The New York Times*, 25 July 1980.

85 Bayard Webster, "Brain Cancer Deaths in Chemical Plants Spur Intense Inquiry," *The New York Times*, 3 March 1981, p. C1.

86 "Gene Watching," *The New York Times*, 13 September 1981.

87 *Malignant Neglect*, *supra*, note 11 at 9, 18.

88 Joanne Omang, "Millions Not Told of Job Health Perils," *The Washington Post*, 24 August 1981, pp. A1, A5.

89 *Ibid.*

90 *The Global Environment and Basic Human Needs*, *supra*, note 3, at 20.

91 "First Annual Report on Carcinogens," U.S. Department of Health and Human Services, Washington, D.C., August 1980, p. 13.

92 *Environmental Quality—1978*, *supra*, note 56 at 180.

93 *Environmental Quality—1979*, *supra*, note 13 at 194-95.

94 "The Asbestos Problem," "The Harvard Medical School Health Letter," Cambridge, Massachusetts, August, 1980; "Asbestos in Schools," The Environmental Defense Fund, Washington, D.C., 1979.

95 Barry I. Castlemen and Manuel J. Vera Vera, "Impending Proliferation of Asbestos," *International Journal of Health Services*, Volume 10, Number 3, 1980, p. 390.

96 *Malignant Neglect*, *supra*, note 11 at 106; Jane E. Brody, "Cancer Found in Asbestos Workers' Kin," *The New York Times*, 19 September 1974, pp. 1, 27.

96b United Press International, "200,000 Delayed Deaths from Asbestos Forecast," *The The New York Times*, 25 December 1981.

97 Fred Graham, "See You in Court," CBS Reports, CBS News, 9 July 1980; "Corporate Crime," Subcommittee on Crime, U.S. House of Representatives, Washington, D.C., May 1980, p. 23; "Occupational Diseases and their Compensation," Part I: Asbestos-Related Diseases, Hearings Before the Subcommittee on Labor Standards, 1, 2, 8 May 1979, p. 152.

98 "Occupational Diseases and their Compensation," *Ibid.* p. 150; Congressman George Miller, "Criminal Penalties for Corporate Coverups," *Congressional Record*, 15 November 1979.

99 Fred Graham, *supra*, note 97; *Environmental Quality—1975*, *supra*, note 4 at 106.

100 Barry Castleman and Manuel Vera Vera, *supra*, note 95 at 401-403.

101 Richard Severo, *The New York Times*, 6 May 1980.

102 "EPA Is Slow . . . ," *supra*, note 55 at i, ii.

103 *Ibid.*, cover, p. ii.

104 *Ibid.*, pp, iii, 1, 33.

105 *Ibid.*, p. 19.

106 *Ibid.*, p. 49.

106b *Chemical Regulation Reporter*, 15 January, 1982, pp. 1089-1093.

107 *Toxic Chemicals and Public Protection*, *supra*, note 7 at 122.

108 United Press International, "Food and Drug Administration: Meat Dye May Cause Cancer," *The Washington Post*, 6 April 1973, p. B6.

109 Kay S. Nelson, "Vegetarianism," Hinsdale Sanitarium and Hospital Health Education Department."

110 Harris, *supra*, note 25 at 8, 11-12; Balbien, Harris and Page, *supra*, note 30 at 10.

111 *Environmental Quality—1975*, *supra*, note 4 at 375.

112 Harris, *supra*, note 25 at 8, 11-12; Balbien, Harris, and Page, *supra*, note 30 at 2.

113 Jim Mason, "Turning Animals Into Machinery," *The New York Times*, 2 September 1980, A23; Mimi Sheraton, "How America Eats: A Nutrition Dilemma," *The New York Times*, 11 June 1980, p. C14.

114 *Ibid.*

115 Jean Carper, "Does Eating Fat Cause Cancer?" *The Washington Post*, 25 November 1979, C1.

116 *Malignant Neglect*, *supra*, note 11 at 23, 126-127.

117 *Ibid.*

118 Philip Shabecoff, *pers. comm.*, December 1981.

Chapter Eight Tris

1 Robert H. Harris, "Potential Health Hazards of the Flame Retardant Chemical Tris," Environmental Defense Fund, Washington, D.C., 9 December 1976, p. 5.

2 Herbert Bishop, The Wall Street Journal, 4 January 1977; "Banning Distribution of Tris," Hearings Before the Subcommittee on Antitrust, Consumers and Employment, U.S. House of Representatives, April and May 1977, p. 3.

3 Ibid.

4 Harris, supra, note 1 at 5.

5 Environmental Defense Fund and Robert H. Boyle, Malignant Neglect, Alfred A. Knopf, New York, 1979, pp. 199, 208.

6 Ibid., p. 196; Harris, supra, note 1 at 3.

7 "Banning Distribution of Tris," supra, note 2 at 14, 132, 135-136, 140-155.

8 Ibid., p. 4.

9 Malignant Neglect, supra, note 5 at 197; Harris, supra, note 1 at 5.

10 Malignant Neglect, ibid., pp. 199, 208; Federal Register, 1 June 1977, p. 28061; "Ban Asked on Children's Wear with Flame Retardant," The New York Times, 9 February 1977.

11 "A Brief Review of Selected Environmental Contamination Incidents with a Potential for Health Effects," prepared by the Library of Congress for the Committee on Environment and Public Works, U.S. Senate, August 1980, p. 302; Consumer Product Safety Commission, "Tris and Fabric, Yarn, or Fiber Containing Tris, Additional Interpretations as Banned Hazardous Substances," Federal Register, 1 June 1977, p. 28062; "CPSC Bans Tris-Treated Children's Garments," News Release, U.S. Consumer Products Safety Commission, Washington, D.C., 7 April 1977.

12 "Banning Distribution of Tris," supra, note 2 at 5-6.

13 Jack Anderson, "Good Morning, America" ABC, 24 May 1977; Robert Rauch, Environmental Defense Fund, Washington, D.C. pers. comm. May 1977 and 1 October 1980.

14 Malignant Neglect, supra, note 5 at 200.

15 Harris, supra, note 1 at 5.

16 Federal Register, supra, note 11.

17 Robert H. Harris, "Estimating the Cancer Hazard to Children from Tris-treated Sleepwear," Environmental Defense Fund, Washington, D.C., 8 March 1977, p. 2.

18 Federal Register, supra, note 11 at 28061.

19 Stewart Brand, "Human Harm to Human DNA," Co-Evolution Quarterly, Spring 1979, p. 19.

20 "Report on Export of Products Banned By U.S. Regulatory Agencies," Government Operations Committee, U.S. House of Representatives, Washington, D.C., 4 October 1978, pp. 29-30.

21 "U.S. Export of Banned Products," Hearings Before a Subcommittee of the Committee on Government Operations, U.S. House of Representatives, July 1978, p. 241.

22 Ibid., pp. 29-32.

23 "Report on Export of Products . . . ," supra, note 20 at 3, 28.

24 Mark Hosenball, Women's Wear Daily, 1 May 1978; Mark Hosenball, "Karl Marx and the Pajama Game," Mother Jones, November 1979.

25 "Report on Export of Products . . . ," supra, note 20 at 10.

26 James Brody, "An Export Trade in Death," Advertising Age, 15 May 1978, p. 99.

27 "U.S. Export of Banned Products," supra, note 21 at 232-237.

28 "Interagency Working Group on Hazardous Substances Export Policy, Draft Report," Federal Register, 12 August 1980, pp. 53754-53768.

29 "Banning Distribution of Tris," supra note 2 at 5-6.

30 Malignant Neglect, supra, note 5 at 200.

31 Elizabeth Jones, Compliance Division, CPSC, Interview, 17 October 1980.

32 Thomas M. Burton, "U.S. Panel Probes Sales of Pajamas Treated with Cancer-Causing Tris," Philadelphia Bulletin, 3 February 1981, p. 1; "Commission Files Complaints Against Two Firms Following Sales of Tris-Treated Children's Sleepwear," News Release, Consumer Products Safety Commission, Washington, D.C., 20 February 1981; United Press International, "More U.S. Sales of Tris-Treated Clothes Reported," The Washington Post, 21 February 1981, p. A6.

33 Martha Hamilton, "3 Firms Accused of Selling Banned Pajamas," Business & Finance Section, The Washington Post, 21 March 1981.

34 CPSC News Release, supra, note 32.

35 Richard Severo, "Chemical Tris is Found in Semen of Participants in Florida Study," The New York Times, 22 February 1981, p. 13.

36 Merrill Brown, "Reagan Wants to Ax Product Safety Agency," The Washington Post, 10 May 1981; "Stockman Moves to Kill Consumer Safety Panel," The New York Times, 9 May 1981.

37 CPSC News Release, supra, note 11.

Chapter Nine PCB's

1 "A Plague on Our Children," NOVA, WGBH Educational Foundation, Boston, 1979; Dolly Katz, "The PCB's Story: Benefits First, then Possible Dangers," Detroit Free Press, 12 April 1981, p. 4B.

2 John Culhane, "PCB's: The Poisons That Won't Go Away," Reader's Digest, December 1980, p. 114.

3 Ibid., p. 79; Stephanie Harris, "Organochlorine Contamination of Breast Milk," Environmental Defense Fund, Washington, D.C., 7 November 1979, pp. 5-6; Joel Balbien, Stephanie Harris and Talbot Page, "Diet as a Factor Affecting Organochlorine Contamination of Breast Milk," Environmental Defense Fund, Washington, D.C., undated, pp. 3-4; "Six Case Studies of Compensation for Toxic Substances Pollution . . .", prepared by the Library of Congress for the Committee on Environment and Public Works, U.S. Senate, June 1980, p. 85.

4 "Six Case Studies . . . ," Ibid., pp. 72-73; "A Brief Review of Selected Environmental Contamination Incidents with a Potential for Health Effects," prepared by the Library of Congress for the Committee on Environment and Public Works, U.S. Senate, August 1980, p. 291.

5 A. K. Bohn; I. Rosenwaike; N. Herman; P. Grover; J. Stellman; K. O'Leary, "Melonoma After Exposure to PCB's," New England Journal of Medicine, 19 August 1976.

6 Balbien, Harris and Page, supra, note 3 at 3-4; Harris, supra note 3 at 5-6.

7 Bill Richards, "Toxic Chemical, Born of PCB's, Found," The Washington Post, 11 September 1979.

8 Ibid.

9 Environmental Defense Fund and Robert H. Boyle, Malignant Neglect, Alfred A. Knopf, New York, 1979, pp. 61-62; "A Plague On Our Children," supra, note 1.

10 Malignant Neglect, pp. 59, 62; "A Plague On Our Children," ibid.; Culhane, supra, note 2 at 113, 115; Toxic Chemicals and Public Protection, A Report to the President by the Toxic Substances Strategy Committee, Council on Environmental Quality, 1980, p. 3.

11 Ralph Nader, Ronald Brownstein and John Richard, Who's Poisoning America, Sierra Club Books, San Francisco, 1981, p. 177.

12 "A Brief Review . . . ", supra, note 4 at 289.

13 Dolly Katz, "PCB's Found in Milk of All Michigan Mothers Tested," Detroit Free Press, 1 February 1981, pp. 1A, 6A.

14 Katz, supra, note 1.

15 "A Brief Review . . . ," supra, note 4 at 289.

16 Malignant Neglect, supra, note 9 at 79, 206-207; Stephanie G. Harris and Joseph H. Highland, Birthright Denied, Environmental Defense Fund, Washington, D.C. 1977, p. 11.

17 Joseph Highland, "Testimony of the Environmental Defense Fund Before the House Subcommittee on Health and the Environment," 1 March 1980, Environmental Defense Fund, Washington, D.C.

18 "Six Case Studies . . .", supra, note 3 at 72-73.
19 Katz, supra, note 1.
20 Toxic Chemicals and Public Protection, supra, note 10 at 3; Malignant Neglect, supra, note 9 at 56.
21 "A Brief Review . . .", supra, note 4 at 290.
22 Toxic Chemicals and Public Protection, supra note 10 at 3.
23 Bill Richards, "Drop in Sperm Count Is Attributed in Part to Toxic Environment," The Washington Post, 12 September 1979. Jane E. Brody, "Sperm Found Especially Vulnerable to Environment," The New York Times, 10 March 1981, pp. C1, C3.
24 "A Plague On Our Children," supra, note 1.
25 Environmental Quality—1979: The Tenth Annual Report of the Council on Environmental Quality, Washington, D.C., December 1979, pp. 11, 99-100, 448-449.
26 Harris, supra note 3 at 8, 10; "Six Case Studies . . .", supra, note 3 at 75.
27 Harold Faber, "Ban on Fishing for Striped Bass Is Reaffirmed," The New York Times, 25 October 1981; Harold Faber, "Hunters Who Eat Ducks Warned on PCB Hazard," The New York Times, 8 October 1981, p. B1; Associated Press, "Great Lakes Fish Warning," The New York Times, 25 September 1981, p. A25; Associated Press, "High Levels of PCB's Found in Waterfowl in Upstate New York," The New York Times, 4 October 1981.
28 Richard Severo, "Hudson: Portrait of a River Under Attack," The New York Times, 9 September 1980, pp. C1, C3.
29 Nader et al., supra note 11.
30 Culhane, supra, note 2 at 113.
31 Nader et al., supra note 11 at 190.
32 Environmental Quality—1976: The Seventh Annual Report of the Council on Environmental Quality, Washington, D.C., December 1976, p. 45; "A Plague On Our Children," supra, note 1; Malignant Neglect, supra, note 9 at 70-73.
33 The Global 2000 Report to the President—The Technical Report, Vol. 2, The Council on Environmental Quality and the U.S. Department of State, July, 1980, p. 306.
34 Severo, supra, note 28.
35 "Six Case Studies . . .", supra, note 3 at 61-69.
36 Richard Severo, "Environmental Protection Agency Slow to Act on 3-Year-Old Law," The New York Times, 6 May 1980, p. C1, C3.
37 "Year End Report, January 1978—December 1979," Environmental Defense Fund, Toxic Chemicals Program, Washington, D.C., p. 9; Philip Shabecoff, "Court of Appeals Acts to Restrict Toxic Chemicals," The New York Times, 1 November 1980, p. 12.
38 Environmental Quality—1979, supra, note 25 at 219.
39 Katz, supra, note 1.
40 Ibid.
41 Malignant Neglect, supra, note 2 at 59, 81.
42 Albert Crenshaw, "Toxic Levels Rising in Great Lakes," The Washington Post, 12 September 1979, A-10.
43 "A Brief Review . . .," supra, note 4 at 281.
44 Bryce Nelson, "PCB Pollution Grave Question, U.S. Says," The Los Angeles Times, 7 October 1979.
45 Philip J. Hilts, "The Deadly Business of Cleaning Up," The Washington Post, 26 February 1981, p. A16; "State to Test On Toxic Exposure," The New York Times, 11 October 1981; "Governor Challenged," The Washington Post, 6 March 1981, p. A10.
46 "Up to 30,000 Flee Gas Leak," The Washington Post, 26 August 1981, p. A14.
47 Nader et al., supra note 11 at 201.
48 The Washington Post, 8 July 1980, A9; Michael Brown, Laying Waste, Pantheon, New York, 1979, 241-247.
49 Sharon Begley, "Toxic Waste Still Pollutes Roadways," Newsweek, 27 October 1980, p. 25.
50 Rod Nordland and Josh Friedman, "Poison at Our Doorstep," The Philadelphia Inquirer; reprint of six part series 23-28 September 1979.
51 Brown, supra, note 48, Begley, supra, note 49.
52 Nordland and Friedman, supra, note 50.
53 Peter Behr, "Goodyear Claims Breakthrough in PCB Disposal," The Washington Post, 22 August 1980, p. E-1; Joanne Omang, "PCB-Disposal Process Nears Approval," The Washington Post, 18 September 1980, p. B3.
54 "A Brief Review . . . ," supra, note 4 at 284; Nelson, supra, note 44; Congressional Quarterly, 6 September 1980, p. 2643; Associated Press, "PCB's Discovered in Foods in West," The Washington Star, 15 September 1979.
55 "A Brief Review . . .," ibid., pp. 284-287; William Longgood, The Darkening Land, Simon and Schuster, New York, 1972, p. 495; Malignant Neglect, supra, note 9 at 77.
56 "A Brief Review . . .", ibid.
57 "Year End Report . . . ," supra, note 37 at 5.
58 "Six Case Studies . . . ," supra, note 3 at 71.
59 Malignant Neglect, supra, note 9 at 220; Harris and Highland, supra, note 16 at 35.

Chapter Ten   Lindane

1 "Lindane: Position Document No. 2/3," Office of Pesticides and Toxic Substances, U.S. Environmental Protection Agency, pp. I-1; II-63-64, 75; V-18, June 1980.
2 Interview with Anne Hollander, Office of Pesticides and Toxic Substances, U.S. Environmental Protection Agency, Washington, D.C., 9 December 1981.
3 "Lindane . . ." supra, note 1.
4 Ibid., pp. II-65-66.
5 Ibid., pp. II-65, 75; V-19.
6 Sandy Rovner, "Health Talk," The Washington Post, 7 November 1980, p. F5.
7 "On Lice and Their Cure", Environmental Defense Fund, Washington, D.C., undated.
8 "Lindane . . . ," supra, note 1 at II-1-2, 18.
9 Ibid., pp. II-23; IV-2.
10 Ibid., pp. II-31, 77-78.
11 Ronald B. Taylor, "Most Cases of Poisonings Go Unreported," The Los Angeles Times, 28 June 1979.
12 "Lindane . . . ," supra, note 1 at II-59.
13 Ibid., II-38.
14 Ibid., page IV-3.
15 Dr. H. Wade Fowler, Jr., Executive Secretary, FIFRA Scientific Advisory Panel, Environmental Protection Agency, Memorandum on Lindane, 9 October 1980.
16 "Lindane . . . ," supra, note 1 at I-1.
17 Ibid.
18 Rachel Carson, Silent Spring, Crest Books, Greenwich, Connecticut, 1962, pp. 159, 176, 280.
19 "Pesticide Oversight," Hearings Before the Subcommittee on Environment, U.S. Senate, 9 and 16 August 1974, pp. 60, 141.
20 "Deficiencies in Administration of Federal Insecticide, Fungicide, and Rodenticide Act," Eleventh Report by the Committee on Government Operations, U.S. House of Representatives, 13 November 1969, p. 7.
21 "Six Case Studies of Compensation for Toxic Substances Pollution . . .", prepared by the Library of Congress for the Committee on Environment and Public Works, U.S. Senate, June 1980, p. 148.
22 "Hazardous Waste Disposal," Report by the Subcommittee on Oversight and Investigations, U.S. House of Representatives, September 1979, p. 9.
23 Ian T. Nisbet, "Pesticide Do's & Don't's," National Audubon Society, New York, (undated); "Getting the Bugs Out: A Guide to Sensible Pest Management In and Around the Home," National Audubon Society, New York, 1981, pp. 9-14.

Chapter Eleven   DBCP

1 "Worker Safety in Pesticide Production," Hearings Before the Subcommittee on Agricultural Research and General Legislation," U.S. Senate, 13 and 14 December 1977, pp. 151-152.

2 Ibid., p. 158.

3 Ronald L. Taylor, "DBCP Still Used Despite Dangers," The Los Angeles Times, 28 June 1979.

4 "Worker Safety in Pesticide Production," supra, note 1 at 152.

5 Bill Richards, "Health Officials Cite Hazards At Pesticide Plant," The Washington Post, 14 December 1977.

6 Taylor, supra, note 3.

7 "Worker Safety in Pesticide Production," supra, note 1 at 151.

8 Ibid., pp. 152, 163.

9 Ibid.: "Involuntary Exposure to Agent Orange and Other Toxic Spraying", Hearings Before the Subcommittee on Oversight and Investigations, U.S. House of Representatives, 26 and 27 June 1979, pp. 163-164.

10 Taylor, supra, note 3.

11 "Wider Sterility Problem With Pesticide Is Found," The Washington Post, 25 August 1977.

12 "Six Case Studies of Compensation for Toxic Substances Pollution . . . ," by the Library of Congress for the Committee on Environment and Public Works, U.S. Senate, June 1980, 159-160.

13 Ronald B. Taylor, "DBCP Found in Water Supplies of 12 Counties", The Los Angeles Times, 5 September 1979.

14 Friends of the Earth, Washington, D.C., Letter to Members of Congress and Attachments, 18 September 1980.

15 United Press International, "Pesticide Found in Md. Possible Cancer Causer," The Washington Post, 22 October 1980, p. B7.

16 Taylor, supra, note 13.

17 "Preliminary Risk Assessment for DBCP", Memorandum from Ray E. Albert, Chairman, Carcinogen Assessment Group, Environmental Protection Agency, to Edwin Johnson, Deputy Assistant Administrator, Pesticide Programs, Environmental Protection Agency, Washington, D.C., 7 September 1980.

18 Christopher Jenkins, "Sterility-Causing Pesticide Slips Back into U.S.," Pacific News Service, San Francisco, California, 29 March 1979; Jeff Kempter, Project Manager, DBCP, Environmental Protection Agency, interview, 20 October 1980; "Revision of DBCP Risks," Memorandum from Lois Rossi, Environmental Protection Agency, to Jeff Kempter, Environmental Protection Agency, 26 July 1979.

19 Bill Peterson and P. Shinoff, "Firms Had Sterility Data on Pesticide," The Washington Post, 23 August 1977; "Worker Safety in Pesticide Production," supra, note 1, p. 184.

20 Statement of Dr. Eula Bingham, Assistant Secretary for Occupational Safety and Health (OSHA), U.S. Department of Labor, Worker Safety in Pesticide Production, ibid., p. 184.

21 "Worker Safety in Pesticide Production," supra, note 1 at 24, 148, 184, 223.

22 Taylor, supra, note 3.

23 "Worker Safety in Pesticide Production," supra, note 1 at 184.

24 Ibid., p. 124, 148-149.

25 Ibid., p. 158.

26 Ibid.

27 Taylor, supra, note 3.

28 Dave Davis and Josh Hanig, "Song of the Canary," Public Broadcasting System, 1980; Daniel Ben-Horin, "The Sterility Scandal", Mother Jones, May 1979, p. 60.

29 Ben-Horin, ibid.

30 Taylor, supra, note 3.

31 "Worker Safety in Pesticide Production," supra, note 1 at 24, 29-30.

32 Barry Castleman, "The Export of Hazardous Factories to Developing Nations," International Journal of Health Services, Volume 9, No. 9, 1979, pp. 590-91; Ronald B. Taylor, "Production of Highly Toxic Pesticide Shifts to Mexico", The Los Angeles Times, 9 September 1978.

33 Jenkins, supra, note 18.

34 Taylor, supra, note 3.

35 Richards, supra note 5; The Global 2000 Report to the President, The Technical Report, Volume Two, Council on Environmental Quality and U.S. Department of State, Washington, D.C., July 1980, p. 343.

36 Robert L. Jackson, "U.S. Bans Use of Pesticide DBCP Except in Hawaii," The Los Angeles Times, 30 October 1979.

37 David Burrington, NBC Nightly News, 10 January 1981.

38 David Weir, "The Boomerang Crimes," Mother Jones, November 1979; David Weir and Mark Schapiro, Circle of Poison, Institute for Food and Development Policy, San Francisco, 1981, pp. 19-22.

39 Ibid.

40 Ibid.

41 David Burnham, "Peach Growers Suggest that Those Who Don't Want Children Handle Pesticide that Can Cause Sterility," The New York Times, 27 September 1977, p. C17.

42 Dale R. Bottrell, Integrated Pest Management, Council on Environmental Quality, Washington, D.C., 1980, pp. 36-37.

43 Ibid., p. 32

44 Ibid., p. 82.

Chapter Twelve   Endrin

1 Rachel Carson, Silent Spring, Crest Books, Greenwich, Connecticut, 1962, p. 34.

2 Tim O'Brien, "Environmental Protection Agency Studies Ban on Pest Killer," The Washington Post, 15 September 1973.

3 Donald H. Davis, "Silent Summer: Spraying of Alabama Cotton Destroys Birds, Fish, Squirrels," The Washington Post, 11 September 1973, p. A3.

4 Frank Graham, Jr., Since Silent Spring, A Fawcett Crest Book, Greenwich, Connecticut, 1970, pp. 29-30.

5 Environmental Defense Fund and Robert Boyle, Malignant Neglect, Alfred A. Knopf, New York, 1979, p. 131; "Bad News for the Birds," Time, 5 October 1981, p. 52.

6 "Suspended and Cancelled Pesticides," Environmental Protection Agency, Washington, D.C., October 1979, p. 8.

7 Ibid.

8 Ibid.

9 United Press International, "Pesticide May Force Duck Hunting Halt," The Washington Post, 16 September 1981; Philip Shabecoff, "Environmental Agency Terms Pesticide Level in Birds Safe," The New York Times, 18 September 1981, p. 8.

9b "Montana Backs Hunting In Dispute on Waterfowl," The New York Times, 26 September, 1981.

10 "Montana Pesticide Spraying Perils Waterfowl Hunting in 17 States," The New York Times, 16 September 1981, pp. A1, A19; National Audubon Society, Letter to EPA Administrator Anne Gorsuch, 29 September 1981.

11 National Audubon Society, ibid.

12 Ibid., "Montana Pesticide Spraying . . .", supra, note 10.

13 "Montana to Permit Use of Controversial Endrin," The New York Times, 19 October 1981.

14 Chris Madson, "An End for Endrin," Kansas Wildlife, November/December, 1981.

Chapter Thirteen   Toxaphene

1 "Toxaphene: Position Document 1," Toxaphene Working Group, U.S. Environment Protection Agency, Washington, D.C., 19 April 1977, p. 19; Bill Richards, "Insecticide Seen Possible Agent in Human Cancer," The Washington Post, (undated) 1979.

2 "U.S. Export of Banned Products", Hearings Before A Subcommittee of the Committee on Government Operations, U.S. House of Representatives, 11-13 July, 1978, p. 55.

3 "Toxaphene . . .", supra, note 1 at 8.

4 Ibid.

5 Ronald B. Taylor, "Cattle Deaths Stir Pesticide Debate", The

Los Angeles Times, 5 November 1979.
6 *Federal Register*, 25 May 1977, p. 26861.
7 Taylor, *supra*, note 3.
8 *Ibid*.
9 *Ibid.*; "Toxaphene . . .", *supra*, note 1 at 12.
10 "Toxaphene . . .", *ibid.*, p. 20.
11 *Federal Register*, *supra*, note 6 at 26862.
12 Jay Sheppard, Office of Endangered Species, U.S. Department of Interior, Washington, D.C., *pers. comm.*, 9 October 1980.
13 *Federal Register*, *supra* note 4 at 26861-26862.
14 Taylor, *supra*, note 3; Ed Bradley, "60 Minutes," CBS News, 22 November 1981.
15 Taylor and Bradley, *ibid*.
16 "Suspended and Cancelled Pesticides," U.S. Environmental Protection Agency, October 1979, p. 19.
17 "Toxaphene . . .", *supra*, note 1 at 3.

Chapter Fourteen   DDT

1 Rachel Carson, *Silent Spring*, Crest Books, Greenwich, Connecticut, 1962, p. 97.
2 *Ibid.*, pp. 97-98.
3 Robert S. Strother, "Backfire in the War Against Insects," *Reader's Digest*, June 1959.
4 "Reproductive Failures in Peregrine Falcons Increase," News Release, U.S. Department of the Interior, Fish and Wildlife Service, Washington, D.C., 1 November 1973.
5 Martin Waldron, "Pelicans Gain on Texas Coast After Drop in Use of Pesticides," *The New York Times*, 5 August 1973.
6 William Longgood, *The Darkening Land*, Simon and Schuster, New York, 1972, pp. 132, 134.
7 Frank Graham, Jr., *Since Silent Spring*, A Fawcett Crest Book, Greenwich, Connecticut, 1970, p. 113; Charles F. Wurster, "DDT Reduces Photosynthesis by Marine Phytoplankton," *Science* 159, 1968, pp. 1474-1475.
8 Longgood, *supra*, note 6 at 137.
9 *Ibid.*, p. 143; *Environmental Quality—1975: The Sixth Annual Report of the Council on Environmental Quality*, Washington, D.C., December 1975, p. 368; Dr. Samuel Epstein, Testimony Before the Senate Subcommittee on Energy, Natural Resources, and the Environment, 15 April 1970; "Effects of 2, 4, 5-T on Man and the Environment," Hearings Before the Subcommittee on Energy, Natural Resources, and the Environment, U.S. Senate, 7 and 15 April 1970, p. 418.
10 *Environmental Quality—1975*, *ibid.*, pp. 368, 375, 387; "A Brief Review of Selected Environmental Contamination Incidents with a Potential for Health Effects," prepared by the Library of Congress for the Committee on Environment and Public Works, U.S. Senate, August, 1980, p. 223.
11 Longgood, *supra*, note 6 at 143.
12 George M. Woodwell, "Toxic Substances and Ecological Cycles," *Scientific American*, Volume 216, No. 3, 1967.
13 Justin Frost, "Earth, Air, Water," *Environment*, August 1969.
14 Longgood, *supra*, note 6 at 136.
15 *Environmental Quality—1975*, *supra*, note 9 at 375.
16 Stephanie G. Harris and Joseph H. Highland, *Birthright Denied*, Environmental Defense Fund, Washington, D.C., 1977, p. 2.
17 Longgood, *supra*, note 6 at 132-134.
18 Graham, *supra*, note 7 at 165-166.
19 Longgood, *supra*, note 6 at 140.
20 Samuel Rotrosen, "Why Are We Exterminating DDT? Ban Called Political Instead of Scientific," Business section, *The New York Times*, 27 August 1972.
21 *Federal Register*, 7 July 1972, pp. 13369-13376.
22 "The Presidential Candidates," League of Conservation Voters, Washington, D.C., April 1980.
23 "True or False," League of Conservation Voters, Washington, D.C., 1980.

24 *Environmental Quality—1974: The Fifth Annual Report of the Council on Environment Quality*, Washington, D.C., December 1974, pp. 161-162.
25 Robert van den Bosch, *The Pesticide Conspiracy*, Anchor Books, Garden City, New York, 1980, pp. 75, 77-79.
26 S. G. Herman, "Douglas Fir Tussock Moth in the Northwest: The Case Against DDT Use in 1974," The Evergreen State College, Olympia, Washington, November, 1973.
27 Linda Greenhouse, "Gypsy Moth Peril Is Seen Vanishing: Westchester Said to Be Rid of Pest After 2 Years," *The New York Times*, 24 September 1972, p. 73.
28 "Safety Factors for Children Employed as Hand Harvesters of Strawberries and Potatoes," prepared by Clement Associates, Washington, D.C., for the U.S. Department of Labor, 18 May 1979, pp. 58-72.
29 Michael Bancroft, "Letter to Organizations and Individuals Concerned with Human Exposure to DDT and Carbaryl", Public Citizen Litigation Group, Washington, D.C., 17 April 1978.
30 E. Michael Kasnia, "DDT Usage at Dulles Airport," Memorandum to Director of Investigations Branch, Washington, D.C., Food Drug Administration, 3 August 1976.
31 "DDT Concentrations Found," *The Washington Post*, 3 August 1981, p. A7.
32 *Environmental Quality—1975*, *supra*, note 9 at 369.
33 Mike Hollis, "The Persistence of a Poison: Effects of Chemical Plant Still Plague Alabama Town," *The Washington Post*, 15 June 1980, p. A2.
34 David Weir and Mark Schapiro, *Circle of Poison*, Institute for Food and Development Policy, San Francisco, 1981, p. 13.
35 Hollis, *supra*, note 33; "A Brief Review . . .", *supra*, note 10 at 224.
36 *The Global 2000 Report to the President—The Technical Report*, Volume II, The Council on Environmental Quality and the U.S. Department of State, Washington, D.C., July 1980, p. 307.
37 R. Jeffrey Smith, "U.S. Beginning to Act on Banned Pesticides," *Science*, 29 June 1979, p. 1393; Erik Eckholm and S. Jacob Scherr, "Double Standard and the Pesticide Trade," *New Scientist*, London, 16 February 1978, p. 443.
38 Better Regulation of Pesticide Exports and Pesticide Residues in Imported Food Is Essential," U.S. General Accounting Office, Washington, D.C., 22 June 1979, pp. 8, 53.
39 *Ibid*.
40 Nelson Bryant, "Wood, Field, and Stream: The Use of DDT Upsets Nature's Balance," *The New York Times*, 10 May 1973, p. 63.
41 "Makers File Suit Against DDT Ban, 26 Concerns Call Substitute Chemical More Harmful," *The New York Times*, 21 October 1972.
42 Elsie Carper, "EPA Bans Nearly All DDT Uses," *The Washington Post*, 15 June 1972; *Federal Register*, 7 July 1972, p. 13374.
43 Dale R. Bottrell, *Integrated Pest Management*, Council on Environmental Quality, Washington, D.C. 1980, p. ix.
44 *Ibid*.

Chapter Fifteen   Aldrin And Dieldrin

1 *Environmental Quality—1975: The Sixth Annual Report of the Council on Environmental Quality*, Washington, D.C., December 1975, p. 381.
2 Joseph Highland, "Corporate Cancer," Environmental Defense Fund, Washington, D.C., undated.
3 George C. Wilson, "Environmental Protection Agency Blocks Manufacture of Dieldrin," *The Washington Post*, 3 August 1974 pp. Al, All; E. W. Kenworthy, "U.S. Bars Aldrin and Dieldrin, 2 Most Widely Used Pesticides," *The New York Times*, 3 August 1974, pp. 1, 38.
4 "A Brief Review of Selected Environmental Contamination Incidents with a Potential for Health Effects," prepared by the Library of Congress for the Committee on Environment and Public Works, U.S. Senate, August 1980, p. 175.

5 Environmental Defense Fund and Robert H. Boyle, *Malignant Neglect*, Alfred A. Knopf, New York, 1979, pp. 122-124.
6 *Ibid.*, pp. 126-127.
7 Frank Graham, Jr., *Since Silent Spring*, A Fawcett Crest Book, Greenwich, Connecticut, 190, p. 41.
8 Rachel Carson, *Silent Spring*, Crest Books, Greenwich, Connecticut, 1962, pp. 33, 88, 150-151.
9 *Ibid.*, p. 98; *Audubon Field Notes*, "Fall Migration—16 August to 30 November, 1958," Volume 13, 1959.
10 Robert S. Strother, "Backfire in the War on Insect," *Reader's Digest*, June 1959.
11 Graham, *supra*, note 7 at 229.
12 Carson, *supra*, note 8 at 33-34, 88.
13 Samuel S. Epstein, *The Politics of Cancer*, Anchor Books, Garden City, New York, 1979, p. 289.
14 Highland, *supra*, note 2.
15 Peter Crane, "Politics and Pesticides: Dieldrin Gets A Reprieve," *The Washington Post*, 28 July 1974, p. C-5; Epstein, *supra*, note 13 at 262-63.
16 *Ibid; Malignant Neglect, supra*, note 5 at 121-122, 127.
17 *Environmental Quality—1975, supra*, note 1 at 381; "A Brief Review . . ." *supra*, note 4 at 173-174; "Aldrin/Dieldrin," Criteria Document, U.S. Environmental Protection Agency, Washington, D.C., 1976.
18 Epstein, *supra*, note 13 at 252-53.
19 *Malignant Neglect, supra*, note 5 at 206-207.
20 "A Brief Review . . .", *supra*, note 4 at 175.
21 "The Environmental Protection Agency and the Regulation of Pesticides," Staff Report to the Subcommittee on Administrative Practice and Procedure, U.S. Senate, Washington, D.C., December 1976, pp. 24-25.
22 *Malignant Neglect, supra*, note 5 at 124-125; *Environmental Quality—1974: The Fifth Annual Report of the Council on Environmental Quality*, Washington, D.C., December 1974, p. 160.
23 Wilson and Kenworthy, *supra*, note 3.
24 Morton Mintz, "Scientist Warns of Cancer Threat in Aldrin and Dieldrin," *The Washington Post*, 15 August 1974.
25 *Ibid.*, Kenworthy, *supra*, note 3.
26 "Pesticide Oversight," Hearings Before the Subcommittee on Environment, U.S. Senate, 9 and 16 August 1974, p. 55.
27 Morton Mintz, "Plan to Ban 2 Pesticides Is Upheld; Cancer Risk Cited," *The Washington Post*, 22 September 1974, p. A-11; *Malignant Neglect, supra*, note 5 at 122.
28 *Malignant Neglect, ibid.*, p. 129.
29 Crane, *supra*, note 15.
30 *Ibid.*
31 "Suspended and Cancelled Pesticides," U.S. Environmental Protection Agency, October 1979, pp. 1, 5.
32 David Weir, "The Boomerang Crime", *Mother Jones*, San Francisco, California, November 1979.
33 "Report on the Export of Products Banned by U.S. Regulatory Agencies," Committee on Government Operations, U.S. House of Representatives, 4 October 1978, p. 8.
34 Loretta McLaughlin, "Concern is Growing in Peru Over Use Of Banned Chemicals," *World Environment Report*, 8 October 1979, p. 5.
35 *Environmental Quality—1974, supra* note 22 at 161; *Malignant Neglect, supra*, note 5 at 128.
36 Associated Press, "Banquet Brand Turkey Products Called 'Suspect'", *The Washington Post*, 27 June 1980, p. A-8.
37 Associated Press, "Banquet Foods Recalls Turkey," *The Washington Post*, 29 June 1980, p. A-2.

Chapter Sixteen   Chlordane And Heptachlor

1 "Environmental Protection Agency, Pesticide Products Containing Heptachlor or Chlordane," *Federal Register*, 26 November 1974, p. 41300.
2 William A. Butler and Jacqueline Warren, "Petition for Suspension and Cancellation of Chlordane/Heptachlor," Environmental Defense Fund, Washington, D.C., October 1974.
3 "Train Suspends Major Uses of Chlordane/Heptachlor . . .", Environmental News, U.S. Environmental Protection Agency, Washington, D.C., 24 December 1975.
4 Robert S. Strother, "Backfire in the War on Insects," *Reader's Digest*, June, 1959.
5 Frank Graham, Jr., *Since Silent Spring*, A Fawcett Crest Book, Greenwich, Conn., 1970, pp. 59-66.
6 Rachel Carson, *Silent Spring*, Crest Books, Greenwich, Conn. 1962.
7 Samuel S. Epstein, *The Politics of Cancer*, Anchor Books, Garden City, New York, 1979, p. 275.
8 Dr. Guy Arnold, interview, 12 December 1981; Adrian Peracchio, "Agent Orange May Be Just One Garnish in a Toxic Cocktail," *The Washington Post*, 20 June 1980, p. A4.
9 Carson, *supra*, note 6, pp. 31-32.
10 *Federal Register, supra*, note 1, p. 41299.
11 "The Environmental Protection Agency and the Regulation of Pesticides," Staff Report to the Subcommittee on Administrative Practice and Procedure, U.S. Senate, December 1976, p. 24.
12 Environmental Defense Fund and Robert H. Boyle, *Malignant Neglect*, Alfred A. Knopf, New York, 1979, p. 135; Peter Milius, "Leptophos Handled at 9 Plants in U.S.," *The Washington Post*, 4 December 1976, pp. A1, A8; Bill Richards, "Health Officials Cite Hazards at Pesticide Plant," *The Washington Post*, 14 December 1977.
13 "Chemical Firm Fined," *The Washington Post*, 19 January 1981, p. A5.
14 Butler and Warren, *supra*, note 2.
15 Epstein, *supra*, note 7 at 272.
16 *Federal Register, supra*, note 1 at 41298.
17 "Train Suspends Major Uses . . . ," *supra*, note 3.
18 "Environmental Protection Agency, Velsicol Chemical Co. Et Al., Consolidated Heptachlor/Chlordane Hearing," *Federal Register*, 19 February 1976, p. 7556.
19 "Train Suspends Major Uses . . . ," *supra*, note 3.
20 *Ibid.*
21 *Federal Register, supra*, note 1 at 41300.
22 *Ibid.*; "Train Suspends Major Uses . . . ," *supra*, note 3.
23 "Report on Export of Products Banned by U.S. Regulatory Agencies," Committee on Government Operations, U.S. House of Representatives, 4 October 1978, p. 8.
24 Herbert Denton, "Contaminated Pork Shipped to Schools," *The Washington Post*, 24 May 1980, p. A1.
25 *Ibid.*, pp. A1 & A8.
26 "Suspended and Cancelled Pesticides," U.S. Environmental Protection Agency, Washington, D.C., October 1979.
27· Associated Press, "Poisoning of Water Perhaps No Accident," *The Washington Post*, 28 November 1981, p. A2; Associated Press, "Officials Say Pesticide Put In Water Supply Purposely," *The New York Times*, 17 December 1981; Associated Press, "Water Supply Contaminated in Areas of Pittsburgh," *The New York Times*, 10 December 1981.

Epilogue

1 Caroline E. Mayer, "Easing of Hazardous Exports Studied," *The Washington Post*, 9 September 1981.
2 Peter Behr, "Regulators Reined," *The Washington Post*, 18 February 1981, pp. A1 & A5.
3 Jeremiah O'Leary, "EPA Will Propose Relaxing Portion of Clean Air Act," *The Washington Star*, March 1981.
4 Peter Behr, "Regulation Target List Being Prepared," *The Washington Post*, 21 February 1981, p. A1.
5 Peter Behr and Joanne Omang, "White House Targets 27 More Regulations for Review," *The Washington Post*, 26 March 1981, pp. A1, A15.
6 Clyde Farnsworth, "U.S. Proposes Eased Car Standards," *The New York Times*, 7 April 1981, pp. A1, D7.
7 Felicity Barringer, "30 More Regulations Targeted for Review," *The Washington Post*, 13 August 1981, p. A27.

8 Philip Shabecoff, "Oil-Saving Plans Face U.S. Attack," *The New York Times,* 27 March 1981, p. D1.

9 "CPSC to Appeal Slash," *The Washington Star,* 25 February 1981, p. E1.

10 Caroline E. Mayer, "Cuts in CPSC Budget, Staff Hit," *The Washington Post,* 2 December 1981, p. D8.

11 Merrill Brown, "Reagan Wants to Ax Product Safety Agency," *The Washington Post,* 10 May 1981; "Stockman Moves to Kill Consumer Safety Panel," *The New York Times,* 9 May 1981.

12 Ward Sinclair, "Chemical Industry to Have Powerful Voice About New EPA Chief," *The Washington Post,* 15 January 1981, p. A8.

13 Joanne Omang, "Denver Lawyer Reagan's Choice to Head EPA," *The Washington Post,* 21 February 1981, p. A4.

14 Carl T. Rowan, "So Many Foxes Loose in the Coop," *The Washington Star,* 1 March 1981.

14b Frank O'Donnell, "Manager of the Year," *The Washington Monthly,* December, 1981.

15 Frank Greve, "Foxes In Charge of EPA's Eggs," Atlanta Constitution, 23 July 1981; Philip Shabecoff, "New Environmental Chief Vows to Ease Regulatory Overburden," *The New York Times,* 21 June 1981; "Executive Notes," *The Washington Post,* 22 February, 1982.

16 Morton Mintz, "EPA Is Shaken as 2 Top Aides Quit Posts," *The Washington Post,* 26 September 1981, p. A4.

17 Joanne Omang, "EPA and Dow Negotiating Settlement on Herbicide," *The Washington Post,* 11 April 1981, p. A4.

18 Philip Shabecoff, "Environment Agency Chief Announces Reorganization," *The New York Times,* 13 June 1981, p. 8; Joanne Omang, "EPA Enforcement Split Up In Agency Reorganization," *The Washington Post,* 13 June 1981 p. A3.

19 Joanne Omang, "EPA Seeks to Bail Out on Water Cleanup," *The Washington Post,* 11 August 1981, p. A13.

20 Caroline E. Mayer, "U.S. Relaxing Enforcement of Regulations," *The Washington Post,* 15 November 1981, p. F1; Philip Shabecoff, "Environmental Cases Off Sharply Under Reagan," *The New York Times,* 15 October 1981.

21 Joanne Omang, "U.S. to Ease Insurance Rules for Hazardous Waste Dumps," *The Washington Post,* 30 August 1981, pp. A1, A15.

22 Joanne Omang, "Firms Urged to Back Air Act Changes," *The Washington Post,* 23 August 1981.

22b "Gorsuch Defends Pollution Plans," *The Washington Post,* 18 February, 1982, p. A8.

22c Robert D. Hershey, Jr., "U.S. May Repeal Rules Limiting Lead in Gasoline," *The New York Times,* 19 February, 1982.

22d "The Federal Triangle," *The Washington Post,* 29 January, 1982.

22e Jonathan Dedmon, "Paraquat's Use Approved On U.S. Marijuana Crops," *The Washington Post,* 23 February, 1982.

23 Philip Shabecoff, "Funds and Staff for Protecting Environment May be Halved," *The New York Times,* 29 September 1981, p. 1.

24 Philip Shabecoff, "Ecology Unit Cut Reportedly Urged by Budget Office," *The New York Times,* 19 November 1981, pp. A1, A22.

25 William Drayton, "Panorama," WTTG-TV, 20 November 1981, "This Ice Queen Does Not Melt," *Time,* 18 January, 1982, p. 16.

26 John B. Oakes, "The Reagan Hoax," Op-Ed Page, *The New York Times,* 1 November 1981.

26b Russell E. Train, "The Destruction of EPA," *The Washington Post,* 2 February, 1982, p. A15.

26c John W. Hernandez, Jr., Letter to Alphons J. Hackl, Publisher, Acropolis Books, Ltd., 1 February, 1982.

27 Warren Brown and Philip Hilts, "Labor Nominee's Firm Has Mixed OSHA Record," *The Washington Post,* (undated) 1981, p. A2.

28 Lance Gay, "Donovan Reviewing Regulations That Hamper Economic Growth," *The Washington Star,* 12 February 1981.

29 *Ibid;* Philip Shabecoff, "Swift Attack on Regulations," *The New York Times,* 13 February 1981.

30 "Safety and Health Director Orders Purging of Booklet He Calls Unfair," *The New York Times,* 27 March 1981.

30b "OSHA Proposes Cotton Dust Rule Review," *The Atlanta Constitution,* 10 February, 1982, p. 1-D.

31 Howie Kurtz, "OSHA Head Accused in Carcinogen Case," *The Washington Star,* 17 July 1981, p. A3; Joanne Omang, "EPA Staff Finds Formaldehyde No Great Cancer Risk," *The Washington Post,* 19 September 1981.

31b Caroline E. Mayer, "Formaldehyde Foam Insulation Ban Ordered," *The Washington Post,* 23 February, 1982; "EPA Now Calls 2 Chemicals No Threat to Human Health," *The Washington Post,* 18 February, 1982.

32 Victor Cohn, "Occupational Safety, Health Chief Fired," *The Washington Post,* 6 March 1981, p. A13.

33 "Sagebrush Rebel at Interior," *Newsweek,* 5 January 1981, p. 17.

34 John B. Oakes, "Watt's Very Wrong," *The New York Times,* 31 December 1980, p. A15.

35 Andy Pasztor, "Watt Tells Senate He Would Balance Use, Protection of Resources as Interior Chief," *The Wall Street Journal,* 8 January 1981.

36 Kathy Sawyer, Dan Morgan, "Reagan Appears to Be Going with Illinois Farm Chief for Agriculture," *The Washington Post,* 21 December 1980.

37 "Block Named to Head Agriculture," *The Washington Post,* 24 December 1980.

38 Caroline E. Mayer, "U.S. Relaxing Enforcement of Regulations," *The Washington Post,* 15 November 1981, p. F1.

39 "True or False," League of Conservation Voters, Washington, D.C., 1980.

40 "The Presidential Candidates," League of Conservation Voters, Washington, D.C., 1980.

41 Oakes, *supra,* note 34.

42 "New Health Plans Focus on 'Wellness'," *The New York Times,* 24 August 1981, p. D1.

Additional Recommended Reading

Recent books on the poisoning of the environment, and what to do about it.

Michael Brown, *Laying Waste*, Pantheon, New York, 1979

Rachel Carson, *Silent Spring*, Crest, Greenwich, Conn., 1962

Barry Commoner, *The Closing Circle*, Bantam Books, New York, 1972

Environmental Defense Fund and Robert Boyle, *Malignant Neglect*, Alfred A. Knopf, New York, 1979

Samuel Epstein, *The Politics of Cancer*, Sierra Club Books, San Francisco, 1978

Lois Gibbs, *Love Canal: My Story*, State University of New York at Albany, 1982

Dudley Giehl, *Vegetarianism: A Way of Life*, Harper & Row, New York, 1979

Frank Graham, Jr., *Since Silent Spring*, Fawcett-Crest, Greenwich, Conn., 1970

Carol Keough, *Water Fit to Drink*, Rodale Press, Emmaus, Pa., 1980

Frances Moore Lappé, *Diet for a Small Planet*, Ballantine, New York, 1975

William Longgood, *The Darkening Land*, Simon & Schuster, New York, 1972

Jim Mason and Peter Singer, *Animal Factories*, Crown, New York, 1980

Ralph Nader, Ronald Brownstein, and John Richard, *Who's Poisoning America*, Sierra Club Books, San Francisco, 1981

Robert Rodale, *The Basic Book of Organic Gardening*, Ballantine, New York.

Robert van den Bosch, *The Pesticide Conspiracy*, Anchor Books, New York, 1980

David Weir and Mark Schapiro, *Circle of Poison*, Institute for Food and Development Policy, San Francisco, 1981

Thomas Whiteside, *The Pendulum and the Toxic Cloud*, Yale University Press, New Haven, Conn., 1979

Brochures, Booklets, and Government Reports

Dale R. Bottrell, *Integrated Pest Management*, Council on Environmental Quality, (U.S. Government Printing Office), Washington, D.C., 1980

Stephanie G. Harris, "Grow It Safely! (Pest Control Without Poisons)," Health Research Group, Washington, D.C., 1975

Stephanie G. Harris and Joseph H. Highland, *Birthright Denied: The Risks and Benefits of Breast-Feeding*, Environmental Defense Fund, Washington, D.C., 1977

Ron Nordland and Josh Friedman, "Poison at Our Doorsteps," *The Philadelphia Inquirer*, Philadelphia, 1979

Helga and William Olkowski, "How To Control Garden Pests Without Killing Almost Everything Else," Rachel Carson Trust for the Living Environment, Washington, D.C., 1977

Bob Wyrick, "Hazards for Export," *Newsday*, Long Island, New York, 1981

"Contamination of Ground Water by Toxic Organic Chemicals," Council on Environmental Quality, (U.S. Government Printing Office), Washington, D.C., 1981

"Getting the Bugs Out: A Guide to Sensible Pest Management In and Around the Home," National Audubon Society, New York, 1981

"The Global Environment and Basic Human Needs," A Report to the Council on Environmental Quality from the World-watch Institute, (U.S. Government Printing Office), Washington, D.C., 1978

The Global 2000 Report to the President, Council on Environmental Quality and U.S. Department of State, (U.S. Government Printing Office), Washington, D.C., 1980

"Groundwater Protection," U.S. Environmental Protection Agency, Washington, D.C., 1980

"A Plague on Our Children," Nova, WGBH Educational Foundation, Boston, 1979

"The Poisoning of America," Parts I and II, *The Los Angeles Times*, Los Angeles, 1979-1980

"Toxic Chemicals and Public Protection," A Report to the President by the Toxic Substances Strategy Committee, Council on Environmental Quality (U.S. Government Printing Office), Washington, D.C., 1980

Index

410   Index